In the fall of 1930, Pauling began work on a determination of the structure of the carbon tetrahedron, implementing a simplified version of the Schrödinger wave equation that had been developed by John C. Slater, an American physicist. ... Pauling worked through the fall without any major breakthroughs until finally, in December 1930, deciding to simplify the mathematics of the project by removing the radial function from his equation. ... Pauling was elated to find that all variables could be accounted for using his new mathematical method, and it was immediately clear that this was a major discovery.

In February 1931, Pauling mailed his results to the *Journal of the American Chemical Society (JACS)*. Confidently titled "The Nature of the Chemical Bond: Application of results obtained from the quantum mechanics and from a theory of paramagnetic susceptibility to the structure of molecules," the paper was published by *JACS* only six weeks after the manuscript arrived at the journal's offices. The incredible speed of this turnaround was abetted by the fact that the paper had not been refereed, as editor Arthur Lamb could think of no individual properly qualified to review the revolutionary content of Pauling's work.

Soon recognized as a classic of 20th-century scientific writing — more than 40 years later, Pauling would call it "the best work I've ever done" — the paper defined six rules for the shared electron bond and presented Pauling's findings in uncomplicated terms, thus enabling peers to examine them without becoming lost in the mathematics. ... As Pauling would recall in 1977,

"It seems to me that I have introduced into my work on the chemical bond a way of thinking that might not have been introduced by anyone else, at least not for quite a while. I suppose that the complex of ideas that I originated in the period of around 1928 to 1933 — and 1931 was perhaps my most important paper — has had the greatest impact on chemistry

THE MANY WORLDS OF
LINUS PAULING

Other World Scientific Titles by the Editor

Visions of Linus Pauling
ISBN: 978-981-12-6075-9
ISBN: 978-981-12-8647-6 (pbk)

THE MANY WORLDS
OF
LINUS PAULING

EDITOR

CHRIS PETERSEN
Oregon State University, USA

World Scientific

NEW JERSEY · LONDON · SINGAPORE · BEIJING · SHANGHAI · HONG KONG · TAIPEI · CHENNAI · TOKYO

Published by

World Scientific Publishing Co. Pte. Ltd.

5 Toh Tuck Link, Singapore 596224

USA office: 27 Warren Street, Suite 401-402, Hackensack, NJ 07601

UK office: 57 Shelton Street, Covent Garden, London WC2H 9HE

Library of Congress Cataloging-in-Publication Data
Names: Petersen, Chris (Christoffer) editor
Title: The many worlds of Linus Pauling / editor Chris Petersen.
Description: New Jersey : World Scientific, [2026] | Includes bibliographical references and index.
Identifiers: LCCN 2025026633 | ISBN 9789819815333 hardcover |
    ISBN 9789819815340 ebook for institutions | ISBN 9789819815357 ebook for individuals
Subjects: LCSH: Pauling, Linus, 1901–1994 | Pauling, Linus, 1901–1994--Knowledge and learning |
    Chemists--United States--Biography | Social reformers--United States--Biography |
    LCGFT: Biographies
Classification: LCC QD22.P35
LC record available at https://lccn.loc.gov/2025026633

British Library Cataloguing-in-Publication Data
A catalogue record for this book is available from the British Library.

For any available supplementary material, please visit
https://www.worldscientific.com/worldscibooks/10.1142/14370#t=suppl

Desk Editor: Shaun Tan Yi Jie

Typeset by Stallion Press
Email: enquiries@stallionpress.com

About the Editor

Chris Petersen is Archivist for Oral History and Digital Projects at the Special Collections and Archives Research Center (SCARC), Oregon State University Libraries and Press, USA, where he has worked since 1996. He is also the founder, editor, and publisher of *The Pauling Blog*, which released original research on Linus Pauling nearly every week for over 13 years. In addition, he administers the SCARC oral history program and has led more than 200 interview sessions comprising in excess of 300 hours of content. He previously co-edited *The Pauling Catalogue*, a six-volume set describing the Ava Helen and Linus Pauling Papers, which are housed at Oregon State University Libraries. His first book for World Scientific, *Visions of Linus Pauling*, recieved an Outstanding Academic Title award from *Choice* magazine in 2023.

Preface

Pauling at the Grand Canyon, circa 1940s.

This book represents the latest chapter of a story that I've been living for nearly 30 years. It's a tale that began in 1996, when I first started working in the Special Collections department at Oregon State University (OSU) Libraries. For the next decade, much of my time was devoted to organizing the Ava Helen and Linus Pauling Papers, an enormous archival collection that had been donated by the famed Dr. Pauling ten years before. After Pauling's death in 1994, his estate was gradually resolved and, by the time I showed up, most of the material had arrived in Oregon. In 2006 we completed the processing work, the product of which was released to the public in the form of a six-volume self-published catalog. During this period, we also became very interested in digitization for the web and ultimately created a series of documentary history websites and digital collections that, at present, consist of about 132,000 html files and supporting images.

In 2008 I created a blog, The Pauling Blog, which we primarily used to publish original research that we conducted on Pauling's life. Generally speaking, I selected the topics, identified the relevant sourcing within the Pauling Papers, and assigned the research and writing of blog posts to student employees whom I had hired to develop content for the project. Once they had completed their drafts, I would review their work, edit their copy, pick out illustrations and publish the posts, usually one per week. We continued in this fashion for more than 13 years, generally employing about 20 hours per week of student labor to support the undertaking. With the onset of the pandemic and the need to devote more energy to other tasks, the initiative gradually began to lose steam. By the time it wound up, we had published more than 865,000 words as authored by 38 different people.

In 2021 I received an email from World Scientific Publishing asking if I would be interested in "turning the blog into a book." I was flattered by the inquiry and most certainly interested but also knew that publishing the full contents of the blog would require around seven books, which neither I nor World Scientific were quite ready to do. But I was confident that we had circulated enough unique stories — either entirely new to the literature or far more in depth than had appeared in previous biographies — that it would be fairly easy to compile many of them into a volume of both scholarly and general interest. The publication that resulted, *Visions of Linus Pauling* (2023), received an Outstanding Academic Title award from the library journal *Choice*, and was commercially viable enough for World Scientific to extend an offer to prepare a sequel.

My first round of selections for *Visions* far outstripped the word count on offer, necessitating some painful decisions about what to cut. Two large batches of material were especially difficult to leave behind: a substantial number of posts that we had written on Pauling's relationship with his home state of Oregon, and a novel deep dive into his decidedly impactful tenure as chair of the Division of Chemistry and Chemical Engineering at the California Institute of Technology. In 2024, when World Scientific expressed interest in a second book, I knew right away that these two collections of writing would provide the foundation for what I wanted to put together.

And so we have *The Many Worlds of Linus Pauling*, a volume that plumbs the same source as *Visions* for its contents but does so in a somewhat different manner. Whereas the first book consists of discrete but thematically unrelated chapters presented in chronological order, *Many Worlds* is comprised of four parts that hint at a unifying concept: Pauling's relationship with physical, geographic and institutional spaces. Importantly, most of the chapters compiled for this volume are also mash-ups of many blog posts written by multiple authors and often published several years apart. As such, while *Visions* might be thought of as a curated collection of writings that generally bear a strong resemblance to their online companions, *Many Worlds* is more singular in its composition and is reflective of choices that I have made to interweave content originally developed at various times and for various purposes. This approach made for a more complicated editorial experience but has, I believe, resulted in a much richer end-product.

***

The Many Worlds of Linus Pauling consists of four parts: 1. Pauling in Oregon; 2. Caltech Administrator; 3. Period of Wandering; and 4. Travels. Its title is meant to evoke the theme that underlies and loosely connects the four parts, but it also works in the context of the many worlds theory that some physicists use to make sense of the quantum universe. By my understanding, the theory essentially posits that any event can and does have an infinite number of outcomes that result in differing lived experiences for all creatures, who themselves live in an infinite number of parallel universes. As a result, while in this universe I find myself writing a book preface in a comfortable space on a lovely spring day, it is a certainty that a different version of me is absolutely not doing this and, indeed, that there are an infinite number of universes in which I never existed at all.

In Pauling's case, certain key moments of chance and fate that proved crucial to his path are easy to identify and fascinating to consider. In Part 1, we tell the story of his early life largely through the eyes of his parents, Herman and Belle, both of whom struggled mightily with physical infirmity, financial hardship and, in Belle's case, debilitating mental illness. Young Linus was an obviously bright boy, but his material circumstances were such that a life working in a machine shop would have seemed far more likely than a career as an academic, to say nothing of the revolutionary impact for which he is known today (in this universe, at least). Likewise, in Part 2, one wonders about the institutional trajectory that Caltech would have followed were Pauling not a member of its faculty writ large and, in particular, not in a position to attract significant funding from the Rockefeller Foundation during the Great Depression.

Aspects of Pauling's peace activism are explored in Part 2, and so too are the consequences that Pauling faced as a result of speaking his mind. In Part 3, we stay connected to that theme by examining Pauling's consistently frustrating attempts to find a suitable organizational home following his resignation from Caltech. Pauling's eldest son, Linus Jr., once characterized this decade of wandering as a time where his father "felt that his creative scientific career was over," spurred by a belief "that creativity and science was a product of youth." Critically, it was during this period that Pauling made the chance acquaintance of a researcher named Irwin Stone, who turned him onto the potential health benefits of Vitamin C. Creatively adrift and institutionally unmoored, the enticing prospect of a high-profile project with major popular implications was precisely what Pauling needed at the time and it ultimately dominated the final third of his life.

Part 4 sheds light on the extent to which Pauling was truly a citizen of the world. In it, we learn more about the exquisite timing of his Guggenheim-funded travels through Europe to learn the new quantum mechanics from the ground up. We likewise uncover another reason why Pauling so spectacularly failed to correctly elucidate the molecular structure of DNA: he was crossing the Atlantic on the *Queen Mary* with his family and many other passengers including Erwin Chargaff, a biochemist who had some important new details to share about the genetic material but was too abrasive to be around for very long.

In addition, and as noted above, the structure of this book has created space for me to splice in a wide array of materials that add depth to the primary posts that serve as the foundation for each of its four parts. For instance, while much of Part 1 focuses on Pauling's early years, its basic premise, a study of Pauling's relationship with the state of Oregon, allows us to tell a few stories about OSU that we are especially well-equipped to share, given our access to the university's records. In one case, an examination of the life of Pauling's protégé Ralph Spitzer, I've also been able to tap into oral history work that was never actually published on the Pauling Blog.

Likewise, in Part 2 we use a survey of the history of the new Crellin Laboratory at Caltech to memorialize the tragic death of one of its staff members, Elizabeth Swingle, the victim of the type of accident that too often occurred for chemists of her era. In Part 3's discussion of Pauling's tenure at the University of California, San Diego we touch on his controversial engagement with eugenicist ideas. And in Part 4, through a study of his contacts with Albert Schweitzer, we see how Pauling's core belief in the minimization of suffering for all of Earth's creatures could have informed that later thinking on eugenics.

***

Because nearly all of the chapters in this book were written by multiple people, many of whom never met, I have decided against including bylines for its contents. (The original titles, dates and authors of all blog posts used in *Many Worlds* are itemized in its sourcing notes.) That said, there is one student author whom I would especially like to highlight here, followed by two other people who have cast a long shadow over all our Pauling activities.

The student author is Andy Hahn — also known professionally as Andre Michael Hahn — who worked for us on two occasions while studying for his Ph.D. in History of Science. Andy's contributions to the Pauling Blog were immense and are credited throughout *Visions* as well as *Many Worlds*. A careful researcher and skilled writer, Andy could be counted on to engage with topics that were largely beyond the grasp of most who worked in support of the project. I could point to many examples where this was the case, but perhaps the most impressive is Part 2 of this book, Caltech Administrator, which is largely a product of Andy's creation.

For some time, I had wondered about how the Pauling Blog might wrestle with Pauling's connection with Caltech, where he studied and worked for more than 40 years. A full accounting of this relationship would be a book in itself and was certainly beyond our capacity, but in thinking more about the question I became intrigued by several boxes of administrative records that had been organized into the Biographical series of the Pauling

Papers and had mostly gone untouched. They largely dated to Pauling's 21 years as chair of the Institute's Division of Chemistry and Chemical Engineering and felt like a potential goldmine for engaging with a major, but understudied, component of Pauling's career. I pointed these materials out to Andy and asked him to review their contents, instructing him to focus on Pauling's experiences as chair and encouraging him to write what he found. A few months later, Andy submitted a 42-page manuscript that we eventually published in 20 parts. In my view, it stands today as one of the most important contributions that the blog ever made and I'm very pleased now to be circulating it in this book.

***

In my Preface to *Visions*, I made note that the Pauling Blog had continuously stood on the shoulders of two giants, one of whom was Robert Paradowski. In 1972 Paradowski defended a mammoth (567 pages) doctoral dissertation, "The Structural Chemistry of Linus Pauling," and not long after began collaborating with Pauling as his preferred biographer. In the years that followed, Paradowski collected hundreds of hours of interviews with Pauling and others, and deeply engaged with both the Pauling Papers as well as the records of the Linus Pauling Institute of Science and Medicine, well before their deposit with the OSU Libraries.

One product of this work was "The Pauling Chronology," which was originally published in an obscure Japanese coffee table book (*Linus Pauling: A Man of Intellect and Action* — OSU's is the only copy cataloged in the WorldCat union database) and later updated and re-released on our website, first in English and then in Spanish and German. If one has an hour or two to spend, there is no better way to engage with Pauling's story. Indeed, Paradowski's chronology was literally required reading for all who wrote for the blog and is a resource to which we still routinely refer for grounding as we engage with the collection's primary sources.

And yet, as I write this in April 2025, there is relatively little else to show for Paradowski's efforts. His dissertation is without question the definitive account of Pauling's hugely influential early work, and a few later essays and lectures hint at his enormous (and elegant) capacity for analysis and interpretation of Pauling's incredibly complicated life. But there is no biography.

In February 2001, Paradowski visited OSU to review components of the Pauling Papers that had been recently arranged (we set up a makeshift office for him in a kitchenette adjacent to the reading room, so that he could continue sifting through boxes after hours) and to participate in a day-long symposium that we hosted to celebrate the centenary of Pauling's birth. At the conclusion of his talk, Paradowski gave a status report of sorts:

"...toward the end of his life, someone had sent [Pauling] a Bible with gigantic print, and he was having trouble with his vision but because this book had such big print, he was reading the Bible again [...] and perhaps he ran into this passage in Ecclesiastes and I'd like to quote it as my finish: 'Let us search then like those who must find and find like those who continue to search, for it is written the man who has reached the end is only beginning.' And that's the way I feel about my work on Pauling, no matter how much I do, it seems like I'm just beginning."

During the question-and-answer period that followed, an audience member returned to this topic and asked directly, when was the long-awaited biography going to be published? Paradowski responded,

"...my book is going to be called *The Alpha Helix*, the first book that I'll be publishing on the material I've gathered. But one reason I'm taking so long is that you people [at Oregon State University] keep getting more and more papers that I haven't been able to look at, and my mind keeps changing. As I read more papers I keep understanding things more deeply, and I don't like to make mistakes, I like to have things very accurate. And when I read some of the stuff that's been written and I notice a lot of errors, it bothers me. I know it's inevitable that there will be errors and mistakes, but I think truth is the daughter of time, and if you keep doing the work that eventually it's going to get clearer and clearer. It still bothers me that there are papers that are not yet here, that I'm not going to be able to see, and I've heard that Linus burned a lot of papers, and that bothers me too: that no one will ever see those papers. So one wonders what he burned and what was in those papers."

As one who has without question introduced his fair share of mistakes into the historical conversation about Pauling, I would encourage you, the reader, to seek out Paradowski's book, should it enter the marketplace. Until then, I find myself of like mind with Derek Davenport (1927–2017), a chemist and historian who authored a review of several works on Pauling for the journal *Chemical Education Today*. In his piece, which was published in September 1996, Davenport offered the following.

> "[Paradowski] has spent the last 20 years accumulating a unique store of knowledge concerning his subject. Since Paradowski writes well and has unusually catholic interests, the book — or rather books, since it will be in at least three volumes — should be well worth waiting for. But for how long? Paradowski tells me he definitely expects completion by the centennial year, 2001. I hope I live to read it!"

***

Regardless of what happens, we are blessed to have Thomas Hager's *Force of Nature: The Life of Linus Pauling* available to us. Published in 1995, the year after Pauling's death, *Force of Nature* is the best biography of Pauling written to date and was so commonly used by us at OSU that we just as often referred to it as "the Bible" as we did its actual title. At 720 pages, *Force of Nature* is detailed enough to provide solid footing for most of Pauling's story while still remaining accessible to the general reader. It's been a professional lifeline for me and many others for the entirety of my career and I'm very grateful that Tom took on the challenge of writing it.

Trained in microbiology, immunology and journalism, Hager spent much of his early career in communications, marketing and freelance writing. He lived in Eugene, Oregon and for many years was an employee of the University of Oregon, at one point serving as director of the university's press. My institution, OSU, is located about 50 miles away from Tom's, and for nearly 150 years the two schools have shared an intense rivalry. As such, Tom was the recipient of numerous Duck jokes from us Beavers and was quick to give as much as he got. Over the course of our many interactions,

Clifford Mead (Oregon State University Head of Special Collections), Linus Pauling and Tom Hager walking on the Oregon State University campus, 1991.

I came to know him somewhat well and considered him a highly valued colleague, mentor and friend.

Force of Nature was the first book that Tom authored on his own and I can only imagine what an intimidating prospect it must have been to wrestle with such a gargantuan figure for his inaugural solo effort. But the Pauling biography was a success and gradually Tom found himself in the rare position of being able to make a living solely through the written word. Subsequent books on synthetic fertilizers and sulfa drugs added both critical and sales momentum to his career, and he enjoyed a big commercial breakthrough with *Ten Drugs* (2019), which has been translated into at least 15 languages.

As Tom's work — and mine, for that matter — moved further away from Pauling, he and I had fewer occasions to correspond, but I always took comfort in knowing that he was there as a resource should the need arise. In February 2022 though, we received an email from his wife, the highly accomplished writer and professor Lauren Kessler, asking if we would be interested in adding a few more of his books to our collection. I've been an archivist long enough to know that messages of this sort are frequently an ominous sign, and that turned out to be the case in this instance too. Enveloped in the fog of the pandemic, Tom had quietly passed away from a rare form of lung cancer. No obituary was published, and it took more than a year and a half for his Wikipedia page to be updated to reflect his death.

Tom Hager (April 18, 1953 — October 15, 2021) was a dapper, generous man who wore an easy smile and was blessed with a twinkle in the eye not dissimilar from Linus Pauling's. I miss him and think of him often. This book is proudly dedicated to his memory.

***

In addition to the talented folks who researched and wrote for the Pauling Blog, many other people have contributed to the publication of this volume. For starters, over the course of my career I've been lucky to work under four department heads who have supported the devotion of significant resources to furthering the Pauling legacy. Like most cultural heritage organizations, slender budgets are a fact of life for university archives and the decision to spend a great deal of staff time on a single collection is not one that would ever be made lightly. As such, I'll be forever grateful to Cliff Mead, Larry Landis, Natalia Fernández and Julie Judkins for backing my continuing connection to the Pauling Papers.

Thanks are likewise due to my wonderful colleagues in the OSU Libraries Special Collections and Archives Research Center who, among other things, put up with my going dark every Friday for nine months to focus on writing and editing. Appreciation is similarly owed to my editor at World Scientific, Shaun Y.J. Tan, for dreaming up the endeavor in the first place, and to OSU friends and fellow Pauling enthusiasts Steve Lawson and

Linda Richards, for providing early support for both *Visions* and *Many Worlds*. A special shoutout as well to the Hallie Ford Center for Healthy Children and Families, where most of the work on this book was done.

More personally, I'd like to thank my parents, Mark and Sue, brother Scott, and mother-in-law Mary Lou for their care and support, and make special mention of my late grandmother Eve, a remarkable woman who was actually the first person to tell me about Linus Pauling. Finally and forever, my deepest affection, appreciation and love belong to Karen and Nora — without them, none of this happens.

Biographical Sketch of Linus Pauling

Linus Pauling with his infant son Peter, 1931.

Linus Carl Pauling was born to parents Herman and Belle in Portland, Oregon, on February 28, 1901. He was the first of three children, preceding sisters Pauline (1902–2003) and Lucile (1904–1992).

Pauling's childhood was in many respects difficult, marked by consistently hard financial times and punctuated by a moment of tragedy (see Chapters 1–4 of this book for more). Fortunately, in his early adolescence Pauling discovered a love for chemistry and became determined to pursue it as a career. When he was 16, he enrolled at Oregon Agricultural College (OAC), where he majored in chemical engineering and, just six months shy of graduating, met the woman who would become his wife, a home economics student and fellow Oregonian named Ava Helen Miller (Chapters 5–10). The two quickly fell in love and by the spring of 1922 Pauling had proposed marriage. Both of their mothers disapproved of this idea, and after completing his OAC degree, Pauling left for graduate studies at the California Institute of Technology while Ava Helen remained in Oregon. Early the next summer the couple wed, and Ava Helen moved to Pasadena with her husband.

Caltech proved to be an ideal location for Pauling. Under the guidance of mentor Roscoe Dickinson, Pauling mastered the techniques of X-ray crystallography, which he continued to advance in subsequent years to solve the crystal structures of a great many compounds. He completed his Ph.D. in 1925 and was immediately offered funding by the Institute to stay in Pasadena. Crucially, Pauling also received a Guggenheim Foundation grant to study in Europe during the 1926–27 academic year. Leaving one-year-old Linus Jr. behind in the care of Ava Helen's mother, the couple spent the year abroad, with the elder Linus focusing intently on making connections and absorbing as much as he could about the emerging field of quantum mechanics (Chapter 25). In 1927 Pauling returned to a faculty position at Caltech and began a period of momentous work, during which he started to apply the new physics to the classical understanding of how atoms form into molecules. In 1931 the first in a series of papers bearing the prefix "The Nature of the Chemical Bond" appeared in the *Journal of the American Chemical Society*. Over the next two years, six more of these papers were published, their findings summarized and expanded upon in a 1939 book, also titled

The Nature of the Chemical Bond. The work was quickly recognized as being of major import, and certain ideas that it put forward — including the theory of hybrid bond orbitals and a novel electronegativity scale — proved fundamental to the development of modern chemistry. In 1954 Pauling would receive the Nobel Prize in Chemistry for this body of "research into the nature of the chemical bond and its application to the elucidation of the structure of complex substances."

In the 1930s, Pauling increasingly shifted his interests toward biological topics, a reorientation that was in part inspired by the Rockefeller Foundation's desire to fund projects focusing on the "science of life." Pauling's initial concentration was the hemoglobin molecule, which he studied using the nascent technique of electron diffraction. As time moved forward, Pauling expanded this program to include groundbreaking investigations in immunology and the structures of other complex proteins. Two major breakthroughs came about in the 1940s, first with Pauling's discovery of the alpha-helix — a major structural configuration found in many protein molecules. (A rushed triple-helical proposal for DNA published in February 1953 proved far less successful.) Pauling also led a group that experimentally proved that sickle cell anemia is a molecular disease, the first condition to be so described. Likewise, as an outgrowth of his immunological investigations, Pauling showed that antibodies and antigens bond as a result of precise complementarity of shape, a trait called biological specificity. It was during this period as well that Pauling published the first edition of his seminal textbook, *General Chemistry*, thus beginning a long and fruitful relationship with the W.H. Freeman & Co. publishing house.

As with many other scientists, the war years forced Pauling to pivot from his established research agenda. Scientifically, Pauling turned his attention away from protein structures in favor of government contracts to improve rocket propellants and secret inks. Pauling was also charged with the development of an oxygen meter for use in aircraft and submarines, and artificial blood that might aid wounded soldiers in the field. The conclusion of the war proved to be of greater consequence to Pauling's story. Shaken by both the devastation and the implications of the atomic bombings of Hiroshima and Nagasaki, Pauling quickly felt drawn by a moral imperative to speak out against the proliferation of these new weapons of mass destruction.

Though generally interested in a great many issues emergent during the nuclear age, Pauling eventually found a niche as a leading voice against the testing of nuclear weapons in the atmosphere. Pauling was among the first to forcefully argue against the practice of above-ground test detonations, pressing his belief that the radioactive fallout unleashed by these tests — ever larger and increasing in number as the Cold War advanced — was contaminating the Earth in ways that would seriously damage human health. In 1957 he and Ava Helen decided to focus their anti-nuclear activism in the form of a petition that would be circulated around the world with the aim of affirming scientific consensus against above-ground nuclear testing. Ultimately signed by more than 11,000 scientists across the globe, the "Appeal by Scientists to Governments and People of the World" was formally presented to the United Nations in January 1958.

Pauling's bomb test petition did much to amplify the debate around weapons testing and, in the estimation of many, helped prompt the Partial Test Ban Treaty agreed to by the United States, the Soviet Union, and Great Britain in June 1963. Pauling received his second unshared Nobel Prize that December, "for his fight against the nuclear arms race between East and West." And though this and other decorations cemented his reputation as a hero for segments of the population, his outspokenness during a tense period in history also exacted a significant personal cost.

As his profile rose, Pauling was routinely dismissed as a Communist sympathizer by critics in the media and government. In 1952 his passport was temporarily revoked, and in 1958 he was twice called before the United States Senate Internal Security Subcommittee with orders to reveal the names of those individuals who had helped to circulate his petition. Faced with threats of imprisonment for contempt of Congress, Pauling refused to comply with the demand, and ultimately the subcommittee stood down. While he won that particular battle, episodes of this sort led to mounting criticism from conservative media commentators, some of which resulted in a spate of libel lawsuits filed by Pauling in the early- to mid-1960s. Pauling also gradually lost his footing at Caltech, as trustees and other donors came to take a dim view of his non-scientific pursuits. Pauling had been appointed chair of the Institute's Division of Chemistry and Chemical Engineering in 1937 and oversaw a period of significant institutional growth until 1958, when growing calls that he step down finally reached a crescendo. Following that, Pauling's office and laboratory space allotments were reduced. The final straw came with the Institute's decision not to formally acknowledge Pauling's receipt of the Nobel Peace Prize. By the time he returned from the award ceremonies in Oslo, Pauling had resigned from Caltech, his institutional home of more than 40 years.

What followed was a period of wandering that began with a move to a think tank, The Center for the Study of Democratic Institutions, located in Santa Barbara, California (Chapter 22). Though invigorating in certain respects, the center lacked scientific apparatus of any kind, and before long Pauling had moved again, this time to the University of California, San Diego (Chapter 23). After two years in La Jolla, Pauling relocated once more, to Stanford University (Chapter 24). It was during this nomadic phase that Pauling developed what would become a famous fascination with vitamin C. Spurred by a chance encounter with a biochemist named Irwin Stone, Pauling became increasingly convinced that chronically low concentrations of ascorbic acid in the bloodstream were a major contributor to suboptimal health for humans who, as with the other primates, are among a small handful of the Earth's species incapable of synthesizing their own vitamin C. As the 1960s progressed, Pauling applied this lens to concurrent interests in mental health, coining the term "orthomolecular psychiatry" to describe the treatment of disorders like schizophrenia with appropriately dosed vitamins and minerals. Later investigations were circulated among the public in books including the best-selling *Vitamin C and the Common Cold* and a later volume titled *Cancer and Vitamin C*.

Pauling's vitamin C work was looked upon with a skeptical eye by much of the medical community as well as many scientific colleagues. As the controversy surrounding his dogged advocacy of orthomolecular medicine mounted, and as his own interest continued to grow, Pauling increasingly felt a need for more concentrated institutional support. In 1973 he began the final chapter of his career by co-founding what would become the Linus Pauling Institute of Science and Medicine (LPISM), originally located in Menlo Park, California, and later in neighboring Palo Alto. For the remainder of Pauling's life, his namesake institute pursued an often sprawling research agenda, a core piece of which remained firmly connected to the original focus on vitamin C. LPISM did so amidst high-profile setbacks to its research objectives, chronic issues securing funds, and a series of personnel conflicts that resulted in costly periods of litigation. Pauling's burden was made all the heavier in December 1981, when Ava Helen succumbed to stomach cancer at the age of 77.

Linus would outlive his dear wife by another twelve-and-a-half years. During that time, he continued to speak out forcefully for peace and against nuclear weapons during the Reagan years, and likewise against the Persian Gulf War in the early 1990s. Scientifically, his vitamin C interests came to include investigations on its possible benefit for heart health. Pauling also engaged deeply with theoretical studies during this period, focusing in particular on quasicrystals, transition metals, and his own theory of nuclear structure, and spending many days in solitude, working through calculations at his secluded ocean-side ranch near Big Sur. At the end of 1991, Pauling was diagnosed with rectal cancer, an illness that gradually advanced over the next two-and-a-half years. He passed away at his home on August 19, 1994, aged 93 and survived by his children Linus Jr. (1925–2023), Peter (1931–2003), Linda (b. 1932), and Crellin (1937–1997). In a career spanning more than 70 years, Pauling published over 1,100 articles in the scientific and popular press, and was granted 47 honorary doctorates. Among many other awards, Pauling remains the only person to have received two unshared Nobel Prizes.

Contents

Part I
Pauling in Oregon

Chapter 1

Herman Pauling: Striving for a Better Life

Herman Pauling with young son Linus and infant daughter Pauline, 1903.

L inus Pauling's earliest documented ancestor was Andreas Pauling, born ca. 1630. Records indicate that Andreas' grandson, Johann Christoph Pauling, married and started a family in Preusslitz, Prussia. The Paulings remained there for at least two generations, until Johann Andreas Pauling (perhaps the grandson of Johann Christoph) move to Golbitz, in what is now western Germany.

In 1842 a son of Johann Andreas', Christoph Friedrich (born 1808), immigrated to the United States with his wife and two daughters. A son, Frederick, was born during the family's passage across the Atlantic, and two additional sons, William Frederick and Charles Henry (whom everyone called Carl), were born in the U.S. The Paulings settled as farmers in Concordia, Missouri, and Christoph Friedrich and all three of his sons would eventually fight on behalf of the Union during the American Civil War.

Charles Henry "Carl" Pauling — Linus Pauling's paternal grandfather — joined his father and brothers during the Civil War by enlisting in Company E of the 45th Regiment of the Missouri Volunteers. After the war, Carl met Adelheit Blanken, who had come to Missouri with her family from Germany. Adelheit gave birth to four children; the youngest, Herman Henry William, was born in 1876.

The following year, the family traveled to Biggs, California where they lived in a predominately German community. It was here where Herman's sister, Anne Charlotte, was born. Herman and Anne would be the only two of Carl and Adelheit's five children to survive into their twenties.

***

In 1882 the Paulings moved again, this time to Oswego, Oregon, near Portland, where Carl worked as an iron monger at the largest foundry west of the Rockies. It was likewise in Oswego that Herman began to attend grammar school. By the tenth grade, in 1890, he had grown tired of school and dropped out, talking his way into an apprenticeship with a local druggist instead.

In the late 19th century, medicine was not well regulated, leaving the door open for basically anyone to come along and make whatever claims they wished about the efficacy of their products. The druggist with whom Herman apprenticed taught him to avoid making wild statements and instead instilled the importance of the druggist's responsibility to his customers. This professional ethic was closely bound to the careful preparation of extracts, compounds, ointments, tinctures, oils and other products. By his 19th birthday, Herman felt confident enough to move to Portland, where he found work at one of the largest pharmacies in the city. This business, Skidmore Drug Company, employed Herman as a traveling salesman assigned to a 100-mile area ringing Portland; he covered his turf by horse and buggy.

After a financial depression hit the country in late 1893, Herman moved back to Oswego to run his own pharmacy. The economic shock of the time had led to the closing of Oswego's massive foundry, pushing many to leave the community, including the doctors and other druggists. Pretty soon Herman found that he was the only person left in town who could care for the sick, and his reputation quickly grew. The locals were grateful for his skills and his caring disposition — as instilled by his mentor — in addition to the low fees that he charged for his consultations. If he was particularly concerned about them, Herman would routinely visit his patients well after their original appointment, making sure that their health had improved.

Herman's profile spread back to Portland as well, and investors there saw him as the perfect person to open a drugstore in the small eastern Oregon town of Condon. By this time, regulations on drug sales had begun to stiffen, creating opportunities for those who were well-situated to meet the new standards and get in on emerging markets.

When he arrived in Condon in the summer of 1899, Herman's first impression was not a good one. The town center consisted of just six blocks along a Main Street that ran straight away into wheat farms and scrub desert. But by the fall, things were beginning to look up. Just as in Oswego, his solid ethics built trust and he was invited to several town functions as a featured guest. Calling himself a "manufacturing pharmacist," his store was founded on a "No Cure, No Pay" policy — that is, if the prescription didn't work for you, you were refunded in full. The town's weekly newspaper, *The Globe*, described "Doctor" Pauling as "a registered, reliable and experienced druggist."

***

Herman also began to attract the attention of some of the young women in town. Notably, Goldie Stephenson, the oldest daughter of one of the area's wealthier families, invited Herman to her house to meet her sister Belle. The two immediately hit it off and soon could be seen talking to each other at community dinners and dances. By Christmas, Herman had proposed to Belle and she immediately accepted. Herman had business to attend to back in Portland soon after, so had to be away from his fiancé for a brief time, but this gave him the opportunity to express his love in letters. In a valentine, he pledged,

"Dear love, when life's storms are raging fiercely, I offer you my arms as your protection, and you can trust in their fond yet firm embrace. When in after years the cares of home and motherhood bear upon your mind you shall find me ever an able assistant and benefactor."

Following that was a poem:

"A maiden fair with jet black hair/Her heart beats kind and true/She confides in me her every care/This maid with eyes light blue."

Herman and Belle married on May 27, 1900, at the Congregationalist Church in Condon. (Herman was raised as a Lutheran but was quite willing to adapt to Belle's Congregationalist upbringing.) Trouble struck quickly however as, only weeks after the wedding, the Portland investors who had backed Herman's initial move to Condon pulled out of the venture, leaving him scrambling for a new job. His best option appeared to be a clerk's position at a pharmacy in Portland and the newlyweds found a small apartment near Chinatown.

Once relocated, Herman worked as hard as he could to support his bride. Meanwhile, the entertainment and shopping options in Portland gave Belle new possibilities to explore that were unavailable in her native Condon. Belle was also pregnant when they arrived in Portland; Linus Carl Pauling, Belle and Herman's first child, was born at the end of February 1901. Herman was thrilled to have a son and worked even harder to provide for his growing family, but this left Belle alone much of the time to take care of Linus on her own. Only 19 years old, the young mother was drawn to the energy of the city, but her feelings of being stuck at home continued to grow as Pauline was born in August 1902 and Lucile on New Year's Day 1904.

When Herman was around, he would take Linus for walks around the neighborhood in an effort to wear out the energetic toddler. The two wandered through the local shops with the charismatic boy drawing the attention of merchants who taught him to count to 100 in Chinese, a talent that garnered little Linus a measure of celebrity as he displayed his skills for passers-by. Belle sometimes joined in on the walks, and the family often visited a nearby water fountain.

Meanwhile, Herman continued to look for a better job and eventually found work as a traveling salesman for the jeweler and druggist D.J. Fry, based out of the capital city of Salem. By October 1904, the family had left Portland and moved into their new home. For his job, Herman traveled up and down the Willamette Valley, sometimes 70 miles in one day, still using a horse and buggy. With Herman away even more, Belle was left to care for the children in an unfamiliar town, and she quickly made it known that she was not happy with their situation. In his correspondence with his wife, Herman continually emphasized a better future; one where he would own a drugstore and provide Belle with more luxury than she had thus far known in their married life.

***

But Salem clearly wasn't working, and Herman soon renewed his search for employment opportunities. In March 1905 he reached out to the Skidmore Drug Company, his old employer, and was disabused of any notion of a return to Portland. It was at this point that Herman once again set his sights on Condon.

With the help of Belle's brother-in-law, Herbert Stephenson, Herman was able to locate a storefront and a place for his family to live in the small eastern Oregon town. The main logistics settled, Herman wrote to Belle back in Salem, asking her to come join him so that they could start their life anew. Herman's letter also revealed a deeper, and indeed prophetic, motivation behind his relentless work ethic:

> "We cannot imagine what it is, but I feel that either ourselves or our children will someday stand before the world as a specimen of a high standard of intelligence."

***

Prior to the Paulings' return to town, The Northern Pacific Railroad had built a spur to Condon to help connect the area's abundant wheat harvests with the rest of the country. The favorable economic conditions that resulted led to a boost in population and, being the only druggist around, Herman benefited from this rising tide. He brought his store to the town's attention by placing large ads in the weekly *Globe*, announcing products like "Pauling's Pink Pills for Paling," "Pauling's Improved Blood Purifier," and "Pauling's Barb Wire Cure." He sold postcards featuring his son dressed up in buffalo skin chaps and captioned, "A Condon Cowboy," and on occasion he again indulged his inner poet, as with this advertisement:

> When sweet Marie was sweet sixteen/She used Pauling's Almond and Cucumber Cream/Tho' many winters since she's seen/She still remains just sweet sixteen.

Condon brought its challenges as well. The summer heat was hard on Herman, and Belle too had grown accustomed to the milder climate in the Willamette Valley. To escape the scorching temperatures, Belle and the children would travel to Oswego for long stays with her in-laws, while Herman remained behind to manage the store. By his second summer in Condon, Herman began conjuring up ideas to get out of eastern Oregon, but it took a few years and more struggles before that would come to pass.

In 1907 a financial panic spread from Wall Street to all corners of the country, including Condon, and soon made a negative impact on Herman's business. To help shore up his income, Herman partnered with a jeweler who promptly died of pneumonia the following year, leaving Herman to take over his role while also expanding into other areas, like selling eyeglasses. The contacts that Herman had made through his many civic organizations, including Woodmen of the World and the Odd Fellows, helped to keep the drugstore going amidst the panic, but staying afloat did not come easy.

Combined with his continuing need to engage in business travel, Herman's long hours at work placed more and more pressure on his marriage. Belle was outspoken about her disappointments, while Herman tried to keep their problems out of the range of the children. Frustrations were mounting though and in one letter, Herman let out what he had been holding back.

"I have quite enough to worry me without asking you to peck, peck, peck at me. But I guess you cannot help it, as that blessing is characteristic of the Darling family... Were it not for trying to get a start financially so you and the little ones may live in an abbreviated form of luxury in later years, I would not stay in this God forsaken hole a moment. You have discouraged me so often in my efforts that I would think you would eventually come to a conclusion to encourage me a little by discontinuing your nonsensical jealousy."

Ground down by the pressures of life, Herman's health began to suffer. He developed insomnia and what he described as a "tummick ake," a condition that would sometimes render him bedridden. More typically though, his stomach problems could be soothed simply by eating something. Armed with an easy method for treating his symptoms, he pushed along as best as he could.

***

Though he was unable to spend much time with them, Herman adored his children and sought to be a good role model, always hoping that the kids would grow up to be "an asset to the human race." He sometimes brought his son to work with him, instilling memories in the boy of watching his father carefully measuring ingredients for medicines, sometimes testing his compounds through various chemical reactions. According to biographer Robert Paradowski, young Linus also "encountered both cowboys, one of whom showed him the proper way to sharpen a pencil with a knife, and Indians, one of whom showed him how to dig for edible roots."

Understandably, Herman felt a need to keep an eye on his son, who could get into trouble when he was not at home or in his father's store. One day Linus was exploring a building that was in the process of being constructed. One of the workmen saw him there and was angered to have him in the way. Linus tried to climb out a window, but the worker caught him before he could escape and ended up giving the boy a beating.

When Linus came crying to his father, Herman immediately went out, found the offender, and punched him to the ground. As Linus later remembered, his father was arrested soon afterwards. Though he had associated the arrest with assaulting the workman, according to biographer Thomas Hager, it was more likely tied to charges of bootlegging. (Herman was eventually cleared of any wrongdoing.)

While most of the town appreciated their pharmacist, some locals felt otherwise. After Herman had organized a Fourth of July celebration and run promotions in the program for the day's baseball game, the other jeweler in town wrote a letter to the editor of *The Globe* attacking his competitor for his advertising tactics. Herman did not stand for this public affront and retaliated by writing a lengthy response, titled "The Truth Will Out," that described how "Sorehead Charlie" was being unfair and noting that, in business, having enemies is helpful, but not at the cost of fair play. This public exchange went on for several weeks until both sides eventually calmed down.

Amidst it all, Herman and Belle became increasingly eager to get out of Condon. Herman was especially tired of having to deal with the recurrent diphtheria and whooping cough that had afflicted a few of the area's children. Belle, as had been the case since they first married, wanted Herman to work less and frequently complained about the amount of time that he devoted to the pharmacy, which often ran to 14 hours per day. In early 1909, the final catalyst for a move came about in an unexpected fashion, when Herman's store caught fire. The local firefighters who responded to the blaze wound up causing even more damage by breaking the store's front window and the glass figurines that were on display. Badly shaken, Herman focused intently on relocating and, by the fall, had saved enough money to move his family back to Portland.

***

Now returned to the big city, Herman spent some time at the Skidmore Drug Company before opening a new store on the growing east side of town. Herman took a different approach with this start-up and got out of most of the extra lines of merchandise that he had sold in Condon — jewelry, phonographs and the like. (He did, however, add a soda fountain.) But business was slow. In his down time, Herman kept his mind occupied by studying German — Belle had been enrolled in classes at the local high school, and Herman thought it would be fun to join her as she learned. It was also a useful tool for looking busy when customers came in.

Herman likewise continued to encourage Linus's growing curiosity by teaching him Latin to help supplement his budding interest in ancient civilizations. In May 1910, Herman also wrote to the editor of *The Oregonian* newspaper asking for advice about books. Titled "Reading for a Nine-Year Old Boy," the letter concluded as follows:

"In order to avoid the possibility, or probability rather, of having someone advise me to have him read the Bible, I will state that it was through reading this and Darwin's theory of evolution that my son became so interested in both history and natural sciences."

The editor responded by suggesting Plutarch, Herodotus, and Thomas Arnold's *The History of Rome*.

Sadly, Herman did not have much time to follow up on the suggestions. On June 11, while the rest of the family, along with Belle's sisters, were at the Portland Rose Festival, Herman was at the pharmacy when he started to get one of his stomachaches. He went home and ate some of the roast that Belle had prepared for dinner which, as usual, helped to reduce his symptoms and allowed him to go back to the store.

But the pain quickly returned and ferociously so. This time, Herman collapsed and had to be carried home where he lay until his family arrived. After seeing his wife, son and daughters one last time, Herman passed away, leaving Belle to care for their three young children. He died at the age of 34, the victim of a perforated ulcer and attendant peritonitis.

Chapter 2

Belle Pauling: Hard Times

Belle Pauling, early 1900s.

The Darling family history is a rough and tumble one, indicative of the pioneer environment in which Linus Pauling was raised. Brothers Dennis and John Darling, the family's earliest-studied ancestors, settled in Braintree, Massachusetts, just south of Boston, sometime around 1660. Records are slight for at least one or two generations, but it is known that John R. Darling, born in 1750, established himself as a farmer in the mid-Hudson Valley region of New York state. We also know that John R. Darling was a Tory loyalist who, in response to the outbreak of revolution in the American colonies, moved his family to the Bay Quinte region of Ontario, Canada, where he remained until his death at age 98.

The Darlings resided in eastern Canada for some time. John R. Darling's son William Darling, born in 1800, married and raised eight children in the Bay Quinte region. One of William's sons, William Allen Darling (1826–1900), would become Linus Pauling's great-grandfather.

William Allen Darling married in 1850 and fathered six children in ten years. He left both eastern Canada and his family behind in 1863, bound for Chicago, apparently for purposes of fighting in the American Civil War on the side of the Union. During this time, he cut off communication with his wife, who eventually presumed him to be dead. Four years later he married again — a bigamous marriage, technically — and settled on the western shores of Lake Huron in Tawas City, Michigan. He and his second oldest child, William Allen III, would later relocate to the Pacific Northwest.

Around this same period, Darling's first wife died and all six of her children were moved into foster homes. One of those kids, Linus Wilson Darling (1855–1910), ran away from his foster parents in New Jersey, worked for a spell as a mule driver on the Erie Canal and, at the age of 15, found his way to Chicago, where he both worked and lodged in a bakery, sleeping in a large bakery barrel. A year later he left Chicago and, for two years, roamed westward. He eventually settled in western Oregon, near Salem, and found work as a schoolteacher. One of his students was Alcy Delilah Neal. The two began courting and were married in 1878. As they moved around Oregon, looking for a place to settle, they began having children. Their second daughter, Lucy Isabelle "Belle" Darling, was born on April 13, 1881, while the family was living in Lonerock, a tiny village in eastern Oregon.

In the years immediately following, the family arrived on hard times and were, in fact, facing starvation. But they were saved when Linus bet his saddle against fifty dollars on Grover Cleveland to win the upcoming presidential election over James Blaine. Funded by those winnings, they moved 20 miles northwest, to Condon, where Linus opened the town's first general store, selling patent medicines and running the post office.

***

Before it was officially settled, the Condon area was known as Summit Springs, a point at which offshoots of the Columbia River converged underground to form a freshwater spring used by Native Americans and white settlers alike. In the mid-1870s, William Potter, a shepherd, built a home at Summit Springs and others soon followed his lead. The town was platted in 1879 and, in 1884, mail service was established by Harry C. Condon, a young attorney from nearby Alkali, Oregon (now known as Arlington) who loaned his name to both the post office and the fledgling community. In 1890, the town became the official seat of Gilliam County and, in 1893, it was incorporated.

As early as the 1880s, the prairie around Condon was used for grazing cattle and sheep. But as the town grew and developed a degree of permanence, settlers began tilling the land for small-scale farming operations. Willing to test their luck with Condon's semi-arid climate, the locals found the richness of the soil — fortified with volcanic ash — and access to fresh water ideally suited to their agricultural pursuits. Wheat and barley quickly developed into staple crops and, during the early 1900s, Condon became known as the Wheat City, shipping more of the grain than any other community of its size.

***

Three years after moving to the town, when Belle was seven, her mother Alcy gave birth to a stillborn son. Badly injured by the traumatic birthing process, Alcy herself passed away one month later.

Linus, however, continued to lead a busy life. Now a beginning law student, he occasionally hired a woman to help take care of the household, but mostly left it up to his daughters, Goldie, Belle, Lucile and Abigail — all born in a span of five years — whom he called the "Four Queens." (A fifth queen, Florence, was at the time too young to pitch in.) The bulk of this work often defaulted to the oldest daughter, Goldie.

Linus eventually remarried, finding a younger widow who owned a large wheat farm and had an extra ten thousand dollars to her name. This newfound financial support allowed him to become a gentleman farmer and begin his law practice.

***

For Belle, this period was one of contrasts. Her mother gone, the household responsibilities mounting, and her father mostly a distant figure, she began suffering from bouts of depression. At the same time, her new family's wealth and large home made her one of the Condon elite.

Indeed, the family's resources offered some measure of protection from the dark economic times plaguing the country, and the Darlings would occasionally take shopping trips to Portland for dresses and other fineries unavailable closer by. In 1895, Belle and Goldie also took advantage of the opportunity to attend boarding school at Pacific University in Forest Grove, Oregon. Belle did well academically, earning marks of 85 in Bible, 91 in grammar, and 98 in arithmetic, but she did not enjoy the boarding school experience and came back to Condon after her first term. Her older sister Goldie was able to make it through the rest of the year before coming home.

Returned to Condon, Belle re-assumed her status as one of the town's most promising young women. One day, in the fall of 1899, Goldie invited the 17-year-old Belle over to her home to meet Condon's new druggist, Herman Pauling. The pair were quickly enamored of one another and by Christmas they were engaged to be married the following May. Their lavish wedding was attended by much of the town.

The fairy tale ended quickly, however, when the investors backing Herman's Condon pharmacy pulled out. Since the area was not big enough to support Herman on its own, he and Belle moved to Portland, finding a residence near Chinatown. Belle was eager to explore the city, but she had discovered that she was pregnant well before she and Herman left Condon and, on February 28, 1901, she gave birth to their first child, Linus, named for Belle's father. Linus was quickly followed by his first sister, Pauline, born on August 2, 1902. Frances Lucile (or "Lucie," named after one of Belle's favorite poems by Owen Meredith) arrived on January 1, 1904.

***

When Herman found a job as a traveling salesman, Belle was frequently left alone to care for the children. Her unquenched desire to enjoy Portland, her absent husband, and her mounting responsibilities as a parent all combined to foster a growing resentment. Frustrated, she repeatedly wrote to her husband while he was on the road, admonishing him for not making enough money and spending so little time at home. Herman usually responded that he was doing all that he could to provide for Belle and the children, and that their future as a family would be brighter.

In 1904, attempting to arrive at this brighter future, Herman took a new job — one that still required travel but was now based in Salem, some

45 miles south of Portland. Herman's hope was that the change would afford him more time at home since Salem was more centrally located in the Willamette Valley. The job did not last long though and, before long, the family returned to familiar territory: Condon.

In April 1905, the Paulings settled into their new home above a general store that belonged to Goldie's husband, Herbert Stephenson. Here, little Linus was free to roam around the town, while Belle stayed at home looking after her daughters. During the harvest period, Belle would also help at the Stephenson wheat farm by cooking for all the temporary workers brought in to process the grain.

Moving to Condon meant that Herman was always close by and that he generally made enough money to meet expenses. Though Condon was not immune to the financial problems afflicting the rest of the country, it was able to bounce back somewhat quickly, in part due to a rising population of homesteaders taking advantage of the 320 free acres being offered by the federal government. Nonetheless, Belle was not happy in her hometown: Herman was still working over 12 hours a day, she missed the culture and excitement of Portland, and the summers were unbearably hot.

The latter two issues were solved, at least in part, by annual summer trips to the milder Portland suburb of Oswego, where Belle and the children stayed with Herman's parents. And though these visits allowed Belle to take in events like the centenary celebration of the Lewis and Clark expedition, she did not like being away from Herman. Letters to her husband sent during this period are full of anxiety over Herman's fidelity and the family's financial situation. By the end of the summer of 1909, the couple agreed that the time was right for a return to Portland.

Once done, Belle's spirits improved, but the happy times were not destined to last. In April 1910, clouds began to gather when Belle's father passed away. While Belle had never been particularly close with her dad, his death was still difficult for her. Three months later, Belle, her sisters, and their children returned home from the Portland Rose Festival to find Herman bedridden and in tremendous pain. Herman's stomachaches, the result of an ulcer, were a recurrent issue for him, but they had never struck so severely. This attack proved fatal; Herman died soon after their arrival.

***

Emotionally devastated, financially imperiled, and widowed with three kids, Belle Pauling found herself at a major crossroads at the young age of 29. After initially seeking out some measure of continuity by hiring a manager for her deceased husband's drugstore, Belle was soon forced to sell both the store and her house in order to buy a larger home. Her intent in doing so was to take on boarders, one of the few income possibilities open to her, given that all three of her children were still under the age of ten.

Having never overseen the family finances, Belle made some initial mistakes by overpaying for the house and, in hopes of offering an attractive space for tenants, outfitting it with several high-end appliances. She also hired someone to cook and clean. Burdened by these expenses, Belle chronically teetered on the edge of being able to make enough to support herself and the children.

Belle's life began to crumble in other ways as well. Her years-long battle with depression only worsened and she was soon diagnosed with pernicious anemia, which left her physically weak. At first, she refrained from telling the children, as she did not want to frighten them with the possibility that they may also lose their mother. Left to care for the children by herself, Belle let them run free around the neighborhood well into the night. She was also unprepared to deal with her maturing daughters. As Pauline later recalled, Belle made her and Lucile feel that menstruation "was a scourge that afflicted only women in their family."

Belle found some relief through weekend visits to Herman's parents in Oswego, times where they could all reminisce about the days when her husband and their son was still living. She also fulfilled one of her lifelong dreams of owning a piano. She and the children began taking lessons and would have singing parties. Lucile, who was the most musically inclined in the family, would play while Pauline joined on ukulele and Linus sang. In 1913 the family was also able to take their one and only vacation, a trip to the Oregon coast where they stayed in a friend's house.

***

Around the age of 12, Linus started helping out financially by taking on an array of odd jobs. He didn't particularly enjoy working, but Belle increasingly came to depend on him for income, so after quitting one job he would soon have to find another.

Belle's reliance on her children, especially Linus and Pauline, began to wear on them. After Belle had to let the cook and maid go, she leaned even more heavily on the kids to take care of the house. Linus later remembered his mother "issuing requests and orders and browbeating the children, often from her bed."

On one occasion, Lucile had taken a boy's bike for a ride without his permission. Once she was finished, the boy punched Lucile to the ground. When she came home crying to her brother, imploring him to go beat up the boy, Linus refused. Belle strongly disapproved of Linus's decision and admonished him as a bad brother. Perhaps unsurprisingly, as he reached adolescence Linus distanced himself more and more from his mother.

As Pauline reached her teenage years, she also came under mounting pressure from Belle. Rather than look for a job to bring in income, Belle wanted Pauline to marry someone wealthy and when Pauline was 17, Belle found just the man for her. Pauline did not take well to her mother's prodding, particularly since the man was 30 years old. Eventually Belle's persistence became unbearable, and Pauline called the police, telling them that her mother was forcing her into marriage with a much older man.

***

In the fall of 1917, Linus enrolled as an undergraduate at Oregon Agricultural College (OAC) in Corvallis, about 90 miles to the south of Portland. Belle's nephew, Mervyn Stephenson, was a junior there and she arranged for Linus to live in the same boarding house. Had her nephew not been there, Belle would not have permitted Linus to attend college, and she continued to believe that he should keep working to help support the family. She rode the train down with Linus to stay with him for his first night at the school to make sure that everything was in order. After she left, any pretense that her nephew would watch over Linus disappeared; Mervyn offered his cousin some advice on getting by as a freshman and after that the two mostly went their separate ways. Linus moved out of the boarding house soon afterward.

With Linus away and not working as much, Belle began a courtship with William Brace Bryden, a lumberman. In a manner similar to her first encounter with Herman nearly 20 years earlier, Belle's sister Goldie had set the two up on a blind date. By June 1918 they were married, but it did not last long; Bryden was not helpful and neither Pauline nor Lucile liked him. When Belle came down with the flu, Bryden offered little assistance, and after Belle recovered the two argued constantly. One day in September, Belle's second husband left to go to the barbershop and never came back.

By the spring of 1922, Linus was ready to step away further from his mother's influence. He and his sweetheart, Ava Helen Miller, were engaged to be married and Linus had solidified plans to attend graduate school in southern California, where he hoped Ava Helen would join him. But Belle and Ava Helen's mother quickly intervened; both thought it best that their children finish their education before getting married. The young couple gave in, and Linus went to California alone.

***

With Linus now 1,000 miles away, Belle kept him up to date on life in Portland, and the news was often not good. As her health continued to worsen, Belle began to see a growing array of doctors. One had her fast for 72 hours and then stayed with her for another day as he administered one quart of olive oil over four doses. Belle told her son, "It is a wonderful treatment, takes all the poisons out of your system. I feel like a new woman. I am weak yet but will soon feel strong." (Pauline also remembered prescriptions for raw liver on bread, which the children called "cannibal sandwiches.") The positive effects rarely lasted long though and Belle continued to seek out different treatments, sometimes against Linus's wishes.

In October 1922, Linus sent Belle a letter trying to get his debts to her in order. This shocked Belle as she could not understand why Linus felt that he owed her, preferring instead that he continue to support the family out of purer motives. Nonetheless, her response could not have helped but place further strain on their relationship:

> "Do you think because I have let you help carry the burden this year that you are repaying me for money I gave you for your education or the cost of your living since you were born or perhaps pay me for the pain I suffered in bringing you into this world? … You are helping the girls and not me personally… I have never worried when I had money to give you but I have worried a lot because I couldn't help you more."

Part of Linus's money was going to support Pauline's college tuition — she had also gone on to OAC, leaving Belle and Lucile alone in Portland. Fortunately, Belle and her youngest daughter got along relatively well and the two sometimes traveled together or spent days shopping downtown. Both were also members of the Order of the Eastern Star; Belle was able to join earlier only because she had made Linus join the Masons. At the meetings, Lucile sometimes sang, which pleased Belle even more.

Linus, however, continued to disappoint his mother. One year, while in Pasadena, he had forgotten her birthday. Belle did not take this well, writing,

"I look around me and I see lots of young men who have mothers (and fathers too) who are lovely to their mothers. I tell myself over and over that you do not mean to be unkind but even so, such a situation is very depressing."

Linus and Ava Helen also began to make new plans to marry, and Belle expressed her hope that the ceremony would be held near to her. Linus, though, did not include Belle in the planning, giving her yet another cause for concern.

Linus made amends by sending his mother a birthday card and a letter that made Belle "feel so much better." Ava Helen also tried to bring her some comfort by offering to visit her, but Belle claimed she was too busy as she had four boarders to attend to at the time. At long last, in June 1923, Linus drove from Pasadena to Salem, where he and Ava Helen were married. The two then went to Portland to stay with Belle briefly before going down to California for Linus to continue his schooling. They returned in the summer of 1925 with their first-born child, Linus Jr.

***

By that time, oldest daughter Pauline had married the athletic director of the local Elks Club and the two had moved to Los Angeles. This devastated Belle, causing her to collapse. From then on Lucile was the only one around to care for Belle, whose health continued to deteriorate. Belle began to suffer bouts of delusion and loss of feeling and movement in her limbs. She left many of the household decisions and responsibilities up to Lucile, something for which the 19-year-old was not well prepared.

In March 1926, Linus and Ava Helen came to visit on their way to Europe. They needed someone to take care of Linus Jr. while overseas and had arranged to leave him with Ava Helen's mother. When Linus saw his own mother, he was shocked by her appearance — her gray hair and poor balance were clear indications that she was not doing well. He was tempted to cancel the trip, but ended up going, leaving some money behind to help pay the mounting bills associated with Belle's care.

Two weeks after Linus and Ava Helen left, Belle sold her home to Lucile for ten dollars, who then rented it out so that she and her mother could move to a smaller apartment. But Belle's condition only worsened as she became increasingly restless and had trouble sleeping. Her moods also grew more volatile, ricocheting from suspicion to happiness to fear. Eventually Belle's behavior became so unmanageable that Lucile called her aunt Goldie for help. The two decided that it would be best if Belle were moved to the state mental hospital in Salem. Her admittance form summed up her difficult life:

"Natural disposition? 'Moral character good. Disposition happy. Lost husband 16 years ago — raised family through great struggles.'"
"First symptoms of mental derangement? 'Worried from illness and too much responsibility.'"

Lucy Isabelle (Darling) Pauling passed away in Salem on July 12, 1926, at the age of 45.

Chapter 3

Pauline and Lucile Pauling

Lucile, Linus, Belle and Pauline Pauling, 1922.

"My name is Pauline Darling Pauling Stockton Ney Dunbar Emmett, and you can see I've had an interesting life..."

— *The Story of Pauline Darling Pauling*,
by Pauline Emmett, December 31, 1994

The sister of one distinguished scientist and later the wife of another, Pauline Darling Pauling, the second oldest of Herman and Belle Pauling's children, led a long and eventful life. Once a record-breaking typist, a women's athletic director, and a successful designer and businesswoman, Pauline found fulfillment in a wide-ranging career and a number of hobbies. Although she remained close to her famous brother over his lifetime, Pauline harbored more artistic aspirations than scientific ones. In addition to her professional successes, she was a seamstress, quilter, painter, and collector of coins and dolls.

Pauline Pauling was born in Portland, Oregon on August 2, 1902. She remembered her childhood as "very stark," remarking that "it was a wonder [the family] survived." An extroverted, energetic and pretty girl, Pauline became something of a socialite as a teenager. She dated a string of boys, frequently attended swimming and singing events, and often arranged social get-togethers. While a student at Franklin High School in Portland, Pauline dropped out for a year to attend the Behnke-Walker Business School. There she learned Pitman shorthand and the touch system of typing. She would later become known for her speed typing, breaking records in an unofficial test.

She met her first husband, Wallace Stockton, while working as a secretary for the Elks Club in Portland. The couple later moved to Los Angeles, where Pauline worked as the Women's Athletic Director for the club. Known as the "Elkettes," the group attracted some of Hollywood's most famous stars and gained considerable publicity for their numerous activities and events.

Pauline and Stockton divorced in the late 1920s, and on October 6, 1932, Pauline married Thomas Ney. After first living in Santa Monica, the two moved to Inglewood, California, where their son Michael was born on December 23, 1934. It was around this time that Pauline took notice of a men's slipper in an issue of *Vogue*. Using a pattern, Pauline refined the design to create a women's slipper. Soon after impressing her friends with the prototype, Pauline began making and selling the slippers from home. Her initially modest business, Paddies, Inc., grew rapidly and before long she was marketing the "Paddy" slipper to upscale department stores like Saks Fifth Avenue, Macy's, Neiman Marcus, and I. Magnin. Unfortunately, Japanese manufacturers were able to copy her design and flood the market with a cheaper model. Pauline eventually lost her big accounts and sold the company.

In 1950 Pauline and Ney divorced. After returning to Santa Monica, Pauline became interested in numismatics and, in 1960, she opened her own coin shop. It was during this time that Pauline became acquainted with Charles "Slim" Dunbar, a fellow coin shop owner from Inglewood.

The two were married on August 25, 1973. Sadly, Slim, in ill health, died just 23 months after their wedding.

Following Dunbar's death, Pauline returned to Oregon. It was there that an old acquaintance, Paul Emmett, re-entered her life. A prominent catalysis scientist, Emmett was a longtime friend, colleague and even roommate of her brother. Once they had made contact, Pauline recalled that Emmett was "underfoot every minute until [she] accepted his proposal." The two were married on May 22, 1976.

Pauline, lively even in her later years, cared for Emmett until he died from Parkinson's disease in 1985. Following her fourth husband's death, Pauline remained in the Portland area until passing away on October 19, 2003. She was 101 years old.

***

Less is known of the youngest Pauling sibling. What can be said is that, of Herman and Belle Pauling's three children, quiet, shy, and warmhearted Lucile was the least hardheaded. Always unsettled when trouble arose between family members, she often took on the role of peacekeeper in the Pauling family.

Frances Lucile Pauling was born in Oswego, Oregon on New Year's Day, 1904. Though both her brother and sister recalled harsh memories of their upbringing, Lucile would remember a happier, more normal childhood, despite circumstances that might suggest otherwise.

After both Linus and Pauline left home when they were just teenagers, Lucile stayed behind to care for their ailing mother, help look after the boardinghouse, and work as a cashier at the Meier & Frank department store. Carrying a heavy burden, Lucile found great comfort in music. An accomplished pianist, she studied and taught music lessons when she wasn't helping her mother.

By 1926 Belle's health was quickly deteriorating, and she was ultimately admitted to the Oregon Hospital for the Insane. When visiting for the first time, Lucile was so overwhelmed by the sight of her mother in the mental ward that she tearfully begged that Belle be removed. Unfortunately, Lucile's request could not be fulfilled and Belle, at the age of 45, died just weeks after being admitted. Overcome by grief, Lucile wrote to her brother, "I left decisions, [Belle's] care, everything, up to others, being absolutely immature and irresponsible, and easily led." Many times after, Lucile would express regret over not having known how to better care for her mother.

In 1937, while working as a secretary in Portland, Lucile married Lemuel Lawrence Jenkins, known to most as "Jenks" and described by Lucile as "a Minnesota man that was just down to Earth." Their son Donald was born soon after — a second child died young — and the family settled in Estacada, Oregon. Lucile cared for the family home while continuing her music studies, teaching lessons and accompanying local musicians on the piano. Due to Jenks' "restlessness," the couple moved eight times within the town of Estacada, finally finding the right place, outfitted with "a farmhouse, 80 acres with a pond," and a nine-hole golf course that Jenks built himself.

Though she was keenly interested in her family genealogy, scant documentation of Lucile's own golden years remains. In a 1990 interview, she reflected that "I didn't stand on my own two feet until I was about 50 years of age." After Jenks passed away in 1963, she did not remarry and was plagued by health problems. A sufferer of heart disease, Lucile died on January 19, 1992, of ventricular fibrillation. At 88 years, her life was the shortest of the three Pauling children — Linus lived to 93 and Pauline to 101 — a noteworthy fact given the short lifespans of Herman and Belle Pauling, who lived to ages 33 and 45 respectively.

Chapter 4

Pauling's Adolescence

A rare photo of Linus Pauling during his adolescence, ca 1910s.

"*When I was [young], I thought that it was proper, something wrong, if I didn't smoke cigarettes; so I smoked a few cigarettes. But fortunately I was so poor that I didn't have money enough to buy them, so I got through the danger period as a result of poverty. It was a fortunate thing; I might well have developed this drug addiction, as the fellows call it.*"

— Linus Pauling, 1960

Early 20th-century Portland, Oregon was largely seen from the outside as an untamed and uncivilized part of the country, full of opportunities and dangers. Aside from an impoverished and rather seedy Japan Town and a thriving red-light district, the city was an industrial center and little else. Due to its convenient access to Oregon's primary rivers, which in turn provided a direct line to the Pacific Ocean, Portland became a hub for the state's shipbuilding and logging sectors. Oregon's booming timber industry single-handedly supported much of the state's economy, providing work in mills, producing lumber for shipbuilders and helping stoke the fires of Portland's fledgling steel industry.

Linus Pauling spent his teenage years in this burgeoning industrial community, immersed in a culture of blue-collar labor and near-poverty. Having lost his father at the age of nine, he spent his youth in pursuits appropriate to his surroundings; dreaming up get-rich-quick schemes and fantasizing about life as a successful corporate chemist. He was the product of his environment — bright, hardworking, and a capitalist to the core.

In the summer of 1913, the Pauling family took their one and only vacation. Mother Belle's chronic illness, pernicious anemia, kept her bedridden much of the time, and to help alleviate the symptoms her doctor suggested a trip to the Oregon coast, where she could get away from the stresses of home and benefit from the sea air. Unfortunately for Linus, the summer by the sea was not to be one of uninterrupted relaxation. Upon arrival at their vacationing spot, he was sent out in search of employment and soon found a job at a local bowling alley where he set pins for other tourists.

After returning to Portland that fall, Pauling was again obligated to earn a wage. This time, he was hired at a local movie theater where he worked as a projectionist, switching out reels and monitoring the film. Pauling, who was too poor to attend the cinema as a paying customer, enjoyed the work because he was able to watch all the newest films. But in time, this too lost its novelty and, as the family slid further into debt, Linus was expected to contribute increasing amounts to the household income.

When Pauling turned 13, his mother purchased him a bicycle so he could work as a delivery boy. He found employment delivering milk, newspapers, and even packages for the postal service. While the jobs offered Pauling a chance to tour Portland and gave him plenty of exercise, they were also dull and, by and large, he resented them. His negative feelings were amplified by the fact that he wasn't allowed to keep his earnings but was instead compelled to hand them over to Belle.

***

Linus's chief refuge from the grim realities of his family life was his intellect. Crucially, at the age of 13, he was first introduced to chemistry by his best friend, Lloyd Jeffress. After watching Lloyd demonstrate a few simple chemical reactions with a homemade chemistry set, his own destiny was set. Linus had previously built a small room in the basement of his mother's boardinghouse to house his mineral collection, and this space quickly doubled as a laboratory. Soon he was collecting chemicals and supplies with which to conduct his own "experiments."

Pauling's Oswego grandparents lived close to the newly built Portland Cement Company, and on weekend visits, the young teenager frequently went to the plant's laboratory and bombarded the chief chemist with questions. Many years later, Pauling remembered this patient individual as "a man who was not very interested in chemistry, but who served as scoutmaster and who was willing to talk with me and to answer my questions."

By this time, Pauling had aspirations of becoming a corporate chemist, as it was well known that companies across the nation were paying high salaries to trained chemists capable of developing saleable products. In a fit of ambition, Pauling and his friend, Lloyd Simon, decided to put their own knowledge of the sciences to work: the two teenagers opened a photo-developing business, purchasing expired materials at bargain prices from a local supplier. Referring to themselves collectively as Palmon Laboratories (they even had business cards made), the boys went door-to-door in search of clients. A few kindly shop owners in the neighborhood agreed to send their business to the young entrepreneurs, but unfortunately prosperity was not close at hand. Between their lack of experience and the poor quality of the materials they had purchased, the developed photos were unusable. The venture was a disaster and, even worse, both boys had sunk most of their savings into it. Linus was poorer than ever.

***

In the meantime, he was starting high school. In February 1914, right before his 13th birthday, Linus entered Washington High (WHS), having finished an accelerated primary school program.

The second oldest school in Portland and formerly known as East Portland High School, Washington was renamed in the early 1900s, and later rechristened Washington-Monroe High School. Though it would eventually close in the early 1980s because of declining enrollment, during Pauling's time it was an excellent choice for a young boy keen on learning all he could.

In his first semester, Linus signed up for a standard course load consisting of elementary algebra, English, Latin, and gym. After the summer, he returned to WHS for the full year and took his first actual science class, physiography, where he learned about minerals. He continued to collect rocks, and although it never grew to be particularly large, he enjoyed analyzing and classifying his stockpile.

Before long, Linus was delving into a more difficult curriculum. On top of his normal classes, he continued with Latin and eagerly enrolled in every science and math course that was offered. Mathematics and the sciences quickly became his favorite subjects, because, as he later recalled:

> "It's like the story of the little boy who, when his teacher asked him, 'Willie, what is two and two?' answered, 'Four.' And she said, 'That's very good, Willie.' And he said, 'Very good? It's perfect!' I liked mathematics because you could be perfect, whereas with Latin, or in studying any language, it's essentially impossible to be perfect."

As his high school career progressed, Linus easily maintained his studies. In fact, high school never presented any significant challenge to him, academically. This was fortunate, because he needed all of his spare time to work his various jobs, and also to feed his growing appetite for chemistry.

Although chemistry quickly became Linus's main interest, he wasn't able to take many classes on the subject. He took first-year chemistry as a junior, which was the only chemistry course that was offered at WHS. Fortunately, the teacher took a liking to Linus, and he was allowed to stay after class to work on additional problems during both his junior and senior years.

In his last semester of high school, Linus took his first physics course. The instructor impressed him, in part because he specifically emphasized the importance of the use of precise terminology in the sciences. This careful use of language was one of the main lessons that Linus took from high school, and it remained with him for the rest of his life. While a student at WHS, he even tried his hand at fiction writing, which resulted in his English teacher encouraging him to write a novel.

At the end of his seventh semester, Linus had run out of math and science classes to take. He had also completed all the requirements for graduation, except for the year of senior-level American history required by the state of Oregon. Upon learning of this requirement, Linus decided to return to WHS for his last semester after summer break, with the intent of taking the two required history courses simultaneously. This decision was quickly vetoed by the principal, and although Linus had been impressed with the breadth of his high school education, he decided to move on to Oregon Agricultural College (OAC, now Oregon State University) in the fall without a high school diploma — none was required by OAC at the time.

Though an expedient decision, the choice to move on lingered as a source of anxiety for the 16-year-old. The following, written on September 5, 1917, is an excerpt from his diary.

"Yesterday and today the feeling has often come to me that never more will I go to school. I think of all the other students beginning their studies, I imagine how I am [sic] member of the graduation class, would appear at Washington, I remember the enjoyment I got out of my studies and school life in general, and I sometimes poignantly regret that I have decided to go to college without graduating from high school. I covet every term of education that I have, and would gladly have more. College still seems so dim and far away that I often forget all about it. In a month and a day from now I will be in Corvallis. I try not to think of College, because of the way it affects me. Why should I rush through my education the way I am?"

Despite his nervousness, Linus stuck with his decision and did not return to WHS. He left for college in early October and ended up thriving at OAC. Many years later, in 1962, he was awarded an honorary diploma from WHS.

***

As always, Pauling continued to work odd jobs, including a brief stint at People's Market and then Apple's Meat Market, making eight dollars a week. Both stores were floundering though, and Pauling was soon laid off due to lack of business.

Desperate to earn money for school in the fall, Pauling took a job as an apprentice machinist under Mr. Schwiezerhoft, owner of the Pacific Scale & Supply Co. Beginning work at $40 per month, Pauling quickly proved to be quite capable and, at the end of his first week, he was given a $5 per month raise. Within a few weeks more, he was making $50 and, by late August, $100 per month. Increasingly dependent on Pauling's skills, Schwiezerhoft did his best to convince him to stay with the company, even going so far as to offer a full 50% salary increase. But despite his mother's pleas and the promise of a living wage, Pauling chose to leave for OAC at the end of the summer.

Chapter 5

Pauling's Freshman Year at Oregon Agricultural College, 1917–1918

Pauling wearing his Oregon Agricultural College 'rook lid,' ca. 1917.

engineering and the School of Mines housed both the mining degree as well as the chemical engineering major.

***

Pauling entered OAC with a particularly keen interest in chemistry, could not major in the discipline as, per state edict, the only School of ence operating at the time was located at the University of Oregon. Ins Pauling chose the next best option, chemical engineering. As he began introductory coursework, Pauling may have found the classroom to be a more crowded than had previous first-term freshmen — the OAC *Baromete* newspaper, reporting on a bump in Engineering majors, hypothesized an invigorated interest "due to the demand for engineers in military service."

As mandated by the Morrill Act of 1862, men attending the college, including those enrolled in shorter vocational courses, were required to participate in military training. A Reserve Officer Training Corps (ROTC) program was officially established during Pauling's freshman year, replacing the Cadet Corps that had existed previously. By this time, the military presence at the college consisted of "one regiment of infantry, a hospital corps, a signal corps detachment, and a band of fifty instruments." Heading into 1918, as the U.S. ramped up its involvement in the Great War, OAC became a military hub of consequence for the state of Oregon, a scenario that repeated itself in the 1940s.

Beyond scholarly inquiry and military training, the college strongly encouraged student involvement in extracurricular activities. This sentiment was echoed in advice that Pauling transcribed into his diary before he started college, wherein it was suggested that he "not take a number of extra [class] hours, but should try to do something for the school."

OAC promoted a vibrant social environment in part by fostering the creation and growth of student clubs and organizations. Most prominently, OAC was home to several Greek letter societies, one of which Pauling joined during his sophomore year. Dormitories, departments and, in certain cases, specific majors, also sponsored their own clubs. As an incoming student studying in the College of Mines, Pauling was naturally a member of the Miner's Club.

More informally, students in Corvallis often entertained themselves with a night out. Popular activities included seeing movies at the Majestic Theater and ending the night with ice cream at Winkley's Creamery. Another hot spot for spending time and grabbing a bite to eat was Andrews & Kerr, which served "Hooverized" waffles and offered a location where "seniors enjoyed high jinks." Pauling was fond of A&K's and frequented it when he could afford to, especially after seeing a show downtown.

In addition to cultivating a culture of student involvement, the college did its best to stoke long-running social traditions. First-year students were colloquially called "rooks" and "rookesses," and were made to wear green caps (for men) or ribbons (for women) to denote their status. Pauling took these rituals seriously and commented on how they had strengthened his school spirit. By the end of his first month in college, he noted in his diary that "I am getting along alright. Have lots of beaver pep."

**

Importantly, tuition for OAC Beavers did not exist. Indeed, in a technical sense, the school was free to all, including out-of-state and international students. That said, there were certain fees dispensed throughout a student's college experience. Some were assessed yearly while others were collected for one-time events, such as graduation. There were also additional costs associated with particular courses. For example, students were required to purchase "gymnasium suits" for Physical Education and other charges were mandated for classes that had a laboratory component. Between yearly fees, semester fees, and a diploma fee, early 20th-century undergraduates could count on spending at least $61 during the time that it took to complete a four-year degree.

With the cost of attendance so low, the main financial concern for OAC's students was room and board. Women, who were required to live with family or on campus, could expect to pay $10 to $20 per semester for rooming, depending on whether they chose a single or a double room. Male students, for whom no campus housing was available, could find lodging in private homes for approximately $16 to $20 a month.

In the diary that he maintained for much of the year, Pauling meticulously tracked his spending habits to the penny, and not long after arriving in Corvallis, he found that he was almost broke. Before the school year began, he estimated that his expenses would sum to $297. However, by late October — just a month into his first term — he revised his initial approximation, noting, "I have spent about $125 already. Board will be $175 more — $300 altogether. Then my numerous expenses will mount up. I do not expect to get off for less than $325."

To keep himself afloat, Pauling worked a series of campus jobs including janitorial duties, chopping wood, and butchering meat for a girls' dormitory. The tasks were dull, hard, and time-consuming; he made $0.25 an hour and worked 100 hours each month. In a 1960 letter to his son Peter, Pauling hinted at the difficulties of this time.

"I have decided that I have a little neurosis resembling the one that affected W.C. Fields. He had had such a hard time in his youth that after he got old and rich he still had trouble to keep from worrying about money. I have decided that the bad three months that I had just after my seventeenth birthday, when I was doing pretty hard physical work but not getting enough to eat because of lack of money, still bothers me to some extent. Of course, Mama and I had some trouble during the first couple of years after we were married, but nothing quite so bad as this earlier three months for me."

When he first arrived in Corvallis, Pauling described his living situation in his diary. He had found lodging in a boarding house close to campus and noted that "I have a nice big room, much larger than two boys usually have. I will share it with a sophomore named Murhard."

Pauling left the boarding house after Fall semester in part because, according to biographer Robert Paradowski, "he made too much noise tramping up the stairs in his heavy military boots." For the second semester he was unable to find a permanent place of residence and often wound up staying with friends. This is the time period that most haunted Pauling, as relayed in his 1960 letter to his son.

***

Academically, the situation was much better than initially expected. Before arriving, Pauling had expressed a great deal of insecurity about his potential to excel as a college student. A month before moving to Corvallis, the 16-year-old wrote this in his diary:

"Paul Harvey is going to OAC to study chemistry — Big manly Paul Harvey, beside whom I pale into insignificance. Why should I enjoy the same benefits he has, when I am so unprepared, so unused to the ways of man? I will not be able, on account of my youth and inexperience, to do justice to the courses and the teaching placed before me."

It did not take long, however, for Pauling to discover that his prowess in the classroom had carried over from the achievements of his high school years.

During the fall semester, Pauling registered for a typical assortment of first-year classes including Modern English Prose, Drill, and Gym. In addition, he began working through the core curriculum for chemical engineering majors, taking courses like General Chemistry, Mining Industry, and Calculus.

As he dug into his coursework, Pauling found that his high school education, though incomplete, was more than satisfactory. So pleased was he with the preparation that he had received that he wrote a letter thanking his high school math teacher, Virgil Earl, for having done an exceptional job. Earl replied to Pauling with gratitude and encouraging words, saying, "you have the ability and the disposition to work so I feel sure that you will succeed in your chosen work." Earl's point of view would later be reflected by waves of admiration from many of Pauling's OAC professors.

Pauling came away from his first semester of college with five A's, two B's, and a D in Mechanical Drawing. (Of Pauling's one notably poor mark during the term, biographer Tom Hager writes, "he wasn't patient enough to let the ink dry on his work [...] and kept smudging it.")

Spring semester revolved around a heavy course load that included Integral Calculus, Descriptive Geology, French, and Qualitative Analysis. It was during this semester that Pauling also received his only failing letter grade, the product of an unsuccessful attempt to circumvent the physical education requirement by joining the school track team. His tryout was evidently a mess, and he did not make the squad. Once this gambit had failed, Pauling chose not to return to the Gym class in which he was enrolled, as his other classwork was quite demanding. He retook, and passed, Physical Education later on during his OAC career.

Though encumbered by significant responsibilities outside of the classroom, Pauling completed his first year with seven A's, one B, and two C's — one in Descriptive Geology and the other in Camp Cookery. The grade scale that OAC used at the time included a letter grade "E," meaning that OAC marks of B, C and D were comparable to contemporary grades of A-, B+, and B-. As such, Pauling's freshman academic record was really quite superb. His marks for military drill were also consistently stellar, and near the end of the year he received a runner-up commendation in the Best Soldier competition at OAC's Military Inspection Day. Though far from easy, Pauling's first year as a college student was a success and concluded with feelings of optimism for what lay ahead.

from semesters in favor of an academic quarter system, which allowed for three-month training periods that dovetailed more readily with the military's needs. OAC also began to heavily promote student enrollment in classes that would support the war effort and tweaked existing courses to fit the needs of the moment.

For Pauling, these adjustments manifested in three specific classes that were new to the college's catalog: Explosives in fall term, Camp Drainage/Trenches Issues in winter term, and Excavation for War Purposes in the spring. As with all other SATC students, Pauling was also committed to a rigorous training schedule, often devoting multiple hours in a day to military drills. These shifts in obligations did nothing to wither his enthusiasm: throughout the war, Pauling remained a steadfast and enthusiastic supporter of the American effort and was later described as "100% for it" by his cousin Mervyn.

World War I concluded midway through fall term and, as might be expected, November 11, 1918 was a memorable day at OAC. Upon hearing the news that the war was over, spontaneous celebrations rocketed through campus, and before long multiple parades and assemblies were organized to honor those who had served. Notably, wartime restrictions on social functions were also temporarily relaxed to allow students to gather in good cheer.

**

Pauling spent fall term rooming with Lloyd Jeffress, his childhood friend who had first introduced him to chemistry. It was through Jeffress that Pauling also met Paul Emmett, a fellow OAC student who would become a close friend, research partner and, eventually, brother-in-law.

Just as Pauling had been academically successful during his freshman year, so too did he excel in the classroom as a sophomore. Taking courses including Engineering Physics, Metallurgy, Analytical Chemistry, and Mining Engineering, Pauling received all A's in his math and science classes throughout the year and tallied a perfect 4.0 grade point average during winter term. In addition to his schoolwork, Pauling remained a member of the Miner's Club, joining the group on field trips to study mine surveying, mining geology, and mining methods.

In the spring, an exciting and important opportunity was extended to Pauling: an offer to join the Gamma Tau Beta fraternity. Pauling eagerly pledged, despite being troubled by the feeling that he had been selected mostly to raise the house grade point average.

Pauling's experience in the fraternity was different than anything he had known before. His upper-class house brothers nicknamed him "Peanie" and expected that he, as with his fellow underclassmen, go out on weekly dates. Those who failed to meet this requirement were submerged and held underwater in a bathtub filled with cold water. This punishment, called "dunking," was a Greek custom, and it wasn't long before it was administered to the shy and short-on-money Pauling.

Before being dunked, Pauling decided that enduring the punishment once would be enough and put his scientific mind to work. He began to breathe deeply to saturate his blood with oxygen and when he was put in the tub, he let the seconds tick by until he had been under for an entire minute. His fraternity brothers soon became frightened and, thinking that something disastrous had occurred, quickly pulled him out. Pauling, of course, was fine, and never again had to worry about being dunked.

Hazing rituals aside, Pauling largely enjoyed fraternity life and came to regard it as having served a crucial role in his maturation. He commented on this in a letter to Thomas D. Hansen, the Executive Director of Delta Upsilon, written in 1988:

"The Oregon State Chapter of Delta Upsilon and its predecessor, the local fraternity Gamma Tau Beta, played an important role in my life. My father had died when I was nine years old, and, up to the time that I became a member of Gamma Tau Beta, there was no one who strove to teach me how to get along with my fellow human beings. As a result, I was rather quiet and withdrawn, to such an extent that I had few friends. My brothers in Gamma Tau Beta and Delta Upsilon helped me to develop my personality and to communicate with other people more effectively."

Over time Pauling became a house leader. One of his main goals in this capacity was to broaden the connections of his fraternity by proposing that they join a nationwide brotherhood, Delta Upsilon. Once the rest of the

In the fall of 1917, a newcomer to Oregon Agricultural College (OAC) would have encountered a lively campus energized by commanding buildings, a bustle of wartime activity, and the promise of significant growth to come. When Linus Pauling first set eyes on his new home, the college's footprint consisted of 349 acres, 91 of them constituting the main campus. The remaining land, located just outside the Corvallis city limits, formed areas for raising and studying livestock, poultry, and crops. At the time, the community boasted of 6,000 permanent residents, "free mail delivery…many churches and no saloons." Living in a town that derived its name from the Latin for "heart of the valley," members of the OAC community took great pride in their scenic surroundings.

For the 1917–18 school year, OAC served more than 3,500 students enrolled either full time or in short courses meant to build specific skills. Long requested by the student body, the number of college personnel had also begun to rise, with the total nearing 200 faculty members. To accommodate this growing corpus of teachers and learners, more and more buildings began to dot OAC's campus. And yet, despite these developments, Corvallis remained a rural community in spirit and Oregon Agricultural College, as the name suggests, was largely agrarian in its focus, with secondary emphasis paid to other disciplines of practical learning.

A land grant college, the OAC curriculum was organized into seven schools: Agriculture, Commerce, Engineering, Forestry, Home Economics, Mining, and Pharmacy. (The college also offered an affiliated but financially self-supporting School of Music, which was created by popular demand in 1908.) To graduate, OAC students needed to earn 136 credits, and though one could not major in any of the liberal arts, all four-year degrees did require liberal arts study in what eventually came to be known as the Lower Division. Pauling fulfilled this largely through foreign language classes, taking coursework in French and German. In addition, all students were required to take gym as well as an additional class in "Hygiene," and pass a swimming test too.

Engineering was a popular course of study for male students, perhaps second only to Agriculture or, in some years, Commerce. In addition, other schools offered engineering-centric degrees that pertained to their respective fields. For example, the School of Forestry advertised a degree in logging engineering and the School of Mines housed both the mining engineering degree as well as the chemical engineering major.

***

Pauling entered OAC with a particularly keen interest in chemistry, but he could not major in the discipline as, per state edict, the only School of Science operating at the time was located at the University of Oregon. Instead, Pauling chose the next best option, chemical engineering. As he began his introductory coursework, Pauling may have found the classroom to be a bit more crowded than had previous first-term freshmen — the OAC *Barometer* newspaper, reporting on a bump in Engineering majors, hypothesized an invigorated interest "due to the demand for engineers in military service."

As mandated by the Morrill Act of 1862, men attending the college, including those enrolled in shorter vocational courses, were required to participate in military training. A Reserve Officer Training Corps (ROTC) program was officially established during Pauling's freshman year, replacing the Cadet Corps that had existed previously. By this time, the military presence at the college consisted of "one regiment of infantry, a hospital corps, a signal corps detachment, and a band of fifty instruments." Heading into 1918, as the U.S. ramped up its involvement in the Great War, OAC became a military hub of consequence for the state of Oregon, a scenario that repeated itself in the 1940s.

Beyond scholarly inquiry and military training, the college strongly encouraged student involvement in extracurricular activities. This sentiment was echoed in advice that Pauling transcribed into his diary before he started college, wherein it was suggested that he "not take a number of extra [class] hours, but should try to do something for the school."

OAC promoted a vibrant social environment in part by fostering the creation and growth of student clubs and organizations. Most prominently, OAC was home to several Greek letter societies, one of which Pauling joined during his sophomore year. Dormitories, departments and, in certain cases, specific majors, also sponsored their own clubs. As an incoming student studying in the College of Mines, Pauling was naturally a member of the Miner's Club.

More informally, students in Corvallis often entertained themselves with a night out. Popular activities included seeing movies at the Majestic Theater and ending the night with ice cream at Winkley's Creamery. Another hot spot for spending time and grabbing a bite to eat was Andrews & Kerr, which served "Hooverized" waffles and offered a location where "seniors enjoyed high jinks." Pauling was fond of A&K's and frequented it when he could afford to, especially after seeing a show downtown.

In addition to cultivating a culture of student involvement, the college did its best to stoke long-running social traditions. First-year students were colloquially called "rooks" and "rookesses," and were made to wear green caps (for men) or ribbons (for women) to denote their status. Pauling took these rituals seriously and commented on how they had strengthened his school spirit. By the end of his first month in college, he noted in his diary that "I am getting along alright. Have lots of beaver pep."

***

Importantly, tuition for OAC Beavers did not exist. Indeed, in a technical sense, the school was free to all, including out-of-state and international students. That said, there were certain fees dispensed throughout a student's college experience. Some were assessed yearly while others were collected for one-time events, such as graduation. There were also additional costs associated with particular courses. For example, students were required to purchase "gymnasium suits" for Physical Education and other charges were mandated for classes that had a laboratory component. Between yearly fees, semester fees, and a diploma fee, early 20th-century undergraduates could count on spending at least $61 during the time that it took to complete a four-year degree.

With the cost of attendance so low, the main financial concern for OAC's students was room and board. Women, who were required to live with family or on campus, could expect to pay $10 to $20 per semester for rooming, depending on whether they chose a single or a double room. Male students, for whom no campus housing was available, could find lodging in private homes for approximately $16 to $20 a month.

In the diary that he maintained for much of the year, Pauling meticulously tracked his spending habits to the penny, and not long after arriving in Corvallis, he found that he was almost broke. Before the school year began, he estimated that his expenses would sum to $297. However, by late October — just a month into his first term — he revised his initial approximation, noting, "I have spent about $125 already. Board will be $175 more — $300 altogether. Then my numerous expenses will mount up. I do not expect to get off for less than $325."

To keep himself afloat, Pauling worked a series of campus jobs including janitorial duties, chopping wood, and butchering meat for a girls' dormitory. The tasks were dull, hard, and time-consuming; he made $0.25 an hour and worked 100 hours each month. In a 1960 letter to his son Peter, Pauling hinted at the difficulties of this time.

> "I have decided that I have a little neurosis resembling the one that affected W.C. Fields. He had had such a hard time in his youth that after he got old and rich he still had trouble to keep from worrying about money. I have decided that the bad three months that I had just after my seventeenth birthday, when I was doing pretty hard physical work but not getting enough to eat because of lack of money, still bothers me to some extent. Of course, Mama and I had some trouble during the first couple of years after we were married, but nothing quite so bad as this earlier three months for me."

When he first arrived in Corvallis, Pauling described his living situation in his diary. He had found lodging in a boarding house close to campus and noted that "I have a nice big room, much larger than two boys usually have. I will share it with a sophomore named Murhard."

Pauling left the boarding house after Fall semester in part because, according to biographer Robert Paradowski, "he made too much noise tramping up the stairs in his heavy military boots." For the second semester he was unable to find a permanent place of residence and often wound up staying with friends. This is the time period that most haunted Pauling, as relayed in his 1960 letter to his son.

***

Academically, the situation was much better than initially expected. Before arriving, Pauling had expressed a great deal of insecurity about his potential to excel as a college student. A month before moving to Corvallis, the 16-year-old wrote this in his diary:

"Paul Harvey is going to OAC to study chemistry — Big manly Paul Harvey, beside whom I pale into insignificance. Why should I enjoy the same benefits he has, when I am so unprepared, so unused to the ways of man? I will not be able, on account of my youth and inexperience, to do justice to the courses and the teaching placed before me."

It did not take long, however, for Pauling to discover that his prowess in the classroom had carried over from the achievements of his high school years.

During the fall semester, Pauling registered for a typical assortment of first-year classes including Modern English Prose, Drill, and Gym. In addition, he began working through the core curriculum for chemical engineering majors, taking courses like General Chemistry, Mining Industry, and Calculus.

As he dug into his coursework, Pauling found that his high school education, though incomplete, was more than satisfactory. So pleased was he with the preparation that he had received that he wrote a letter thanking his high school math teacher, Virgil Earl, for having done an exceptional job. Earl replied to Pauling with gratitude and encouraging words, saying, "you have the ability and the disposition to work so I feel sure that you will succeed in your chosen work." Earl's point of view would later be reflected by waves of admiration from many of Pauling's OAC professors.

Pauling came away from his first semester of college with five A's, two B's, and a D in Mechanical Drawing. (Of Pauling's one notably poor mark during the term, biographer Tom Hager writes, "he wasn't patient enough to let the ink dry on his work [...] and kept smudging it.")

Spring semester revolved around a heavy course load that included Integral Calculus, Descriptive Geology, French, and Qualitative Analysis. It was during this semester that Pauling also received his only failing letter grade, the product of an unsuccessful attempt to circumvent the physical education requirement by joining the school track team. His tryout was evidently a mess, and he did not make the squad. Once this gambit had failed, Pauling chose not to return to the Gym class in which he was enrolled, as his other classwork was quite demanding. He retook, and passed, Physical Education later on during his OAC career.

Though encumbered by significant responsibilities outside of the classroom, Pauling completed his first year with seven A's, one B, and two C's — one in Descriptive Geology and the other in Camp Cookery. The grade scale that OAC used at the time included a letter grade "E," meaning that OAC marks of B, C and D were comparable to contemporary grades of A-, B+, and B-. As such, Pauling's freshman academic record was really quite superb. His marks for military drill were also consistently stellar, and near the end of the year he received a runner-up commendation in the Best Soldier competition at OAC's Military Inspection Day. Though far from easy, Pauling's first year as a college student was a success and concluded with feelings of optimism for what lay ahead.

Chapter 6

Pauling's Sophomore Year at Oregon Agricultural College, 1918–1919

Pauling in military dress, 1918.

On September 23, 1918, 350 sophomore students returned to Oregon Agricultural College (OAC) to resume their classes after a three-month summer break. Among them was 17-year-old Linus Pauling who had been very busy during his time away. Pauling began his summer with an intensive six-week military training course at the Presidio Army post in San Francisco, where he was joined by his cousin Mervyn Stephenson and many other OAC cadets enrolled in the college's required Reserve Officer Training Corps (ROTC) program. Following that, Pauling and Stephenson found work on the Oregon coast at the Tillamook shipyard, where the pair helped build a 4,000-ton wooden-hulled freighter. The job was grueling, but Pauling earned enough to meet his living expenses and save for the coming school year.

Once on campus, Pauling worked in OAC's chemical stockroom, where he was tasked with tracking inventories of supplies and mixing compounds for student use. The job was easy and allowed him to interact with faculty members in Chemical Engineering. During this time, Pauling began sending any surplus money that he made to his mother, Belle, for safekeeping.

***

Two disruptions of global concern defined much of Pauling's sophomore year: the outbreak of the Spanish Influenza pandemic and U.S. involvement in World War I. The arrival of the flu quickly made an impact on campus; within the first month of fall, four students had fallen ill and a handful of new cases were being reported each week. In response, OAC converted its YMCA/YWCA facility into a hospital and brought in two nurses and a physician to handle the situation. Several football games were cancelled, and the basketball team was hampered during the winter. More than 800 cases were reported overall, including four deaths, but mostly the campus avoided the horrors seen elsewhere around the world.

During this period, a far greater shift was taking place on campus as a result of American entry into World War I. Notably, in 1918 a Student Army Training Corps (SATC) unit was established at OAC. Created to allow young men to enroll in the military while furthering their technical education, the SATC effectively replaced the ROTC and quickly became a major presence. Indeed, nearly half of all male students at OAC were enrolled in the SATC during the school year, and Pauling was among them. Following the conclusion of the war Pauling remained active in the ROTC as well, and by the time he graduated he had risen to the rank of Cadet Corporal within Local Company H.

The Great War made a tremendous impact on students at OAC in other ways. For one, the urgency of a war-time curriculum prompted a shift away from semesters in favor of an academic quarter system, which allowed for three-month training periods that dovetailed more readily with the military's needs. OAC also began to heavily promote student enrollment in classes that would support the war effort and tweaked existing courses to fit the needs of the moment.

For Pauling, these adjustments manifested in three specific classes that were new to the college's catalog: Explosives in fall term, Camp Drainage/ Trenches Issues in winter term, and Excavation for War Purposes in the spring. As with all other SATC students, Pauling was also committed to a rigorous training schedule, often devoting multiple hours in a day to military drills. These shifts in obligations did nothing to wither his enthusiasm: throughout the war, Pauling remained a steadfast and enthusiastic supporter of the American effort and was later described as "100% for it" by his cousin Mervyn.

World War I concluded midway through fall term and, as might be expected, November 11, 1918 was a memorable day at OAC. Upon hearing the news that the war was over, spontaneous celebrations rocketed through campus, and before long multiple parades and assemblies were organized to honor those who had served. Notably, wartime restrictions on social functions were also temporarily relaxed to allow students to gather in good cheer.

***

Pauling spent fall term rooming with Lloyd Jeffress, his childhood friend who had first introduced him to chemistry. It was through Jeffress that Pauling also met Paul Emmett, a fellow OAC student who would become a close friend, research partner and, eventually, brother-in-law.

Just as Pauling had been academically successful during his freshman year, so too did he excel in the classroom as a sophomore. Taking courses including Engineering Physics, Metallurgy, Analytical Chemistry, and Mining Engineering, Pauling received all A's in his math and science classes throughout the year and tallied a perfect 4.0 grade point average during winter term. In addition to his schoolwork, Pauling remained a member of the Miner's Club, joining the group on field trips to study mine surveying, mining geology, and mining methods.

In the spring, an exciting and important opportunity was extended to Pauling: an offer to join the Gamma Tau Beta fraternity. Pauling eagerly pledged, despite being troubled by the feeling that he had been selected mostly to raise the house grade point average.

Pauling's experience in the fraternity was different than anything he had known before. His upper-class house brothers nicknamed him "Peanie" and expected that he, as with his fellow underclassmen, go out on weekly dates. Those who failed to meet this requirement were submerged and held underwater in a bathtub filled with cold water. This punishment, called "dunking," was a Greek custom, and it wasn't long before it was administered to the shy and short-on-money Pauling.

Before being dunked, Pauling decided that enduring the punishment once would be enough and put his scientific mind to work. He began to breathe deeply to saturate his blood with oxygen and when he was put in the tub, he let the seconds tick by until he had been under for an entire minute. His fraternity brothers soon became frightened and, thinking that something disastrous had occurred, quickly pulled him out. Pauling, of course, was fine, and never again had to worry about being dunked.

Hazing rituals aside, Pauling largely enjoyed fraternity life and came to regard it as having served a crucial role in his maturation. He commented on this in a letter to Thomas D. Hansen, the Executive Director of Delta Upsilon, written in 1988:

"The Oregon State Chapter of Delta Upsilon and its predecessor, the local fraternity Gamma Tau Beta, played an important role in my life. My father had died when I was nine years old, and, up to the time that I became a member of Gamma Tau Beta, there was no one who strove to teach me how to get along with my fellow human beings. As a result, I was rather quiet and withdrawn, to such an extent that I had few friends. My brothers in Gamma Tau Beta and Delta Upsilon helped me to develop my personality and to communicate with other people more effectively."

Over time Pauling became a house leader. One of his main goals in this capacity was to broaden the connections of his fraternity by proposing that they join a nationwide brotherhood, Delta Upsilon. Once the rest of the house accepted the proposition, Pauling almost single-handedly took care of moving the transition forward and the Corvallis chapter opened during Pauling's senior year in 1922.

By the end of spring term, there were at least 25 fraternities and 13 sororities active at Oregon Agricultural College. Not surprisingly, Greek life on campus was a potent force, prevalent in many spheres including the world of competitive speaking. Speech and debate were very popular at OAC and students from across campus enjoyed taking on topics both serious and comedic. In one instance, competitors were asked to wrestle with the following argument: "Resolved, that an alligator is a better pet than a rhinoceros." Pauling participated in an inter-class competition that spring and eventually developed a reputation for his oratorical skill.

Still just 18 years old, Pauling concluded his sophomore year with excellent grades and an improved social standing. Growing more confident and the slightest bit more financially secure, he had ably navigated the difficulties of an unusual year and was excited to continue his studies in the fall.

Chapter 7

The Boy Professor, 1919–1920

After completing his sophomore year, Linus Pauling returned to Portland and began the summer working a job that he detested. Employed as a milkman by Riverside Dairy, Pauling was asked to take the night shift for 60 hours a week, and before long he was looking for a new job.

Quickly he found a more desirable post with the Oregon State Highway Department as a blacktop pavement inspector. Based in southern Oregon at the Wolf Creek-Grave Creek section of the Pacific Highway, the position focused on monitoring the quality of the bitumen-stone mixes used in the pavement, making sure that the material would hold together under heavy use. Though far from glorified work, and at times very boring, Pauling did enjoy his time working outdoors and living in a tent. In his correspondence from the time, Pauling wrote of his love for the sun, the benefits of spending a substantial portion of the year outside of a laboratory, and the value of working with chemicals in a real-world setting.

That is not to say that the summer was without conflict. During this period, Pauling worked under the partial jurisdiction of a man named E.W. Lazell, a chemical and efficiency engineer stationed in Portland. As the summer moved forward, Lazell issued a series of written reprimands expressing dissatisfaction with Pauling's work. In early September Pauling replied to department official Leland Gregory, regarding a complaint that had been lodged about his handling of paving material temperatures. The "misinformed informant," as Pauling referred to the unnamed complainant (Lazell), could apparently have been better informed had he actually read Pauling's reports.

***

With the end of summer nearing, Pauling was preparing to return to Oregon Agricultural College (OAC) for his junior year when he received devastating news: his mother, who was in dire financial straits, had used all of his savings to keep the family afloat in Portland. As such, there was no money available to fund his continued schooling. Lacking any other option, Pauling continued on with the Highway Department and, once September rolled around, found himself debating paving techniques with other inspectors rather than attending classes in Corvallis.

Fortunately for Pauling, OAC was experiencing a major post-war boost in enrollment and, with the onset of fall term, was confronting a crisis as faculty struggled to meet the needs of a much larger student body. In late October, OAC chemistry professor John Fulton reached out to Pauling, who was by then well-known to the faculty as being an exceptional talent. Fulton relayed the news that a member of the chemical engineering faculty was unexpectedly not available and offered Pauling a temporary position teaching analytical chemistry. Pauling was 18 years old at the time.

Though OAC's proposed $100 per month wage would mark a 20% pay decrease, Pauling readily accepted the job, returning to campus on November 14 and officially beginning work as an instructor in chemistry on November 20. While not a student during the 1919–1920 academic year,

Pauling still lived at the Gamma Tau Beta house as a faculty member. This was a regular practice at the time, especially for those who were unmarried.

Despite the reduction in his salary, Pauling found that his new job did come with a few perks. Nestled in Science Hall on the east edge of campus, Pauling worked primarily on the second floor, which was dedicated to quantitative analysis, and was assigned his own office. He also enjoyed the services of an assistant, Mr. Douglas, who helped prepare solutions for the courses that Pauling instructed (the same job that Pauling had held the year before).

In winter 1920, his debut as a collegiate instructor, Pauling taught three courses: two sections of Quantitative Analysis for mining engineers, chemical engineers and pharmacy students; and one section of General Chemistry for agriculture, home economics, and entry-level engineering students. In these three classes combined, Pauling taught 83 students, one of whom was his close friend Paul Emmett, who received an A in Quantitative Analysis.

As the year progressed, Pauling also developed a consequential friendship with John Fulton, the professor who had hired him and the director of OAC's chemical laboratories. It is likely Fulton who later referred Pauling to a series of papers that became very influential in his research. Following Pauling's graduation from OAC, Fulton also supplied loans of $100 and then $200 to support Pauling's graduate research at the California Institute of Technology.

With one term of instruction under his belt, Pauling's scientific horizons were expanding and his enthusiasm for the opportunities offered by an OAC education started to wither. In particular, Pauling believed that the curriculum's attention to theory was lacking, and increasingly he found himself drawn to the California Institute of Technology and its new Gates Chemical Laboratory. As he considered a transfer, Pauling initiated a brief correspondence with Caltech's Chemistry head A.A. Noyes and went so far as to secure a letter of recommendation from Fulton. In the end though, Pauling was simply unable to financially commit to a move to southern California and decided to stay on at OAC for another term as an instructor.

***

In his teaching and independent study during this time, Pauling deepened his love for the mathematical rigor required by many technical tasks.

In doing so, he developed a stickler's personality, one that reveals itself time and again in his correspondence both then and later.

One such early example is dated March 15, 1920, in a letter from Pauling to Dr. George Smith, the author of a chemistry textbook. In it, Pauling — who had just turned 19 years old — points out a tiny technical mistake. On page 11 of the textbook, Smith talked about errors in weight measurement. Using the tools of the day, Smith pointed out that weighing the same sample of a given substance twice would always yield two results that differed very slightly, the moral being that there is always a built-in percentage error in weight measurements.

Smith went on to explain how observers calculated errors of this sort in weighing mixtures, using as an example the measurement of the weight of a clay sample which contained in it 0.2% of magnesium oxide. Here Pauling found a point of disagreement, believing that the calculation of the percentage error of the magnesium oxide was problematic. In his text, Smith stated that if the measuring error was 0.1% in total, then for the impurity in the sample, which was 0.2% of the total clay, the percentage error would be 0.0002%. Conversely, Pauling thought the error in the weight of the impurity should be compared with the impurity itself, and that the percentage error would thus be 0.1%, instead of 0.0002%.

On March 22, Smith wrote a congenial letter back to his young correspondent, saying that he did not agree that this was an error but admitting that the example was confusing and needed further clarification. At the end of the letter, Smith also enthusiastically (and presumptively) noted that he was looking forward to meeting Pauling at the upcoming meeting of the Pacific Division of the American Association for the Advancement of Science.

***

For spring term, Pauling was assigned two new courses to teach: Chemistry 242 and Chemistry 245. Both were quantitative analysis surveys designed for engineering students, and each included at least one lecture, one recitation, and anywhere from 3–12 hours of lab work per week. Every morning, Pauling blocked out several hours to prepare his new slate of lectures.

Pauling's research notebook for that year — emblazoned with a handwritten "Keep Out! No Admittance" across the front cover — is riddled with student grades, calculations and notes on experimental methods. Pauling was compelled to consult this journal when, after finishing his position at the end of spring term, Professor Fulton contacted him at his Wolf Creek address. In his letter, Fulton requested that Pauling decipher some of the notes on quizzes that he had administered to his students during the previous term. Fulton also needed clarification on unknown solutions that he had produced and used in his classes.

***

Pauling's appointment ended with the conclusion of the academic year and also coincided with his first publication, "The Manufacture of Cement in Oregon," which appeared in *The Student Engineer*, a journal sponsored by the Associated Engineers of Oregon Agricultural College. In his three-page article, Pauling specified the process by which cement was produced, from crushers cutting large rocks as a first step to the kilns yielding the final, small round particles used for cooling in the finishing mill.

As OAC's students were wrapping up their final examinations in mid-June, "the boy professor" was returning to his position with the Oregon State Highway Department. He did so having also applied for a job as an assayer at Mountain Copper Company in Keswick, California, but ultimately deciding not to make the move so far south. During the summer months that followed, Pauling worked especially hard to accrue enough savings to support a true junior year at OAC. Fortunately, he was able to do so and returned to campus the following fall, eager to resume his studies after a one-year hiatus.

Chapter 8

Pauling's Junior Year at Oregon Agricultural College, 1920–1921

Pauling at the Oregon Agricultural College Interfraternity Smoker, 1920.

The 1920–21 academic year at Oregon Agricultural College (OAC) was, in many respects, a period of growth and change. With World War I now concluded, the school expanded, many of its programs became better known, and the campus was buoyed by a sense of optimism. One signal that the Roaring Twenties had reached Corvallis was the aptly named "Inter-Fraternity Smoker" contest. The cheeky competition — in which Linus Pauling participated — saw fraternity members "v[ying] with one another to produce the best characterization of womankind."

Pauling's junior year was also the last one to be offered tuition-free for OAC students. As before, all female students were required to stay in the dorms unless their parents lived in Corvallis, or they received special permission from the Dean of Women. At a cost of $18 per term for a single or $9 for a double, plus $5 for deposit and incidentals, all rooms had access to "pure mountain water, both hot and cold," lights, heat and "other modern conveniences" including a bed, pillows, linens, towels, sheets, and a wardrobe. Men also now had the option to live on campus, and plenty of private rooms could still be found around town.

Despite a great deal of positive momentum, many at OAC recognized that it could be difficult for recent graduates to find work outside of the Pacific Northwest, except for those from the School of Agriculture, who had "no such hardship." Some argued that a reason for this was the name of the school, and that if OAC were to be rechristened as Oregon State College, a "handicap" that was "neither fair nor equitable" to graduates outside of the School of Agriculture would be removed. As it happened, the school was eventually renamed, but not until 1927 and even then, as Oregon State Agricultural College. It became Oregon State College in 1937 and, at long last, Oregon State University in 1961.

***

When Pauling moved back to Corvallis in September 1920, he rejoined the Gamma Tau Beta fraternity. While Pauling's involvement in the house had been crucial to his social development, his year out of school had also provided ample opportunities for personal growth. Partly as a result, when Pauling returned to student life he was no longer so strictly interested in chemistry and instead began to dabble in a number of additional pursuits, excelling at most.

One of the most significant diversions attracting Pauling's interest was competitive public speaking. When he first came to OAC, Pauling was, by his own account, a shy individual lacking in self-confidence. But later in his undergraduate career, Pauling developed a keen capacity for oratory that he sharpened over his final two years in Corvallis.

Competitive speech was so widely followed on campus that an insert in the 1907–1908 Rooter's Club booklet featured a cheer specifically created for OAC's orators and the student newspaper would routinely publish up to six columns reporting on forensics competitions. This culture was boosted when, in 1920, OAC established a speech department and began building a club team that was stronger than ever before. This new department was a major asset to the college in part because speeches were used not only for competitions, but also to enhance the experience of nearly all campus events. Speakers commonly addressed the general student body or the college's athletes before an athletic event to raise confidence and excitement in competitors and spectators alike. Once the game had started, oratory was also used to engage in "verbal combat" with students from other institutions.

Pauling had gained significant experience speaking to groups while teaching entry-level chemistry to fellow OAC undergraduates. During this same time, Pauling's close friend Paul Emmett became quite active in the OAC Forensics Club and helped introduce Pauling to the thrills of presenting on stage. Emmett represented OAC in the 1920 triangular debate, an annual competition involving three colleges. The following year, both Emmett and Pauling were featured in the forensics section of *The Beaver* yearbook.

***

OAC's Forensics Club was led by Professor George Varney, a new arrival to the college who had experience training orators at different institutions, including a state champion. Despite this, when Pauling decided to compete in the annual inter-class contest, he instead sought the help of his own personal coach, an English professor whose past experience as a preacher qualified him to guide students in the art of public speaking.

Representing the Juniors in OAC's internal competition, Pauling's speech offered a grand assessment of the 20th century thus far and speculated as well on its future. Titled "Children of the Dawn" (which he meant to refer to members of his generation), the piece opens with a poetic description of a dream, one in which humanity and Earth are described as specks within the greater universe. In Pauling's dream though, humanity has developed so effectively as to reach beyond Earth to understand the entire universe. This dream, Pauling reveals, is an allegory for the possibilities that lay ahead for his classmates.

From there, the speech chronicles the development of science and thought since ancient times and culminates with Pauling's main argument: that Darwin's theory of evolution can be applied to society, science and civilization. In this, Pauling describes the developments of the past as necessary steps to completing a "Great Design" in which the entire universe progresses in accordance with the principles of evolution. The speech concludes on a hopeful note by suggesting that unimaginable achievements were on the horizon. "It is impossible for us to imagine what developments in science and invention will be witnessed by the next generation," Pauling wrote. "We are not the flower of civilization. We are but the immature bud of a civilization yet to come."

Impressive as Pauling's speech was, he wound up tying for second place in the OAC competition, losing top honors to William Black, a senior and three-time participant in the event. Contrary to Pauling's idealistic and relatively simple premise, Black's oration, titled "Our Tottering Civilization," presented an elaborate and frankly racist view of the era. At the time, China and Japan had undergone periods of rapid modernization and immigrants from east Asia were very well established as active participants in the U.S. economy. The arguments issued in "Our Tottering Civilization" largely stemmed from a fear that further development of these cultures, both in and beyond the United States, could eventually lead to the subjugation of Western ideals. Black's manuscript concludes by exhorting white nations to join forces against the further development of "colored" nations. In addition to claiming top prize in the OAC event, Black's submission eventually won second place in the statewide intercollegiate oratory contest.

***

Throughout his junior year, Pauling found himself compelled to advance his studies in chemistry through somewhat unorthodox means. In part because OAC emphasized the practical and applied sciences, Pauling had begun to find some of his classes to be a bit rote in their emphasis on solving problems of interest to engineers rather than academic chemists. But having spent the previous year teaching, Pauling re-enrolled at OAC with a willingness to seek out opportunities in non-traditional ways. Fortunately, the school year reciprocated.

As his knowledge base grew, the young Pauling often found himself questioning aspects of what he was learning and seeking to uncover more. For example, as a chemical engineer in training, he was learning that measurements of magnetism differed from material to material, but he had no insight into why. Prior to his junior year, Pauling may well have been resigned to the notion that these were unanswerable questions. However, more satisfactory solutions soon emerged with the help of a few influential professors.

Though he had saved up enough money to return to school, Pauling still needed to earn a wage to pay for ongoing expenses, so he took a job as an assistant to Samuel Graf, a professor of mechanics and materials at OAC. Even though the job consisted mostly of working through computations, it also allocated time for Pauling to engage with the scientific literature. The college's chemistry head, John Fulton, helped facilitate this by giving Pauling a few of his own chemical journals, and during his shifts as Graf's assistant, Pauling consumed these journals with relish.

It was in this setting that Pauling first encountered the work of G.N. Lewis and Irving Langmuir, both of whom were exploring some of the most exciting questions in subatomic chemistry. While their publications did not answer all of Pauling's questions, reading Lewis and Langmuir made Pauling realize that this new field of subatomic chemistry *could* solve problems, many of which he had not even known existed.

While the history of the field of subatomic chemistry is quite complex, many of the ideas that Lewis and Langmuir were developing emerged because of breakthroughs that the Danish chemist Niels Bohr made with the formalization of his quantum theory in 1918. At OAC all of the chemical engineering courses were physical and practical in their orientation. The kind of theoretical work that Bohr, Lewis, and Langmuir were doing was novel — and not being taught at OAC — but making its acquaintance equipped Pauling with new tools to explore some of the questions that he was pondering as a 19-year-old undergraduate. This breakthrough renewed Pauling's fervor for chemistry and his determination to pursue it as a career.

Pauling's moment of inspiration was especially well-timed as he soon learned that graduate fellowships at the California Institute of Technology were available and that they offered the chance to study with A.A. Noyes, one of the country's leading physical chemists and a mentor to several promising young scholars.

***

During this period, Pauling was further honing his ability to think independently and pursue a problem persistently, a set of traits that informed much of his later academic work. His course reports from a metallography lab in the spring of 1921 provide a nice glimpse into his growing scientific acumen and overall confidence. The reports are not written in the formal and impersonal manner that one might expect to find. Instead, quite often Pauling uses plain terms and interjects many of his own thoughts in the writeups. From item to item, a personal voice is easily identified, and the reports make for engaging reading. One interesting example is the concluding paragraph to Pauling's report on "Preparation and Examination of Specimens" (April 25, 1921), which is typically lighthearted and even boastful. Presumably addressing his professor, Pauling writes,

> "I have made free use of technical terms throughout on the assumption that you would understand them, but in case you do not, I refer you to my experiment on metallography the first quarter of this year, in which complete definitions are given. I have also attempted to use words of one syllable to as great an extent as is practicable in order to prevent any mental strain. Let me repeat that, for a really good article, you should read my previous experiment."

When approaching a lab topic, Pauling developed the habit of consulting all the relevant publications that he could find. And in analyzing them, he sometimes issued opinions that reflected both growing acuity and intellectual restlessness. See, for example, this aside included in a 1921 report titled "Heat Treatment and Tests of Specimen and Case-Carburizing":

> "Quite often in reading, I wonder where people find all the things they do to write about. Just about as often I wonder what the idea of writing so much is, and why it is necessary to really do it. Then again, I find the secret. It is this. All this writing is necessary because we are acquiring so much knowledge that we are behind in writing it down as it is, and there is still room for more books. I wish that someone would prepare (or rather, had already prepared) a short concise article on heat treating steels covering about five such pages as this. His work at least would be considered useful by me."

***

As the year progressed, Pauling also garnered increasing attention from the school's honor societies. First and foremost, Pauling was elected to Forum, OAC's most prestigious and academically stringent honorary. Created six years earlier and akin to Phi Beta Kappa, Forum was comprised of juniors and seniors who were elected by current members on the basis of their "scholastic attainment and leadership."

Pauling was also a member of Sigma Tau, the national honor society for engineering, and served as president of Chi Epsilon, the chemistry honor society. He remained a member of the Chemical Engineering Society, which he had joined as a freshman and now served as treasurer. He also helped out a bit with the production of the *Beaver* yearbook, and was tasked, along with fellow student Ernest Abbot, to create a page documenting Forensics activities.

And as usual, Pauling earned stellar grades. Over the course of his junior year, he received all A's, except for a B in Military Drill during the first quarter and, interestingly, a B in inorganic chemical engineering in the third quarter. The 1920–21 academic year also marked the moment where Pauling was awarded the elusive A in track and field that he so desired.

But more than anything, students and professors alike were recognizing his superior academic talent. Tellingly, OAC's yearbook from Pauling's junior year pictures him as a member of the Class of 1922, notes his major, hometown, and fraternity nickname, and includes this brief description: "a prodigy, yet in his teens."

Chapter 9

Ava Helen in Oregon

Ava Helen Miller (far left) with her sisters and mother, 1914.

Ava Helen Miller's maternal lineage traces back to Philip Edmond Linn, a third-generation American born on November 25, 1811, in Mercer County, Pennsylvania. According to *Genealogy of the Philip E. Linn Families* (Albert B. Shankland, compiler; 1965) Philip's father was a "minute man" who fought in the War of 1812 and then, following the conclusion of hostilities, moved his family to Kentucky, where Philip was raised. In 1831 Philip married Mahala McDannald and the couple eventually moved to Illinois where Philip worked a series of jobs, mostly in timber mills, and Mahala gave birth to 14 children.

Shortly after Mahala's death in 1859, Philip remarried and, in 1865, set out west for the young state of Oregon. He, his second wife, and many of his 16 children settled in Clackamas County, near the present-day community of Eagle Creek. Evidently a man of reasonable means, Philip donated the land and materials needed to build the area's first Southern Methodist church as well as the first bridge to span the Clackamas River at what is now Estacada, Oregon.

The seventh of Philip's children was a daughter, Mary Ellen, born on October 8, 1844. Little is known about Mary's youth except that, during her family's westward migration, she walked from Missouri to Oregon (at the age of 17) to help save the strength of her father's oxen. On April 11, 1867, Mary wed John Jay Gard, an Illinois native who had also traveled to Oregon with his family via covered wagon. John and Mary would have eight children, the eldest of whom was Elnora Ellen Gard, born February 19, 1868. She would eventually become Ava Helen's mother.

**

Ava Helen's father was George Richard Miller, born on March 5, 1856, in Hindern, Germany. In an interview with biographer Thomas Hager, Linus Pauling shared these memories of his father-in-law:

> "[George] had come from Germany, Hamburg, when he was in his teens, perhaps. ... And he was a schoolteacher in the elementary school, primary school, in the Willamette Valley, and met Nora Gard in the school. She was a student. ... So they were married and I suppose that he homesteaded a 160-acre farm."

It was on this farm near Oregon City that Ava Helen and her 11 siblings were raised. In his *Force of Nature*, Hager describes the hugely influential setting in which the Miller children grew up:

> "Politics was a part of life in the household. Ava Helen's mother had been a suffragist, and her father was a liberal Democrat with leanings toward socialism. Her parents divorced when Ava Helen was eleven, and she and her younger brothers and sisters were raised by their mother; the combination of socialist discussions around the dinner table and the example of her self-sufficient mother engendered in Ava Helen a lifelong concern for social justice and a strong feeling that women were capable of anything they set their minds to."

Following her parents' divorce, Ava Helen remained close with her mother but seems to have fallen out of touch with her father. In his discussions with Hager, Linus Pauling again notes that,

"The father seems to have been rather...dictatorial, working everyone hard on the farm. ... I don't know what his history was after the divorce, but he was living in Chicago in...March of 1926. Ava Helen and I were on our way to Europe and we managed to see him for 15 minutes or 20 minutes perhaps."

***

Ava Helen was born on her family's farm on December 24, 1903. When she was in grammar school, she, her mother, and some of her 11 siblings moved to Canby, Oregon. In 1918 she moved again to Oregon's capital city, Salem, to live with her sister Nettie Spaulding and attend Salem High School. It is not clear why this decision was made, though one might speculate that it related to family finances.

Much of what we do know about Ava Helen's high school experience is contained in a journal that she kept during the period. As historian Mina Carson writes in her biography, *Ava Helen Pauling: Partner, Activist, Visionary*,

"Her notes suggest a lively, flirtatious disposition. There were plenty of boys to write down...'He joined the navy during the war,' she wrote of Claire Gaines of Canby, 'and I have ever felt happy to think I refused to kiss him good-by which perhaps took a bit of conceit out of him. [He] married in 1921.' On the back of a photograph of Haines she wrote in retrospect: 'my heart's first flutter.'"

Her lifelong interest in politics was also evidenced during her Salem years. Again from Carson:

> "The Spaulding household…was probably lively and certainly close to the state's political heartbeat. Her sister Nettie was secretary to one of the Oregon Supreme Court justices, so there was a direct link to affairs in the capitol, and 1630 Court Street, the Spaulding home, was just a few blocks' stroll from the Supreme Court building and the State Capitol. Ava Helen carried her father's Democratic politics into her adolescence; of a family friend, an admired physician, she wrote: 'We quarreled about politics. He is a Republican.'"

One senses that the Salem years were mostly happy for the precocious young woman.

> "[She] graduated from Salem High School in three years. She was class president her senior year. For her senior class picnic at Silver Creek Falls that spring she helped organize the food for a class of one hundred twenty-five. She dared kiss a boy for the camera. She was a girl of fun and will, as well as a sense of duty."

***

When the time came to begin considering a higher education, Oregon Agricultural College (OAC) proved a natural choice in part because, by the time of her high school graduation, three of her siblings were already there. These siblings were brother Milton, a senior majoring in agriculture; brother Clay, a junior also studying agriculture; and sister Mary, a senior in home economics. Like Mary, Ava Helen entered OAC as a home economics major. And instead of living in a women's dormitory like most first-year students, Ava Helen stayed with her siblings in a Corvallis rental home that her mother Nora also resided in.

Ava Helen entered OAC in the fall of 1921 and spent a total of five quarters there, one term shy of two full academic years. During her time as an undergraduate student, she took a number of English, French and Spanish courses. She also took a chemistry class every term. Courses in home

Ava Helen Miller around the time of her enrollment at Oregon Agricultural College, fall 1921.

economics were more varied and included Clothing and Textiles, Child Care, and Food Preparation and Selection. Overall, she did quite well, finishing with a 89.21 grade point average, a figure that would have been even higher had she not received an F in an English course during winter term of her freshman year.

Ava Helen also pursued interests outside of academics during her stint in Corvallis. During her first year she joined the school's drama club, the Mask and Dagger. The club produced several plays while she was a student, including larger productions like Shakespeare's *A Midsummer Night's Dream* and a modern "light comedy" by Booth Tarkington called *Clarence*. Near the middle of the year, Ava Helen hit the stage in a one-act offering titled *Pierro by the Light of the Moon* by Virginia Church; Ms. Miller was cast as Columbine, the second lead.

During her second year at OAC, Ava Helen likewise served as secretary of the Lyceum Club, which consisted of "musicians, readers, and lecturers." The club focused on coordinating cultural events for the local community and, during Ava Helen's tenure, hosted several prominent artists and musicians, including the New York-based Flonzaley String Quartet.

Chapter 10

Pauling's Senior Year at Oregon Agricultural College, 1921–1922

Linus Pauling on graduation day, June 1922.

During the summer of 1921, Linus Pauling again worked with the road crew, zigzagging across Oregon. By now, he had established seniority and was bringing in good money. In the process, he had also become a sort of kid brother to the rest of the group, who admired his intelligence and seemed to enjoy his company.

Pauling returned to OAC in 1921 as a senior. In the 12 months to come, he would finish his coursework, become engaged, and move to Pasadena, California to begin his graduate studies. Needless to say, it was a year of great change for Pauling, but one that he embraced in full.

***

In the fall of 1921, seniors at Oregon Agricultural College (OAC) were welcomed back through a series of "Get Acquainted" dances, which aimed to help them become more comfortable with their apex position in the social hierarchy. Though these dances were a running tradition at OAC, each senior class approached them in their own way. During this particular year the dances were themed, with one especially memorable event, the Goof Dance, challenging participants to wear the craziest outfits they had.

For Pauling, the start of the year marked a continuation of his effort to earn solid marks and gain entry into a highly regarded graduate program. Throughout his time at OAC, he had always applied himself, and by his senior year, those efforts were evident.

The OAC *Beaver* yearbook traditionally included basic information on all its seniors, as well as additional details documenting their participation in extracurricular organizations and clubs. These blurbs often consisted of a handful of words, but Pauling's was, not surprisingly, several lines long. As per OAC custom, Pauling's entry lists his major (chemical engineering), hometown (Portland), and fraternity (Delta Upsilon). Decorations included his membership in Sigma Tau, the engineering honor society into which he was inducted during his junior year and served as secretary during his senior year. His participation in the Scabbard and Blade, a military honor society that he joined during his junior year, is also listed. So too are his memberships in the Chemical Engineering Association (junior year treasurer) and the Chi Epsilon civil engineering honor society (junior year president). As a senior, Pauling also served as a Captain in the Reserve Officer Training Corps and his efforts in competitive speaking were noted as well.

***

Academically, Pauling continued to excel — he earned 18 A's and 3 B's in his last year in Corvallis — and aspects of his promise are noted in a notebook simply titled "Phy Chem Data Book" that was kept by Fred Allen, Pauling's physical chemistry professor during his final year at OAC. The notebook contains results compiled by Allen from the experiments that his students ran, as well as annotations and short biographical notes made by Allen later in life. While the data aren't of especial interest, the annotations, compiled in 1962, do provide insightful details on the class in general, and on Pauling in particular.

On one of the first pages, Allen writes, "The 14 men named on next page were in a Phy. Chem. Course under FJ Allen the school year 1921–22. It was a remarkable group," adding that half of the individuals enrolled in the class would eventually be listed in the 1961 edition of *American Men of Science*.

Pauling is obviously the most notable person among those listed, and another familiar name is Paul Emmett who, along with Pauling, would go on to receive his Ph.D. from Caltech. But for Allen, Pauling was in a league entirely his own:

"Except for Pauling, Emmett would have been top man in the class. No censure is intended when I say that the gap from Pauling to the others in the class is akin to the hardness gap from diamond to corundum."

He added:

"Pauling is the only student I have encountered who showed definite qualities of genius as an undergraduate."

***

Mostly due to his excellence in the classroom, Pauling was invited by members of the OAC faculty to apply for a Rhodes Scholarship, an opportunity that offered three years of paid study at Oxford University, an institution whose scientific facilities were, in Pauling's own words, "not excelled in the world."

Outside of general qualifications including age, class standing and citizenship status, certain additional factors were weighed by the Oregon state scholarship committee. These characteristics, as defined by the Rhodes Scholarship Memorandum, included:

(1) Qualities of manhood, force of character, and leadership.
(2) Literary and scholastic ability and attainments.
(3) Physical vigor, as shown by interest in outdoor sports or in other ways.

Should a candidate who was qualified in all three areas fail to emerge, state committees were to select those who showed "distinction either of character and personality, or of intellect, over one who shows a lower degree of excellence in both."

In his initial application letter, Pauling referenced his engagement in campus honor societies, his impressive scholastic record, his status as junior class orator and his (somewhat dubious) involvement in track and field as relevant personal qualifications. Pauling also acquired seven letters of recommendation from faculty members as well as his summer employer. Each of the letters shed light on perceptions of Pauling as a student. Notably, though the recommendations are overwhelmingly positive and illuminating, emphasizing his competence, character and intelligence, some of the faculty clearly felt the need to overcompensate for his lack of established athletic prowess.

In the end, Pauling's application was denied by the Rhodes committee, perhaps in large part because of weakness in the "physical vigor" category. Though naturally disappointed by the decision, he was able to see a silver lining. As he recalled years later, Oxford's chemistry department, where he would have studied, was stuck in the past. In Pauling's view, the department's faculty were not interested in some of the new innovations emerging within the discipline, and had he attended Oxford at an impressionable age, he may well have been steered down a less vibrant path.

More tangibly, Pauling's application drew the attention of several professors who provided support to the prodigious student for the remainder of his undergraduate career. Floyd Rowland, a chemical engineering professor at OAC, noted that Pauling "possesses one of the best minds I have ever observed in a person of his age, and in many ways is superior to his instructors." Likewise, the English professor who helped Pauling with his speech earlier in the year observed that Pauling "does not expect results without hard work but seems to delight in digging hard." German professor Louis Bach followed suit with a keen observation later affirmed by untold others:

"[Pauling] is endowed with a remarkable memory in combination with good judgement, sound analytical and synthetic discrimination: a brilliant mind."

***

Ava Helen Miller and Linus Pauling with two Oregon Agricultural College classmates, 1922. This is potentially the earliest photograph of Ava Helen and Linus ever taken.

During fall term Pauling again worked as Samuel Graf's assistant, but as he was preparing to travel home for Christmas vacation, OAC offered him a new job teaching freshman chemistry for home economic majors. Thinking the extra money would be useful, he decided to accept the offer.

On January 6, 1922, Pauling entered a classroom in Science Hall. He was nervous, but basically ready to teach. The era being what it was, this class of home economics students consisted entirely of women. Feeling a need to establish his authority from early on, Pauling decided to ask a tough question. He ran his finger down the registration sheet, looking for a name that he knew he would not mispronounce. The words that came next

would change his life: "What do you know about ammonium hydroxide…
Miss Miller?"

Ava Helen Miller responded with a quite satisfactory answer. Chemistry
was always one of her best subjects (an organic chemistry instructor named
Mr. Quigley remarked that she was among his best students) and the class had
studied the compound during the previous term. Of this initial encounter,
Ava Helen later wrote, "In recitation room #211. Chemistry O.A.C. He was
my teacher — a student assistant. His curls are lovely."

Following their first meeting, a few weeks passed before the two spent
any more time together. An OAC instructor had recently been severely criticized for the attention that he had paid to one of his students, and though
Pauling was quite taken with Ava from the beginning, he was determined
not to endure the same fate.

One day, however, a note came back to Ms. Miller in her chemistry notebook stating that if she waited after class, her instructor would walk across
campus with her. She waited and the two walked, and then went for more
walks, becoming better and better acquainted. Their relationship quickly
blossomed from there, and just before the end of the term, Pauling asked
her to marry him. She said yes, and he promptly lowered her final grade by
one letter to avoid any suspicions of favoritism. This decision remained a
source of good-natured ribbing between the couple for years to come and
clearly did little to alter the trajectory of their growing affection. By the end
of the school year, Ava Helen had written Linus a check for the amount of
"My heart, my life, my love, my all."

***

During his senior year, Pauling also began thinking about graduate school.
The talented young man was well-aware of the need to move beyond OAC to
continue his learning, and throughout the year the decision of where to go
for graduate studies weighed heavily on his mind. Always keen on a future
in chemistry, Pauling stayed as current as possible on recent developments
in the field and knew that there were a handful of institutions equipped to
provide him with an advanced education that could keep up with the rapidly
changing discipline.

When the time came to apply to graduate programs, Pauling narrowed
his list to four possibilities: Harvard University, University of California

at Berkeley, University of Illinois, and California Institute of Technology. Of these schools, Pauling was perhaps most attracted to Berkeley because it was headed by G.N. Lewis, who had discovered that electron bonds are shared. Harvard was also enticing, in part because its program was led by Theodore Richards, who was America's only Nobel Laureate in Chemistry at the time. Richards had attended the University of Illinois for his graduate work, and this connection had helped to boost its program. Caltech, by comparison, was the smallest and newest of the options in which Pauling was interested.

But Pauling ultimately chose Caltech, a decision that was made, in part, because of a fortunate sequence of events. All of the universities that Pauling wanted to attend were interested in him, and Harvard offered an attractive fellowship that would cover his tuition. But shortly after receiving this offer, Caltech's letter arrived. Like Harvard, the Pasadena school promised a full-ride fellowship, but Caltech's package also included a $350 stipend to work as a teaching assistant in undergraduate chemistry courses. Importantly, Caltech had also accepted Pauling's close friend, Paul Emmett, and the two would wind up living together for their first year as graduate students. These two factors tilted the scale in Caltech's favor, and Pauling would remain there for more than 40 years.

***

As graduation day neared, Pauling was asked to deliver the senior class speech. He was a likely choice to fill this role, given his strong academic standing and his success in the junior year debate contest. But unlike past years, where speakers tended to offer fairly generic observations, Pauling's speech was notably more pointed.

Delivered on May 31, 1922, six days before commencement, the speech that Pauling prepared urged his fellow classmates to use the knowledge that they had gained at OAC to attack the problems facing society. Where his junior year oration, "Children of the Dawn," might come across as simplistic and perhaps overly optimistic, Pauling's senior class talk was characterized by its emphasis on personal responsibility and the "problems of the state," a term that referred to the social and political issues that had emerged from the destruction of World War I. "Our lives are to stand as testimonials to the efficacy of the work that our college is doing," Pauling said. "Education, true education, such as our own college gives us, is preparation both for a life of appreciation of the world and for a life of service to the world."

Another point that Pauling stressed to the senior class centered on the idea of "repaying OAC." In this, one might surmise that Pauling was speaking both of value gained from OAC and from the system of higher education as a whole. It is also important to point out that the systematic killing of troops that characterized World War I had fractured the public's feelings about research in the sciences. As noted by biographer Thomas Hager, a common argument at the time was that science was the cause of the war's deadly nature, and out of this experience, numerous questions lingered. Should education work to propel science and technology? Was further development of science potentially harmful to society?

Given this context, Pauling's calls for individual responsibility and service to society can be viewed as a reaction against the negative connotations then being ascribed to certain educational pursuits. He may have had this in mind when pointing out that OAC, Oregon's land grant institution, "has contributed in a wonderful way to solving the multitude of problems arising in the state." He reiterated this idea near the conclusion of his talk:

"This, then, is the way we can repay OAC — by service. Our college is founded on the idea of service, and we, its students, are the representatives of the college. It is upon us that the duty falls of carrying out that basic idea. We are going into the world inspired with the resolution of service, eager to show our love for our college and our appreciation of her work by being of service to our fellow men."

By emphasizing the idea that knowledge acquired at OAC was a tool that could be used for the benefit of society, Pauling's speech makes the argument that the development of knowledge in any field cannot be intrinsically evil. Rather, each educated individual can apply their knowledge in either beneficial or harmful ways. In Pauling's view, those same individuals bear a responsibility to use their talents for the improvement of society.

***

Pauling's early relationship with Ava Helen Miller is of notable importance in part because it bridges two critical periods in his life: the end of the OAC chapter and the beginning of his long run at Caltech. Linus graduated from OAC in June 1922 and moved to Pasadena while Ava Helen stayed in Corvallis for more schooling, the couple's desire to wed temporarily squelched by both of their mothers. Separated for one year, the two wrote to each other nearly every day, and in these letters, Linus expressed his true self to Ava Helen in a way he had not done — and never would do — with anybody else.

Prior to his move to Caltech, Pauling spent one final summer in Oregon as a paving inspector, and the 94 (!) letters that he wrote to Ava Helen during this time offer an intimate view of a rapidly maturing point of view.

The elements that generally defined Pauling's correspondence with his future wife were a) their wish to be engaged, and b) the strong maternal opposition to marriage that the two currently faced. Always the romantic, Pauling was accused by some of Ava's friends as being "too mushy," and indeed there is much written between the two about love, children and an idyllic life spent together.

However, over the course of their exchanges, Pauling likewise discussed his evolving sense of self and both suggested reading materials to one another, with the bulk of Ava Helen's picks tending toward the philosophical or metaphysical. Inspired, Pauling reflected in great detail on his perceptions of the soul, his conflicted feelings between animism and materialism, and his predisposition towards pacifism.

Money, an all-too-common theme for the duration of his undergraduate experience, also makes its presence felt throughout their correspondence. At times Pauling secretly mailed cash to Ava Helen to help finance trips to see him. He also devoted substantial energy to garnering adequate funds to support their marriage after the summer's end, with or without help from their parents. Other themes such as Ava's influence on Pauling's diet, as well as his developing fascination with fruits, hint at patterns that would come to define important periods of his future life.

Pauling also read from his own selection of books and took quite a liking to *David Copperfield* among other fiction titles. Far and away though, a major defining activity for Pauling during his summer off hours was working through proof sheets of the first nine chapters of a newly revised chemistry textbook, *Chemical Principles*, that had been sent to him by A.A. Noyes, the head of the Division of Chemistry and Chemical Engineering at the California Institute of Technology.

Completed while he was stationed near the Pacific coast at Astoria, Pauling devoured all 500 of the listed problems. After discussing his other interests with Noyes by mail, Pauling also began reading books on X-ray crystallography, a new technique being used to delve into the structure of molecules. (One of these texts was *X-rays and Crystal Structures* by a father-son team, W.H. and W.L. Bragg, the latter of whom would eventually become a chief scientific rival of Pauling's.) Having completed his reading, and prompted by some nudging from Noyes, Pauling would begin his graduate career as an X-ray crystallographer under the direction of major professor Roscoe Dickinson.

By the end of his final stint with the highway commission, Pauling had clearly grown weary of his summer occupation. In an August 1922 letter to Ava Helen he confided, "I really hate working in a paving plant. I do it just because I earn more than I would elsewhere." Bored, lonely and finished with the problem sets given to him by Professor Noyes, Pauling was hungry to launch into a new chapter of life.

***

Meanwhile, Ava Helen resumed classes at OAC the following autumn. Though now separated by 900 miles, the two kept in close contact — in addition to near daily letters, Ava sometimes sent Linus candy, and Linus would reciprocate with flowers. More commonly, they discussed the prospect of their marriage and their eager anticipation of the time they would be spending together in the future. Until then, and though they missed each other greatly, both managed to stay well-occupied. Linus was working intently on his crystallographic research while overseeing classes and labs. Ava was busy keeping up with schoolwork and friends.

Eventually, Ava Helen and Linus decided to marry following the completion of Linus's first year of graduate studies, no matter their mothers' feelings. In due course they informed their relatives and purchased their rings; after the completion of winter term, Ava Helen dropped out of school to plan her wedding.

To make the trip up from southern California for the ceremony, Linus purchased a Model T Ford from his mentor, Roscoe Dickinson, and headed north for Oregon. Unfortunately, Pauling's driving experience was limited to just a few minutes of practice and, come nightfall, he crashed into a roadside pit in the Siskiyou Mountains, resulting in an injured leg and a wrecked car. After waiting all night for help, Pauling was able to get his car repaired and arrived at the wedding on time.

More than a year later than originally intended, Linus and Ava Helen married in Salem, Oregon on June 17, 1923. The two would spend a brief honeymoon in Corvallis before moving to Portland over the summer, where Linus worked for the Warren Construction Company. With the onset of autumn, the newlyweds returned to Pasadena, Linus renewed his studies and the couple began a new life together.

Chapter 11

An Honorary Doctorate from Oregon State Agricultural College

Honorees at the 1933 Oregon State Agricultural College Commencement. Included in the photograph (left to right) are Dr. Marvin Gordon Neale, Commencement speaker; David C. Henry, honorary degree recipient; Linus Pauling; Dr. William J. Kerr, Chancellor of the Oregon State System of Higher Education and Oregon State Agricultural College president from 1907–1932; and Charles A. Howard, honorary degree recipient.

The early years of Linus Pauling's academic career were marked by a dizzying array of accomplishments. Offered a faculty position in 1927 at the conclusion of his Caltech graduate studies, he was promoted to full professor just four years later and, by 1933, was supervising twice as many graduate students and post-doctoral fellows as any other professor at the Institute. His Caltech salary also increased substantially during this time, leveraged by his receipt of numerous offers from other institutions trying to pry him away from Pasadena. Since he was usually asked to teach only one seminar per term, he was generally left with plenty of time to conduct research, often as a visiting professor at peer universities.

Notably, Pauling was also the first recipient of the Langmuir Award, granted by the American Chemical Society in 1931 for his structural chemistry work. A.C. Langmuir, the brother of Nobel laureate chemist Irving Langmuir, established the award for "outstanding chemical research" conducted by an individual in the beginning stages of their career. In his presentation of the award to Pauling, Langmuir recognized the recipient as a rising star and predicted that he would one day win the Nobel Prize. In some respects, the award marked the start of Pauling's emergence as a public figure.

Around this time, Pauling gave a seminar at Caltech on the quantum mechanics of the chemical bond that famously baffled Albert Einstein, who was in attendance. As recounted in a 1931 *New York Times* article,

"Last winter Dr. Einstein was an interested listener while Dr. Pauling discussed his chemical bond research. After the lecture, reporters noted that the German sage asked Dr. Pauling a number of questions. 'I'm afraid I'm not up on the chemical bond,' Dr. Einstein was heard to say. 'I shall have to brush up on the subject before taking more of your time.'"

Not long after, Pauling became the youngest individual ever invited to join the National Academy of Sciences. It is no wonder that Caltech's chemistry chief A.A. Noyes described Pauling as "the most promising young man with whom I have had contact in my many years of teaching."

***

Pauling's talent was clearly coming to the attention of the broader scientific community, but his undergraduate alma mater, now called Oregon State Agricultural College (OSAC), had been exposed to his potential much earlier. In May 1933, perhaps seeking to strike while the iron was hot, OSAC sent Pauling a telegram offering him an honorary doctorate of science, which would be his first decoration of this sort. Despite the short notice, Pauling promptly and eagerly agreed to be present for the commencement ceremony, which would take place on June 5. Not long after, he hopped in his car and drove from Pasadena to Corvallis to partake in alumni events scheduled for the preceding weekend.

Recent bureaucratic changes made this honorary degree all the more impressive. In 1932 the Oregon State System of Higher Education was created to manage the affairs of colleges and universities across the state, an arrangement that remained in place for more than 80 years. Though longtime Oregon Agricultural College (OAC) president William Jasper Kerr subsequently became the first chancellor of the system, the decision to award Pauling his doctorate required the agreement of several others. Specifically, Pauling was recommended by the state system's administrative council, approved by Chancellor Kerr, and then endorsed by the state board of higher education.

Pauling was one of three alumni to receive an honorary degree from OSAC that year. The others were David C. Henry, a consulting engineer in Portland who received an honorary doctorate of engineering, and Charles Howard, the state superintendent of public institutions in Oregon, who received an honorary doctorate of education. Marvin Gorden Neale, president of the University of Idaho, gave the commencement address. In his speech, delivered in the early years of the Great Depression, Neale spoke of the need to fight against critics of the education system and to work to ensure that support for land grant colleges and universities didn't slip away.

When the moment came to introduce Pauling, Chancellor Kerr listed a string of accomplishments amassed since Pauling's graduation from OAC in 1922. In addition to the Langmuir Prize and the National Academy of Sciences admission, Kerr also emphasized Pauling's achievements during his two years as a Guggenheim fellow, his authorship of over 50 scientific articles, and his appointment as a full professor at Caltech.

The evening report published in the Corvallis *Gazette-Times* newspaper leaned heavily on Pauling's local roots and agreed that his future was bright. The paper also reported that 486 degrees were conferred at the gathering: 418 bachelor's degrees, 52 graduate degrees, and 13 pharmaceutical chemistry diplomas.

***

OSAC Executive Secretary W.A. Jensen wrote to Pauling following the ceremony to confide that his award had been one of the most heartily endorsed he had ever seen. He also conveyed the high hopes that had been relayed by many on campus, concluding with an increasingly common idea: "The Nobel Prize is just ahead!"

When *Science* published news of Pauling's diploma, Fred Allen, another Oregon State alumnus and Pauling's former professor, wrote to extend congratulations. In his letter Allen joked, "I am proud that our alma mater could break away from the precedent which has stumbling over one's beard a prerequisite to an honorary degree."

Pauling was only 32 when awarded his first honorary doctorate, and just 11 years removed from the completion of his undergraduate program! The two other recipients of honorary degrees at the 1933 graduation ceremony were both decades older. Over the course of his life, Pauling would ultimately accumulate 47 honorary degrees from institutions around the world, but he took great satisfaction in knowing that the first came from Oregon State.

Chapter 12

The Story of Ralph Spitzer

Ralph and Therese Spitzer at Oregon State College, 1949.

B orn in Brooklyn, New York on February 9, 1918, Ralph William Spitzer was blessed with a prodigious intellect, beginning first grade at age 5 and graduating from Abraham Lincoln High School not long after his 16th birthday. Interested in chemistry, physics and math, Ralph enrolled at Cornell University in fall 1934 and, as in high school, encountered little difficulty in achieving at a high level.

From September 1937 to February 1938, Linus Pauling, a budding star of the academic world, took residence at Cornell as George Fischer Baker Lecturer. The talks that he gave during this period proved to be hugely important as they formed the nucleus of 1939's *The Nature of the Chemical Bond*, one of the great scientific books of the 20th century. The bright young Spitzer attended Pauling's lectures and at their conclusion approached the speaker to express his admiration and thanks. A conversation emerged from there that ultimately led to Spitzer choosing the California Institute of Technology for his graduate studies.

Ralph arrived in Pasadena in fall 1938 and immediately set to work with Pauling as his advisor. In addition to course studies and teaching assistant duties, Spitzer began researching Pauling's new ideas on structural chemistry, with an eye toward improving a specific component of the theory. Pauling initially dismissed Spitzer's initiative as "crazy" but in later years admitted that his protégé had been correct in his thinking. In various letters of recommendation, Pauling referred to his graduate student as a "pleasant" and "unusually able man" who ranked "well above the average of our Ph.D.'s in physical chemistry."

Spitzer completed his doctorate in three years. Wishing to remain in Pasadena, he accepted a post-doctoral fellowship in Caltech's biology division, working with Hugh Huffman on a project to improve a cooling apparatus designed to attain temperatures approaching absolute zero. From there, Spitzer took on the first of several scientific positions related to the war effort. The job that he most enjoyed was at Woods Hole, Massachusetts and involved experiments meant to hone techniques for sinking submarines.

It was there that Spitzer's interest in politics and world affairs began to emerge. Hailing from a largely apolitical family, Spitzer appears to have had little engagement with social activism until the mid-1940s, but at Woods Hole he led discussion groups focused on current events. In doing so, it is likely that he was inspired by the examples of both Linus Pauling as well as his wife Therese. As with Ava Helen Pauling, Therese Spitzer seems initially to have harbored a more instinctive and vigorous interest in social and political activism than did her science-focused husband.

After the conclusion of the war, the Spitzers moved back across the country to Berkeley, California where Ralph spent a second post-doc year working with Kenneth Pitzer, motivated partly by a desire to publish a paper authored by Pitzer and Spitzer. Their research together focused on determining the structure of cyclopentane by the measurement of gaseous heat capacities, a technique that Spitzer had learned at Caltech. In Berkeley he also participated in the founding of a local chapter of the National Council of Atomic Scientists.

Months later, the end of his fellowship in sight, Spitzer began looking anew for an academic job, asking Pauling to keep his name in mind if he happened to hear of anything. Coincidentally, Pauling soon received a letter from Oregon State College (OSC)'s School of Science, asking for ideas on individuals suitable to fill the position of Assistant Professor of Physical Chemistry. In response, Pauling endorsed Spitzer as "a first-rate man" and he was subsequently offered the job. Spitzer began work in Corvallis on September 16, 1946, devoted full-time to chemistry instruction and working with students enrolled in elementary and advanced physical chemistry, as well as chemical engineering. He also taught advanced classes in chemical theory for graduate students.

Spitzer was pleased with the size of his new school, the small town and the quality of education. In turn, Pauling was happy to have helped a friend and was glad to know that Ralph was enjoying his time in Oregon and that Therese was taking undergraduate courses at OSC. In the months that followed, Pauling continued to encourage Spitzer to do research, as he felt there were many great opportunities that lie ahead for him at the college.

***

In the meantime, Spitzer was becoming increasingly interested in social problems, particularly concerning the atomic bomb, and was thinking more about work he could do to "preserve peace and civilization." One outgrowth of this was a visit to Reed College that may have been partly responsible for the formation of the Portland Association of Scientists. Spitzer also planned to apply for a fellowship abroad in which he could study economics and philosophy, as well as physical chemistry.

The 1948 presidential election was likewise beginning to play a large role in Spitzer's life as he became an active supporter of Henry Wallace. In Spitzer's view, "a whopping big vote for Wallace, whether he wins or not, would serve notice that our bipartisan foreign policy of preparing to win the next war was not what the American people wanted." It was at this point that Spitzer asked Pauling to nominate him for the overseas fellowship, expressing his hope that he would be back home in time to participate in the presidential campaign. Pauling complied, recommending Spitzer as:

"...nearly an ideal man for such a job, combining as he does a sound understanding of the physical sciences and a keen interest in social sciences. He is just the sort of man that we must interest in the social sciences."

Unfortunately for Spitzer, the President of OSC disagreed.

***

On February 8, 1949, OSC President August Strand called Ralph and Therese into his office. The purpose of the summons was to inform Ralph that his contract would not be renewed because "he had become much more interested in 'other matters' than he was in teaching chemistry." In this meeting, Strand also told Therese that the Progressive Party group on campus, of which she was a member, would have to cancel an upcoming meeting because he disapproved of their guest speaker.

Spitzer quickly wrote to Pauling to communicate what had happened. In his letter he emphasized that there was no question of his competency, nor had he been accused of being delinquent in his departmental duties. Encouraging Pauling to get involved, Spitzer added,

"I think if we can smash these attacks on academic freedom and out their democratic rights in the next few years, we can fight off fascism permanently. I am sure you are working hard on this problem and hope that it is possible for you to lend a little assistance."

Pauling responded with shock and added that he would do everything he could to get to the bottom of the matter. He would later write to Strand in Spitzer's defense, specifying that he did not agree completely with Spitzer on questions relating to politics, but that he did support him in his right to hold his beliefs.

***

Within days of the firing, stories were published with headlines reading, "Strand Lashes at Commie Professors" and "Dismissed Educators Just 'Not Wanted,' Says OSC Head." Both articles referred to Spitzer as well as L.R. La Vallee, Assistant Professor of Economics, whose contract had also not been renewed despite reassurance that his academic work had been satisfactory. OSC was soon described in print as "a battle ground" for the heavily debated topic of academic freedom, with the author explaining that, in the minds of many, any alliance with the party of Henry Wallace was synonymous with being a communist.

Meanwhile, President Strand's rationale for dismissing the professors began to clarify. In an address to the college's Faculty Committee dated February 23, 1949, Strand offered that Spitzer's dismissal was not motivated by his Progressive Party membership, but rather because he had followed the Communist party line through his support of an untenable scientific thesis. In this, Strand was specifically pointing to Spitzer's interest in the Lysenko theory of genetics, which de-emphasizes the role that biology plays in heredity and suggests instead that environmental factors are more prone to shaping individual characteristics. While Trofim Lysenko's work was focused mainly on plant genetics, leaders in the Soviet Union used his ideas to push the notion that life in a socialist state could cleanse the proletariat of certain bourgeois tendencies. In making his case, Strand pivoted away from uncomfortable questions of academic freedom by asking a question of his own: "How about freedom from party line compulsion?"

Strand's evidence for Spitzer's alleged Lysenkoism was a letter to the editor of *Chemical and Engineering News* that Spitzer had written in response to an H.J. Muller editorial claiming that science was being destroyed in the Soviet Union. Strand felt that the letter demonstrated Spitzer's support for Lysenko and disregarded what he must have known to be scientific truth.

Spitzer ridiculed Strand's line of reasoning as both preposterous and a clear infringement of academic freedom. In a one-page typewritten response, Spitzer made his case:

I did not support Lysenko in my letter; in any case, it is absurd to reason that agreement with a Soviet scientific theory is evidence of adherence to a party line [...] I did not stir up controversy, but rather commented on an editorial on Soviet genetics. The editorial was by a chemist, in a chemical journal, and was discussed by two other chemists in the same issue.

***

On February 28, 1949, his 48th birthday, Pauling wrote a letter to Strand, stating that he was "greatly disturbed" by the termination of Spitzer's appointment. Speaking as a friend of Spitzer, as an alumnus of OSC, and as the current president of the American Chemical Society (which declined to intervene in the case), Pauling emphasized that Spitzer was an American citizen and that it was his civic duty to take an active interest in politics. Pauling concluded by urging Strand to reconsider his actions.

In his reply of March 4, Strand responded that Spitzer's letter to *Chemical and Engineering News* "showed beyond question that he was devoted to Communist party policy regardless of evident truth." He then continued,

"How far need we go in the name of academic freedom? How stupid need we be and just how much impudence do we have to stand for to please the pundits of dialectical materialism?"

From there, the president concluded fatefully, stating that,

"If by this action, Oregon State College has lost your respect and support, all I can say is that your price is too high. We'll have to get along without your aid."

Pauling's letter, as well as Strand's flinty response, were both published in the OSC student newspaper, *The Daily Barometer*, and later reprinted in *Chemical and Engineering News*.

***

Author Suzanne Clark, in her book *Cold Warriors: Manliness on Trial in the Rhetoric of the West*, wrote of what followed next:

> "Spitzer defended himself vigorously, if with a degree of innocence about the growing power of those who would finally be enlisted to anticommunism. He pointed out that cases such as his own served to damage academic freedom in hundreds of invisible ways as faculty members learned to be afraid. Spitzer immediately turned to the AAUP [American Association of University Professors] on campus, which declared itself without jurisdiction, and asked the Appeals Committee of the OSC Faculty Council to investigate. He made four points: the head of the chemistry department was not consulted; the acting head had no complaints about his work; he had been promised a leave for a fellowship; and he had been promoted to associate professor."

But Spitzer's attempts to save his job did not bear fruit and, as it turned out, Strand was hardly operating on his own. In the report issued by the Appeals Committee, it was revealed that the desirability of reappointing Spitzer or of granting him a leave of absence during the 1949–1950 academic year had been in question since the previous October. Further, the decision against tendering reappointment was a culmination of various consultations on the departmental, school, and institutional levels that had taken place over several months, none of which officially pertained to political party affiliation.

The committee concluded that, in deciding not to renew the appointments of the dismissed junior faculty members, President Strand had acted entirely within his administrative rights, since Spitzer and La Vallee were Associate Professors who had not yet earned tenure. The committee also argued that the President had legal authority to refuse renewal of these contracts without any reasons given, so long as political activity was not specifically identified as the cause for firing.

The Spitzer and La Vallee controversy raised awareness among students at OSC, some of whom expressed their feelings in letters to *The Daily Barometer*. One student wrote,

"[Strand's decision] means that compliance to 'accepted' political thought is required of our college professors. It means that any person who disagrees with either Democratic or Republican party platforms is not a fit person to teach in this institution. It means that Dr. Einstein wouldn't be allowed to teach in our physics department since he has been active in supporting the Progressive Party. For the same reason, Dr. Linus Pauling, OSC graduate and present head of the American Chemical Society, would be considered unfit to teach here."

The conflict also caught the attention of the national press, including a piece by John L. Childs in *The Nation*, titled "Communists and the Right to Teach." Among other details, the article noted that a recent meeting of the National Commission on Educational Reconstruction had determined that "membership in the Communist Party is not compatible with service in the educational institutions of the United States."

Spitzer and La Vallee both made one final return to OSC on May 26, 1949, to speak about "Your Stake in Academic Freedom." The event was promoted on campus as "the story the *Barometer* didn't print."

***

For Pauling, the Spitzer incident was a source of lasting damage to his relationship with his alma mater. In a letter written to an OSC colleague in April 1959, Pauling summed up his feelings at the time:

"I wish that I could accept your invitation to me to participate in the symposium that you are planning, but I have decided, a number of years ago, that I would not return to the Oregon State College so long as the last word that I had from President Strand was his statement, published in the *Barometer*, that Oregon State would get along without me in the future."

And so it was that Pauling made no official visit to the Oregon State campus from 1937 to December 1966, when he returned to lecture on "Science and the Future of Man." Pauling's talk was delivered more than five years after the retirement of August Strand from the presidency of what was by then known as Oregon State University.

***

Prior to his firing, Spitzer had received a National Research Council fellowship to work in Boston on a book that would discuss the social responsibilities of scientists. However, the consequences of his dismissal from OSC appear to have followed him to New England as his grant, traditionally a two-year support package, was not renewed for the customary second year.

Their stay in Boston truncated, the Spitzers spent the summer of 1950 traveling in Europe, where Ralph delivered a talk on Isaac Newton to a history of science group and another on academic freedom at a meeting of the International Union of Students, which was held in Prague. Throughout their travels the Spitzers were observed from afar by secret police until, shortly before a scheduled departure to England, Ralph was arrested in Rotterdam. Incarcerated and held in isolation for a week without explanation, Ralph was finally placed on a ship bound for New Jersey, his passport cancelled, an experience that he later described as a kidnapping. Therese had been sent home one day prior; the couple were allowed only minimal communication with one another during the whole of the ordeal.

Back stateside, Ralph found himself unemployed for several months until accepting a job at the University of Kansas City in 1951. Two years later he was again terminated from his academic position; this time, a disagreement within the department prompted action from an aggrieved colleague, who used Ralph's past at Oregon State as ammunition. Not long after, Spitzer was summoned to Washington, DC to testify before the Senate Internal Subcommittee on Subversive Influences in the Educational Process. Heeding the advice of his attorney, Spitzer cited his Constitutional Fifth Amendment rights in refusing to answer questions about federal jobs and Communist Party membership.

***

At this point the Spitzer family, which now included two children, was ready for a change and a chance encounter with a newspaper advertisement prompted them to look north in their search for new opportunities. In 1954, having learned of the program in passing, Ralph entered an accelerated M.D. program at the University of Manitoba, where he focused on the application of chemistry to medical problems. Spitzer completed this training in three years and then immediately assumed a research position at Winnipeg General Hospital.

Around 1958 Spitzer began seeking work outside of Winnipeg. Before long he met Cam Coady, a renowned pathologist based at Vancouver's Royal Columbian Hospital. With Lindsey Sturrock, the trio created a firm called Coady, Sturrock and Spitzer, later renamed C.J. Coady Associates, and in due course the Spitzers moved to New Westminster, a suburb of Vancouver. In the early 1960s, the family moved again to a property within the University Endowment Lands, adjacent to the campus of the University of British Columbia (UBC), where the Spitzer children were raised and where Ralph and Therese spent the remainder of their years together.

C.J. Coady Associates proved very successful, and it was there that Spitzer spent the majority of his career, working as a chemical pathologist. Spitzer's unit specialized in the analysis of blood samples extracted from individuals suffering from any number of maladies. By examining the chemical composition of a given sample, the unit would offer counsel on medical diagnosis. In so doing they filled a niche that hadn't been occupied to any great degree province-wide, and business boomed.

***

While the move to Vancouver proved a terrific boon to Ralph's career, life in British Columbia was not without its hardships. In 1960 Therese was diagnosed with breast cancer. Treated aggressively, the cancer was kept fully at bay and Therese would live for another 39 years. More difficult was the situation of the Spitzers' son, Matthew. A brilliant boy diagnosed as schizophrenic at the age of 17, Matthew suffered mightily with mental illness, a battle that ended with his tragic death at 34.

As parents the Spitzers were profoundly impacted by Matthew's struggles, which they had attempted to combat using a number of approaches, including the ideas on orthomolecular psychiatry then being espoused by Linus Pauling. Their experiences also informed a book project titled

Psychobattery: A Chronicle of Psychotherapeutic Abuse, which was published in 1981. The book passionately argued against the psychiatric practices of the day using case studies researched by Therese, the publication's lead author, and medical explanations penned by Ralph. The primary target of what the book described as "psychobattery" was the Freudian argument, then current, that mental illness is caused by dysfunction between mother and child, and that treatment necessarily include separation of the patient from their nuclear family.

For the remainder of her life, Therese stayed engaged with social causes, at one point becoming actively involved in the picketing of a neighborhood liquor store that sold South African wine during the era of Apartheid. A strong feminist, Therese also returned to school later in life and met all but the dissertation requirements for a doctorate in Women's Studies at UBC. Therese's passions rubbed off on her daughter as well — Eloise Spitzer attended law school and made a career, in part, out of advocating for the rights of Canada's indigenous peoples.

In the early 1980s, Therese and Ralph met Hisako Kurotaki — a native of Japan who emigrated to Vancouver in the 1970s — through mutual acquaintances in the city's mental health community. Therese and Hisako quickly became dear friends. In 1998 Therese was diagnosed with gallbladder cancer and died just five days after her illness was confirmed. By then Hisako, who was also a paralegal, had been specified as co-executor of Therese's will. The duties that this required meant that Hisako and Ralph, the other co-executor, spent a great deal of time together. Ralph was struggling mightily with the death of his wife of 59 years and Hisako helped to rehabilitate him, both physically and emotionally. Gradually, Hisako and Ralph fell in love and eventually married. As Ralph recalled in a 2013 interview, "I've dated two women in my life and married both of them."

Ralph Spitzer passed away on October 17, 2018, at the age of 100. A terrific athlete, prodigious traveler and accomplished bonsai gardener, Spitzer also built an organ in his living room that eventually came to include 2,000 pipes; the project necessitated raising the roof of his home, with certain pipes running from the basement to the top of the heightened ceiling. Stenciled at the base of the living room pipes, which stretched high above, was a Latin phrase: *Ars Longa Vita Brevis* — "Art is long, life is short."

Ralph and Therese Spitzer in the living room of their Vancouver home, ca. 1970s.

Chapter 13

The Black Student Union Walkout

Rich Harr speaking during the Oregon State University Centennial lecture walk-in, February 25, 1969. Linus Pauling is seated at the back.

The racial tensions that escalated across America's college campuses throughout the 1960s were evident at Oregon State University (OSU) as well. One high-profile event that served as a pivot point for race relations at Oregon State occurred in late February 1969, when members of the OSU Black Student Union (BSU) interrupted a convocation hosted by President James Jensen at OSU's Gill Coliseum. The convocation, which was part of a series of events marking the university's centenary, was to feature a speech by Linus Pauling, Oregon State's most prominent alum.

There was an uneasiness surrounding the talk from the outset, both with respect to Pauling's presence on campus as well as the way in which he was introduced to the large crowd that assembled for his lecture.

These tensions dated back many years, stemming from a schism that developed between Pauling and the school in 1949, when Pauling's former graduate student was fired from his faculty position at Oregon State College because of his political beliefs.

Though the ice between Pauling and OSU had been broken a couple years prior, the situation remained awkward as he arrived on campus for the centenary lecture. Pauling's eyebrows were raised in particular by a decision that university president Jensen would not introduce his talk. Instead, Bert Christensen, who was chair of the OSU Chemistry department, was asked to fill this role. This choice was far from customary for a visitor of Pauling's magnitude and was viewed by many as an affront. Pauling's surprise at the breach in protocol was, however, eclipsed by what happened once the event began.

***

The BSU's decision to interrupt the centenary convocation and stage a subsequent walkout was sparked by reports that OSU football player Fred Milton, an African American, had broken team rules by refusing to shave his goatee. Although this conflict occurred during the off-season, Oregon State football coach Dee Andros — an ex-Marine affectionately known as "The Great Pumpkin" — believed that he still maintained authority over his players and their appearance. Andros gave Milton a 48-hour deadline to comply with the team rule, warning that, if he continued to resist, he would be dismissed from the team and forfeit his OSU scholarship.

Though the immediate implications for Milton were significant, the conflict was also symptomatic of a longer history of racial tensions on campus, as well as concurrent protests related to tuition hikes and the escalation of the Vietnam War. As a subsequent editorial in *The Daily Barometer* student newspaper put it, "this issue is clearly much more than that of a little beard."

Promptly after learning about Andros's ultimatum, the BSU began planning peaceful measures to publicly express their solidarity with Milton and to bring awareness to the struggles that African American students were facing on the OSU campus. The actions that they agreed to initiate included a sit-in at a public event, a class boycott, and a campus walkout. The group also began publishing an underground newspaper, *The Scab Sheet*, which "sold briskly for a nickel" and served as an alternative to *The Barometer*, whose editorial stance was generally sympathetic to Andros's position.

The BSU believed the Milton case to be an infringement of a student's rights to individual self-expression. The group also pointed out that this was not the first instance of a Black student athlete coming into conflict with Andros's policies; in the past, others had been told to keep their hair short and to not wear medallions. BSU President Mike Smith explained that, although the policies were extended across the athletic department, they were based on standards set by white society, and that Black student athletes were pressured to conform to them.

***

The first of the BSU's peaceful protest actions began with a sit-in at a campus lecture, Pauling's lecture, which was to be given on the morning of February 25, 1969. The speech, titled "Advancement of Knowledge: Ortho-Molecular Psychiatry," was one of seven presentations scheduled over three days as part of the OSU centenary celebration. The lecture series was meant to mark the first hundred years of Oregon State by looking toward the future through the prism of "The Second Hundred Years." Each talk was to be followed by a discussion period in which students would be given the opportunity to dialogue with the invited speaker. To encourage campus participation, the university cancelled all classes that conflicted with the seven convocations.

As Pauling was being introduced, an estimated 70 BSU members and supporters filed into Gill Coliseum, the school's 10,000-seat basketball arena. The BSU students subsequently took control of the dais, while Pauling remained seated, and the environment quickly became tense. As later reported by *The Barometer*,

"Before the public address system was turned on, several catcalls were hurled toward the students. 'Go home you [goddamn] nigger,' one student shouted."

The Oregon Stater alumni magazine likewise noted this ugly moment but reported too of "a mathematics professor who stood up, rapped his umbrella sharply on the balcony railing, and cried, 'Let them be heard! Let them be heard!'"

Undaunted, Mike Smith, the BSU president, and sophomore football player Rich Harr explained the group's reasons for staging the interruption and then announced a boycott of athletic events that would start that weekend. The speakers also called for white student support, noting that this was not just about the treatment of Black student athletes, but of all students on campus. These sentiments were repeated later at a rally held in front of OSU's Memorial Union.

After about 20 minutes, the protesters left the gym. The large crowd that had assembled for Pauling's talk gave a mixed response, though most students applauded the BSU's statements. In an oral history interview conducted in 2011, OSU chemistry professor emeritus Ken Hedberg, a close friend of Pauling's, remembered the participants as having been "very well behaved."

In an interview with *The Barometer* published three days later, Pauling made clear that he had not received a straight answer from university administration concerning the rationale for Andros's policy. When asked if he supported the BSU's efforts, Pauling replied, "I support all demonstrations for individual rights." He also reflected on his own student experiences at OAC in the late 1910s and early 1920s, recalling that:

> "I was such a conformist. We as students didn't even ask if the rules were proper ones. We should have demonstrated, and I should have been the leader. We should have been aware of world affairs as students are today... We should have been concerned with justice."

Referring to Milton as "a very good-looking young man," Pauling noted that he too had worn a beard for a time and expressed his belief that young people must be given the freedom to "explore and be different" and that "we would have an increase of world happiness with people allowed to be individuals."

Pauling also expressed his belief that the sentiment displayed on his alma mater's campus mirrored trends across higher education and that, as with many other institutions, the roots of these conflicts lay mostly with the administration's inability to recognize the larger problem and to take measures to fix it.

***

Multiple rallies and counter-rallies were held in the days following the sit-in, and President Jensen took steps to reconcile with the BSU but did not meet with success. On March 1, the BSU announced the next in its series of actions: a class boycott scheduled for March 4. Hundreds of students, faculty, and staff joined in support. Athletic events were also boycotted at OSU and elsewhere, and Black athletes in the PAC-8 Conference joined the protest by refusing to participate in games against OSU.

Amidst this, tensions in Corvallis continued to rise. As documented by the *Oregon Stater*,

> "BSU adviser and acting minority affairs director Karl Helms reported threats on the lives of his small children. A bomb threat cleared Education Hall and another threat was made on McNary Hall, a student residence. A group of Black Panthers appeared on campus for a day, were closely watched by state policemen and left. That day was the only time police were visible on campus in numbers.
>
> "[…] At home, threatening phone calls began. Blacks were told in foul language to leave town or die. … 'If racism was latent before,' Helms said, 'it certainly was overt now.'"

The protests reached their crescendo on March 5, when 47 Black students — essentially the entire African American student population at the school — marched out of the main gate on the east side of OSU's campus. The walkout began with a rally held in the Memorial Union that included a talk delivered by BSU President Smith. Speaking to a gathering of more than 1,000 faculty and students, Smith stressed that Black students could no longer accept "the plantation logic" upheld by the administration and athletic department at OSU, a "hallowed institution of racism." Commenting on these events the next day, the *Scab Sheet* suggested that the OSU walkout was the first of its kind at an American college or university.

The Oregon State students were joined in the walkout by over 100 members of the University of Oregon's Black Student Union, who chartered buses and drove up to Corvallis to participate in solidarity. There was also a rally held in sympathy at Portland State University (PSU) to support the actions on OSU's campus. The president of PSU's Black Student Union spoke at that rally and condemned OSU's "policy of tradition" as being "not in accord with what's going on today."

On March 6 the OSU Faculty Senate convened and passed an amended version of an "Administrative Proposal" that had been drafted by the school's Office of Minority Affairs. This proposal included the creation of a Commission on Human Rights and Responsibilities. The following day, three students withdrew from the university to seek education elsewhere and several others expressed their intent to transfer at the conclusion of the term. In response, the Faculty Senate met again to declare an emergency, an action that allowed the students who withdrew to receive an incomplete on their transcripts rather than a failing grade. Faculty members at the University of Oregon also met around this time to consider a proposal that would allow Black students from OSU to be admitted as expediently as possible, should they choose to transfer.

In an editorial, the *Scab Sheet* expressed a lack of surprise that the Black students had left. After all, "Arrayed against them was a coalition of the University administration, the Athletic Department, the various athletic supporters, white athletes, Chamber of Commerce and alumni."

Furthermore, the university had earned a reputation for being ill-prepared to assist students of color and had compiled a record of inaction in handling problems of this sort. In fact, as explained by the president of the Associated Students of OSU, *inaction* seemed to be the university's formal policy. Indeed, one year before, the university had turned back over $100,000 in unspent federal funds that had been earmarked for the recruitment of historically underrepresented students. In the view of the *Scab Sheet*, the situation had deteriorated "to the point that Blacks could maintain pride and self-respect only by disassociating themselves completely from Oregon State University."

***

The departure of many Black students from OSU's campus in 1969 altered student demographics for years to come and exacted lasting damage to the university's reputation. With respect to athletics, Dee Andros was not able to convince a single African American player to join his 1970 recruiting class, and from 1971 to 1998, OSU's football teams posted consecutive losing records, still the longest run of futility in the history of Division I football.

Fred Milton, whose refusal to shave his beard brought decades of tensions to a head, ultimately transferred to Utah State University. Milton later enjoyed a successful career at IBM and Liberty Mutual Insurance, before moving into the public sector as a civil servant working for the city of Portland and Multnomah County. He passed away in 2011 at the age of 62.

Though a painful moment in university history, the actions taken by the BSU in winter 1969 led to direct and meaningful changes on the Corvallis campus. Later that year, OSU established the Educational Opportunities Program, which was designed to help recruit and retain students of color. Three cultural centers were also established on campus, each devoted to creating community spaces for underrepresented groups and platforms for sharing their experiences across the university.

Chapter 14

A Sentimental Trip

Linus Pauling with his sister Lucile at the 60th reunion of the Oregon Agricultural College class of 1922. Standing behind him are sister Pauline and her husband Paul Emmett.

In the final months of 1981, Ava Helen Pauling was slowing down and making her final public appearances. Diagnosed with a form of inoperable cancer, Ava Helen had decided against the use of chemotherapy and knew that her life was coming to an end.

According to his family, Linus chose to believe otherwise, in large part because he was convinced that he could save her through the use of vitamin C and other supplements. He was unable to talk about her final arrangements, so preparations including Ava's memorial service preferences and her desire to be cremated were discussed with daughter Linda over a long weekend. After surgeries and a number of medical complications, Ava Helen passed away in her home on December 7, just a couple weeks shy of her 78th birthday.

Following the death of his wife of nearly 60 years, Pauling was, understandably, badly adrift. His children helped guide him through the immediate aftermath and, though he gladly accepted their help, he was resistant to other offers of assistance in everyday life. Instead, he stayed as busy as he could and, over the course of 1982, published three papers on the nucleus of the atom — a highly technical and abstract project that afforded him some measure of escape from his grief.

He remained very lonely, however, was often lost in thought, and was having trouble accepting the reality of his wife's death. Of this period, biographer Thomas Hager wrote,

"He still talked to her, holding phantom conversations as he spooned his vitamin C powder into his juice in the morning. He still looked for her, expecting to see her in the doorway, asking him to stop and take a walk, to come to lunch. He would cry and look out to sea. Then he would get back to work."

Though he was managing to maintain his health and daily affairs, Pauling clearly needed an outlet for his sadness. An opportunity to do so came about when he received an invitation to attend the 60th reunion of the Oregon Agricultural College class of 1922. He decided to attend and set off on what would become a meaningful journey.

***

As he thought about his travel itinerary, Pauling identified a desire to revisit several locations that held meaning for him and Ava Helen. His first stop, on June 6, was Dayton, Washington, where he had worked for the Warren

Construction Company and lived with Ava Helen for one month just after their wedding. As he walked around the little town, he noted the spot where they had lodged and also found the location where Ava had outscored him on an IQ test they had both taken.

The following morning, he drove across the border into Oregon, traveling to Arlington and then Condon, where he visited the grave of his grandfather Linus Wilson Darling for the first time. He spent the next day on the Oregon coast, seeking out former vacation and employment spots in Gearhart, Tillamook and Seaside. He then drove to Corvallis, where he stayed for a few days before attending the Oregon State University Golden Jubilee, a joint reunion of the Classes of 1922, 1927 and 1932 that was attended by more than 600 alumni.

The day after his reunion, Pauling spoke on the capitol steps in Salem at a nuclear weapons freeze rally. He spoke again that same night, once again on peace topics, at the First Methodist Church in Portland. The next day he met with his sisters, a cousin and the director of the Oregon Historical Society to donate two diaries that Linus Wilson Darling had kept in the late 19th century.

After lunch with his relatives, he began his drive back home, stopping at a portion of highway along Grave Creek where he had lived for five months in 1919 working with a road crew and sleeping in a tent near a covered bridge. At the time of his visit, the covered bridge was still in existence, but the highway was partially destroyed by the construction of Interstate 5.

Pauling finally made it home two days before his wedding anniversary, having driven a total of 2,400 miles. It appears that the trip was just what he had needed, providing a frame of reference and partial relief from his loss. In a letter to an old friend, Pauling described his travels simply and decisively, offering that "I went on this trip mainly to visit places where I had lived long ago."

Following his return, Pauling decided to move out of the Portola Valley, California house that he and his wife had shared together. His youngest son Crellin moved in with his family, while Pauling bought a condominium on the Stanford University campus. He moved some of his belongings to his ranch at Big Sur, others to Stanford, and decorated his new space with pictures of Ava and himself, framed awards, and mementos from their world travels. The changes helped, but only to a degree. In September he wrote to his best friend, Lloyd Jeffress, "I am getting along pretty well, but I still feel quite lonesome. I have been working hard."

Pauling also became involved once again with his institute, and in early 1983 settled a lawsuit that had been consuming valuable time and resources. He spent half of his time at his ranch, the other half in Palo Alto, and developed a routine that worked for him, waking up before five in the morning and reading himself to sleep at night. Despite his loneliness, Pauling would live for another 12 years, continuing to pursue his scientific work, speak on world peace and manage his affairs.

Part II
Caltech Administrator

Chapter 15

Introduction

Linus Pauling at about the time he became Division Chair in 1937.

In 1937 Linus Pauling took on an administrative appointment that came with two titles: Chairman of the Division of Chemistry and Chemical Engineering, and Director of the Gates and Crellin Laboratories of Chemistry at the California Institute of Technology. In doing so, he succeeded long-time head A.A. Noyes. Prior to this moment, Pauling had assumed administrative responsibilities that prepared him for the position and demonstrated to his superiors that he was a suitable candidate. By 1937 Pauling had also long since proven himself to be a world-class researcher and his rank had advanced accordingly: hired as Assistant Professor of Theoretical Chemistry in 1927, he was promoted to full professor just four years later. Importantly, Pauling's research interests also led to the fostering of a strong working relationship with the Rockefeller Foundation during a key moment in institutional history.

***

From the very beginning of Pauling's tenure as chair, the need for and allocation of space ranked high as an ever-present concern, and this is a topic with which Pauling was well acquainted. A member of a 1929 sub-committee charged with exploring ways to improve graduate instruction and research in physical chemistry, Pauling found that space devoted to graduate research was a pressing need and advocated that the division act accordingly. Later, Pauling himself dealt with shortages in space when compelled to move his laboratory to the astrophysics building beginning in 1932. Once Pauling became chair, these problems continued to linger, if softened somewhat by the construction of two new facilities, the Crellin and Church Laboratories.

In addition to gross square footage, the organization of available space was a regular topic of discussion. During his years as chair, A.A. Noyes sought to address the issue by assigning spaces according to research program, with areas for inorganic, organic, physical, and applied chemistry designated within the newly occupied Gates Laboratory. Pauling took issue with this approach, writing to Noyes in 1931 that the compartmentalization served "no useful purpose and would seriously weaken the Division by the introduction of artificial barriers."

It was Pauling's opinion that the division ought to continue promoting its well-regarded physical chemistry program instead, rather than incur the risk that organic chemistry, a more emergent program, begin to overshadow an existing area of strength. "I am not opposed to the development of work in organic chemistry," Pauling hastened to add, "but I feel that the work in physical and inorganic chemistry is one of the Institute's strongest assets, and that development of organic chemistry should not be made at the expense of physical chemistry."

Pauling even went so far as to propose a floor plan, one in which the sub-basement, basement, and first floor would be devoted to physical chemistry, and the second and third floors to organic. Pauling further suggested that, as the division continued to grow to the point of overcrowding, a new building devoted to organic chemistry could be built, leaving physical and inorganic chemistry to occupy all of Gates. As it turned out, Pauling's vision proved accurate: a new building did come very soon, with construction of the Crellin Laboratory of Chemistry first proposed in 1935 and completed in 1938, not long after Pauling took up the chairmanship.

***

In the years prior to his taking charge, Pauling also developed a reputation as an advocate for his fellow faculty, a stance that sometimes put him at odds with the Institute's upper administration. In 1932, Robert Millikan, a Nobel Laureate who was then the Chairman of the Caltech Executive Council, asked that the faculty vote to take a 10% pay cut amidst the economic depression then gripping the United States. Pauling strongly opposed this request, noting that only the Institute's Board of Trustees could invoke such an action.

Three years later, Pauling voiced his support for raises that were due to newly tenured colleagues Richard Badger and Don Yost but had been delayed because of continuing budget woes. Pauling argued that the bumps in pay would help the division maintain its position as a leader within the profession by rewarding the successes of deserving researchers. As Pauling told Noyes, "I feel that in university administration, just is to be esteemed above expediency, and a satisfied staff above a balanced budget." Pauling's attention to faculty pay remained a hallmark of his tenure as chairman. Indeed, one of his final gestures as division leader, put forth in 1957, was the donation of a $1,500 gift earmarked for Caltech faculty salaries.

***

Another issue with which Pauling would grapple as chair was the imperative that the division be properly equipped, a problem that he had encountered in his own research. In 1930 Pauling spent part of his summer at Arnold Sommerfeld's Institute for Theoretical Physics in Munich, and upon his return to Pasadena, he requested institutional support for an electron-diffraction apparatus that was similar to Sommerfeld's. As with his advocacy of faculty raises, Pauling's request was in keeping with his ambition that the division maintain a position of prominence, this time in structural chemistry research.

In making his case, Pauling argued that the research infrastructures at campuses like the University of Chicago and the Massachusetts Institute of Technology were beginning to leave Caltech in their wake. He expressed this concern to Noyes, writing,

"I should not like to have this laboratory, which has played a significant part in the development of crystal structure since the early days, fall far behind the other and newer crystal structure laboratories in this country."

Pauling likewise believed that researchers themselves, rather than administrators, were in the best position to determine what sort of laboratory equipment was needed to carry out cutting-edge work. And though an admittedly risky proposition, he felt that each researcher should be given their own funding to do with as they pleased. Again to Noyes,

"The most interesting experiments are the least safe — those which might give a surprising result, but which might fail. It is difficult to use these as an argument for buying new apparatus, inasmuch as success cannot be guaranteed. I feel nevertheless that these experiments are fully important as the routine ones."

In the years that followed, Pauling continued to exert influence on decision-making related to the division's general equipment needs and became a formal member of its Equipment Laboratory Committee in 1935.

***

Without doubt, a major factor behind Pauling's elevation to chairman was the strength of his relationship with the Rockefeller Foundation and, more specifically, its Director of the Natural Sciences, Warren Weaver. Pauling had been cultivating ties with the Foundation for at least five years prior to his appointment as chair. In July 1932, he secured Rockefeller funding under what he later described as a "small grant" for $10,000 per year (more than

$220,000 in 2025 dollars) for crystal structure work. This grant was renewed twice and served as a lifeline during difficult economic times. After those three years had passed, Weaver told Pauling that the Foundation would no longer fund his current line of research, but that they would be interested in its biological applications.

Following Weaver's guidance, Pauling began to redirect his focus towards topics in biochemistry. In doing so, he requested $5,000 per year from Caltech's Executive Council to supplement a potential $10,000 annual award from the Rockefeller Foundation, an amount that he ultimately received for three years beginning in 1935. Following the Foundation's approval of this grant, Pauling wrote a thank you letter to Weaver in which he confided that he had already begun preliminary investigations on the structure of hemoglobin. "As I have read about the problems of biochemistry," he added, "I have become more and more enthusiastic about the possibilities of the application of our methods."

For Weaver, Pauling was not only a key player in advancing biochemical research but also a valuable confidant in evaluating proposals submitted by other labs. One case involved a researcher named O.L. Inman, who worked at Antioch College in Ohio. Inman had requested Rockefeller support for studies of chlorophyll that were similar to Pauling's work on hemoglobin, and had asked specifically if he could bring in someone who had worked on hemoglobin in Pauling's lab. When asked for his input, Pauling told Weaver that Inman's idea was doomed to failure since chlorophyll lacked paramagnetic atoms, and Weaver promptly rejected Inman's proposal.

✳✳✳

In an undated note likely written in the mid-1940s, Pauling reflected on his relationship with the Rockefeller Foundation and the role that it played in influencing his research trajectory. "Perhaps," Pauling wrote, "the remark from Weaver that my grant for molecular structure was all right, but that the main support was going in another direction, and the hint that application of m.s. [molecular structure] to biological problems might interest the Foundation greatly" had indeed made an impact on his decision-making.

That said, Pauling did not agree with the notion that Weaver's encouragement had diverted him from more focused attention on purely chemical subjects. Rather, Weaver's suggestion had opened up vital new research pathways of which Pauling had been unaware and that he subsequently became eager to explore. Pauling further described his relationship with the Rockefeller Foundation by likening it to a joke he had read in the *Saturday Evening Post*.

"A young man and young woman were saying goodnight at her door. She said, 'I'll give you a kiss — I owe it to you for bringing me all the way out to 155th Street, and next week I'm going to move out to 242nd.'"

Regardless of its impact on his research agenda, Pauling's willingness to follow Weaver's suggestions, and the reliable funding that flowed from there, would prove to be the tipping point in Pauling's ascent to division leadership.

Chapter 16

Becoming Division Chair

Pauling in lecture, 1935.

In 1935 A.A. Noyes was diagnosed with colon cancer, a serious turn of events that would require him to step down from his position as Chairman of the Division of Chemistry and Chemical Engineering at the California Institute of Technology. In thinking about who might next take the reins, Noyes favored the idea of promoting a strong researcher rather than an experienced administrator and was likewise keen to continue strengthening the division's ties to the Rockefeller Foundation. With these criteria set, Noyes quickly settled on Linus Pauling as his favored successor.

Pauling was aware of Noyes's preferences and gradually began to press the issue himself. When July arrived and little movement had been made toward appointing a new chair, Pauling approached Robert Millikan, professor of physics and Chairman of the Executive Council at Caltech, to make his case more aggressively. As a close friend of Noyes, whose health was clearly on the decline (he would die less than a year later), Millikan was infuriated by Pauling's apparent insensitivity to the circumstances. But this did not stop Pauling: within two weeks, after thinking the situation over, he addressed Noyes directly by letter, claiming that he was considering leaving Caltech since the promised chairmanship had apparently been taken away.

***

For his part, Noyes still wanted Pauling to succeed him, and he passed Pauling's letter on to astronomer George Ellery Hale, who had been central to shaping Caltech into a prestigious institution over the previous two decades. Noyes also met in person with Hale, Millikan, and physicist Richard Tolman to discuss the question of the division's leadership.

Millikan favored Tolman for the position, in part because he was concerned that Pauling's modest upbringing would impact his ability to engage with and woo wealthy donors. Noyes also harbored concerns about Pauling's leadership style, noting that, in the laboratory, Pauling was inclined to delegate specific tasks to his students and staff rather than allowing those under him the opportunity to think through problems for themselves.

Ultimately the four decided that the best course of action was to split the leadership of the division in half. Pauling would be anointed as chairman but on condition that he work with a new Chemistry Division Council, to be comprised of a selection of five fellow faculty. The group also decided that Tolman would represent the division to the Caltech Executive Council and would retain primary responsibility for interacting with donors.

The creation of the Chemistry Division Council, which was modeled on the Institute's existing Executive Council, reflected — and attempted to institutionalize — the inclusive management approach that Noyes had fostered during his tenure. In a letter requesting that the Executive Committee establish the Division Council, Noyes and Tolman emphasized that the chairman would work with the council in a "spirit of cooperation," and that part of the chair's role would be to bring matters before the council and make recommendations.

A separate memo further clarified the roles to be played by the governing parties and suggested that the chair would represent the division to the broader Caltech community but would be constrained by certain restrictions. The memo also envisioned the council as having "final authority and responsibility" for making recommendations to the Executive Council concerning budgets and major expenditures, staffing and promotions, and decisions on the usage of laboratory space. In doing this work, the council would be tasked with meeting every month during the academic year and also when called by the division chair.

***

Pauling received a copy of this memo and, from his annotations, it is clear that he was not in agreement. Pauling highlighted in particular the passage specifying that the chairman would

> "personally decide all administrative questions, except that he will refer matters upon which a consensus of Division opinion is desirable to the Council or to the Committee of the Division, or to the Division as a whole, as indicated in the statement given below of their respective functions."

The various restrictions outlined in the memo were unacceptable to Pauling and he refused to sign off on its contents. Instead, he replied to the memo with a written rejoinder addressed to the Executive Council. In it, Pauling expressed his feeling that the Division Council approach would prove inefficient and stagnate the progress of the unit. "The more reactionary and less ambitious members of the group," he worried, "will determine its policy, inasmuch as to move ahead is harder than to stand still." More specifically, Pauling was concerned that the council would be ruled by those who were most out of touch with current trends in research and instruction, and that the quality of the division would suffer accordingly.

Hesitations about trying to work within this structure, compounded by the difficult financial times being endured nationwide, were such that Pauling chose to decline the chairmanship under the terms offered.

"I would not accept appointment as Chairman of the Division with authority vested in a Council, inasmuch as it would be impossible or difficult to build up the Division under these circumstances. With someone else as Chairman, I would not feel called on or justified in making any effort to build up the Division, this being then the responsibility of the Chairman. Professor Morgan [geneticist and Caltech biology head T.H. Morgan] says that there is no chance of building the West Wing of Gates for five years, no chance of increasing the Chemistry budget, no chance of getting new staff members, no chance that the Institute would promise an increase in budget at some definite time in the future. With no prospect of developing the Division, I would not accept its Chairmanship."

Ignoring Pauling's objections, the Executive Council approved the creation of the Division Council on November 2, 1935, the day after Pauling authored his letter. From that point, it would take more than two years to resolve this disagreement between Pauling and upper administration.

**

Central to the healing process was Warren Weaver at the Rockefeller Foundation. In March 1936, Weaver informed Pauling of the Foundation's interest in supporting "an attack on cancer from below (structure of carcinogenic substances, etc.) but not from above." The following month, further details about the Foundation's proposed level of support were shared at a Division Council meeting, where it was conveyed that the grant could fund research in organic chemistry at a rate of $250,000 over five to seven years, with an additional $50,000 going to the Division of Biology. The Division of Chemistry and Chemical Engineering was asked to submit its grant application by August. Needless to say, a potential windfall of this magnitude served as powerful motivation for the unit to shift its attention toward biochemistry and also provided Pauling with significant leverage in his pursuit of the chairmanship.

This leverage first began to manifest when Noyes put Pauling in charge of identifying three research fellows to attach to the grant. The previous year, Pauling had conducted a similar search and was unsuccessful. During this first attempt, Pauling had sent out letters to chemistry departments and medical schools at the University of Chicago, the University of Michigan, Columbia University, Washington University, and Harvard University, describing the ideal candidate as "original and energetic" but not in need of plum facilities to carry out effective research. This second time around, Caltech's relatively meager facilities would be less of a problem. The potential Rockefeller grant was partly responsible for this, as was a plan to begin construction on a new building the following year.

In November 1935, retired Pasadena steel magnate Edward Crellin, along with his wife Amy, informed A.A. Noyes of their wish to provide majority funding for a new facility to be used by the Division of Chemistry and Chemical Engineering. Planning for this space promptly commenced and, by the next month, had advanced to the point where Noyes could tell the Division Council that construction would begin in spring 1937. In the meantime, fundraising and design work continued, a process that was aided by Crellin's forgiveness of a $60,000 annuity owed to him by the Institute.

By spring 1936 the final plans appeared to be coming together for the new space. It would be called the Crellin Laboratory, while the division's existing facility would remain the Gates Laboratory. Though good news all around, one issue of concern was the square footage to be made available for biochemistry research supported by the Rockefeller Foundation. Another key need was finding a person to organize the laboratory's operations.

Pauling was tasked with steering the search for the head researcher position but, as had happened the previous year, his inquiries yielded few leads. In his correspondence with Thorfin Hogness, a close colleague at the University of Chicago, Pauling learned that they too were looking for someone similar. This shared difficulty in recruiting fresh faces reflected a growing nationwide interest in biochemical research that had been catalyzed, in no small measure, by priorities set forth by the Rockefeller Foundation.

More promising suggestions came from Moses Gomberg at Johns Hopkins University and also from the Rockefeller Foundation itself, once again in the form of Warren Weaver. Gomberg suggested Edwin Buchman and

Weaver suggested Carl Niemann, both of whom struck Pauling as being too early in their careers to satisfactorily fill this position. Making a trip back East, Pauling began to search in person, interviewing "young bio-organic chemists" and getting "advice from several older ones."

As it turned out, the most important conversation that Pauling had during this trip was with Weaver. In it, Pauling learned that the Rockefeller Foundation would not commit to even a "small grant" for "preliminary investigations," if the division was not able to meet an April 1937 deadline for submitting a "well worked out plan" for initial implementation that following September. Weaver also shared his sense that the Foundation's trustees would likely not consent to supporting a program headed by "only young and relatively untried men at the beginning of their careers."

This guidance in hand, Pauling proposed that Caltech hire a mid-career candidate who had already made significant contributions and then add young men who could be groomed by this individual. The only person whom Pauling had met so far who approximated the leadership description was Hans T. Clark, a member of the biochemistry faculty at Columbia University, but Pauling was concerned that Clark's research was not "outstanding." Another possibility was Samuel Gurin, a National Research Fellow at the University of Illinois and Pauling's favorite among the "young men" whom he had interviewed.

In the end, neither Clark nor Gurin was hired. Instead, Edwin Buchman and Carl Niemann — the two candidates suggested by Moses Gomberg and Warren Weaver — were brought on as temporary junior appointments. Though young, both were brimming with promise: Niemann had already published ten papers by the age of 25, and Buchman had secured stable funding for his research on vitamin B1. Ultimately both would stay at Caltech for the remainder of their careers.

**

On June 6, 1936, with the Crellin facility still in its planning phase and a succession plan for division leadership not yet in hand, A.A. Noyes passed away. For many at the Institute, sadness at Noyes's passing was amplified by feelings of resentment toward Pauling, the likely new chair, who was widely seen as having pressed the issue of succession too aggressively during

Noyes's final days. For this and other reasons, several colleagues favored the appointment of physicist Richard Tolman instead, forcing him, just four days after Noyes's death, to clarify that he would not accept the position if offered.

Tolman's preemptive refusal fell on deaf ears with the Executive Council, who recommended him anyway. Tolman, who was already Dean of the Graduate School and whose research did not comfortably align with the division's broader work, responded to the council once more at the end of June, explaining,

> "I am very appreciative — and indeed quite touched — by this expression of confidence on the part of the members of the Executive Council. Nevertheless, both from the point of view of my own work and from that of the welfare of the Institute, I do not think that it would be wisest for me to accept."

Tolman still favored the plan that he had devised the previous year with Noyes, George Ellery Hale and Robert Millikan, in which it was recommended that Pauling take over as leader but work in tandem with the Division Council. Tolman noted that as "an outstanding chemist, who is actively engaged in chemical research, who has a good knowledge of the chemical work being done in this and in other countries, and who is himself recognized as a man who is now making important contributions to chemistry," Pauling was the perfect candidate. Tolman also hoped that the Division Council structure could be maintained and stressed that he would continue to serve on the group, since it was a position from which he could effectively benefit the division.

Despite this vote of confidence, Pauling's strong objections to the Tolman group's proposal in general, and the Division Council structure in particular, continued to hold. Feeling that the window for advancing on acceptable terms at Caltech was nearing its close, Pauling also began to seriously entertain the idea of working elsewhere.

***

In 1929 Harvard University had recruited Pauling for a position as Professor of Physical Chemistry, an offer that Pauling ultimately refused. Seven years later, the division chairmanship seemingly out of reach, Pauling wrote to the new president of Harvard, James B. Conant, asking if the 1929 offer might still be a possibility. Conant replied that the position had since been filled by one of Pauling's former students, E. Bright Wilson, Jr., and that Harvard was not presently in a position to create a new job for Pauling. This news came as a disappointment, but other opportunities were soon to arrive.

In mid-July 1936, Pauling received a thought-provoking letter from Christopher Kelk Ingold of University College London. In it, Ingold wondered if Pauling might be tempted to replace F.G. Donnan as chemistry chair at the university. For Ingold this was a bit of a fishing expedition, and he made it plain that he did not really expect Pauling to take the idea seriously.

As such, it likely came as a surprise when Ingold received Pauling's response a month later. In it, Pauling revealed that he had been devoting a great deal of thought to the offer and divulged his motivations for doing so, writing,

"In general I have been in the past very well pleased with my opportunities here for teaching and carrying on research. Subsequent to the death of Professor Noyes, however, the affairs of our Chemistry Division have been in some confusion, and the problem of administration has not yet been solved."

Pauling also relayed his intrigue at the possibility of working with Donnan and living in such a radically different location.

Though adjusting to life in England would surely be a shift for Pauling and his family, the challenge of doing so did not appear to be an issue of pressing concern. Rather, the most significant obstacle impeding Pauling's acceptance of the offer was its base salary of £1,200 per year, which he calculated to be less than his current Caltech salary of $6,300. Pauling also calculated that the taxes on his University College earnings would amount to £360, which equated to about $1,000 more per year than what he was paying in California.

Pauling communicated these concerns to Ingold, perhaps hoping that University College would reply with a counter. Instead, Ingold merely repeated the original offer with feelings of resignation that it might not be enough to attract a talent of Pauling's magnitude. In his next letter, Pauling took up their correspondence on chemical matters but also described how the experience of meeting biophysicist Archibald Hill, who was visiting Pasadena, had made him "realize more keenly than before" that he would not miss coming to London. The courtship had reached its conclusion.

Not long after declining the University College offer, Pauling was similarly recruited for a one-year stint at Princeton's Institute for Advanced Study, and also for the George Fischer Baker Lectureship, a six-month visiting appointment offered by Cornell University's Department of Chemistry. While neither of these opportunities would have taken him away from Caltech permanently, they clearly signaled that Pauling had reached a new level of stature within the profession.

Pauling's first inclination was to choose the Baker Lectureship, though he made inquiries at Caltech to determine whether he might receive enough leave to fill both visiting positions. Upon learning that he would be afforded only one semester's worth of paid time away, he chose Cornell.

***

One of the people whom Pauling had consulted about his leave options was Robert Millikan, the chair of Caltech's Executive Council. Millikan was pleased with the decision that Pauling made and, as a byproduct of their recent conversations, once again broached the idea of Pauling moving into the position of division chair. In raising this subject, Millikan notably asked if Pauling might still have the "desire to suggest changes in the organization of the Division of Chemistry and Chemical Engineering such as would make your continued association with and leadership of that Division satisfactory to you." With plans for the new Crellin Laboratory finally set and construction soon to begin, Millikan knew that resolving the question of division leadership was becoming more and more important.

Now aware that Millikan was open to negotiation and sensing that he had the upper hand, Pauling waited for the new year, 1937, to respond. In doing so, Pauling made it clear that the changes that he would desire remained consistent with the perspective that he had put forth in his November 1935 letter to the Executive Council. Pauling also communicated his willingness to directly discuss the points he had made in a subsequent letter meant for the Executive Council. As it turned out, Millikan had never forwarded that letter on, so a discussion of its contents would prove essential to satisfying Pauling's demands.

Around this time, Warren Weaver, who had favored Pauling as division chair from the start, came to Pasadena to help mediate the situation. Weaver chose to interject himself in part because Millikan, who had little interest in biochemical research, had by necessity taken over management of a Rockefeller grant that had been worked out with A.A. Noyes prior to Noyes's death.

Once arrived, Weaver met with Pauling and listened to his concerns about the shackles that would be placed on the chair under the stipulations put forth by the Executive Council in fall 1935. Pauling also discussed his salary ambitions and his desire for an appropriate title. Using the sway that he had accumulated as a major source of funding for the division, Weaver thus began negotiating on Pauling's behalf. It did not take long for Pauling's salary to be increased to $9,000 and agreement reached on a new title: Chairman of the Division of Chemistry and Chemical Engineering and Director of the Chemical Laboratories. It would still take several months, however, before all the details were worked out.

***

As Pauling and Millikan mended fences, Pauling began thinking in concrete terms about the future of the division. In mid-March 1937, he wrote a note to self about the unit's financial standing which, in the midst of the Depression, depended heavily on the Rockefeller Foundation. Be it by choice or necessity, Pauling was clearly on board with the Foundation's involvement, writing that it "would help the Division as a whole both scientifically and financially."

By now, the Foundation's influence on division affairs had become increasingly evident. As noted, one particular instance involved the hiring of Carl Niemann to help staff the brand new Crellin Laboratory. At Weaver's urging, Niemann was invited to visit Pasadena with the understanding that he would be offered a $3,000 assistant professorship starting that September. This offer was extended even though Niemann was obligated to spend much of the next year in Europe.

***

On April 14, 1937, Millikan conveyed to Pauling that the Executive Council had informally agreed to increase his salary to $9,000 in 1938 and that he would receive his new title at the next Executive Council meeting. In his documentation of that conversation, Pauling wrote that

"Professor Millikan intimated that I might desire the action to be postponed a little beyond that time in case that [historian William Bennett] Munro and [Richard] Tolman could not be raised simultaneously to $9000 from $8100, and I replied that I preferred a definite understanding in order that I might incur certain obligations."

As promised, Pauling was recognized as chair for the first time at the Division Council meeting that followed and, on May 4, he was informed that the governing body had officially approved his appointment. By mid-August 1937, all agreements had been formalized and approved by the Board of Trustees. At long last, Pauling was in charge.

Chapter 17

First Years in Charge

WORK REWARDED

DR. LINUS PAULING
Whose appointment as head of
the chemistry and chemical engineering departments of Caltech,
was announced yesterday.

As a newly anointed division chair, Linus Pauling was obligated to weigh in on details both large and small. One major and ongoing task was to monitor the division's budget and to think hard about the best ways to direct funding, and this too manifested in decisions with varying degrees of impact. More often than not, Pauling's bias was clearly in favor of devoting funds to research. In one instance, when colleague Howard Lucas requested support to attend a conference on the East Coast, Pauling replied that because Lucas was not presenting a paper, the division could not provide funding. In the future, Pauling suggested, Lucas should arrange to give talks when travelling east. Pauling did, however, agree that it would be a good idea for Lucas to hire an assistant to help him with his investigations on bean pod hormones, and subsequently arranged funding for a six-month temporary position.

Though administrative responsibilities now occupied much of his time, Pauling continued to teach, including the graduate courses "On the Nature of the Chemical Bond" and "Introduction to Quantum Mechanics with Chemical Applications." As chairman, Pauling also held more sway in shaping what was taught, both within the division and across the Institute. Before his first year in leadership had been completed, Pauling used his new title to push for the development of broader coursework in organic chemistry across the campus. This effort yielded quick results, in no small measure because of lingering momentum from A.A. Noyes's activities as the previous division chair and the influx of money coming from the Rockefeller Foundation. Indeed, one might accurately intuit Pauling's satisfaction in writing "Carried!" next to an agenda motion stipulating that, for seniors in physics, applied physics and astronomy, Caltech remove required courses in statistics and replace them with organic chemistry.

Within the division, Pauling was obligated to work with the Division Council before proposing any changes to the curriculum. In spring 1938, the council approved an optional second year of organic chemistry for seniors, a request that the students themselves had been making. By 1942 the organic chemistry requirement became uniform across the division, with applied chemistry majors taking the coursework as juniors alongside chemistry majors. In 1955 Pauling suggested that the organic chemistry requirement be moved to the sophomore year and expressed his belief that there was too much physical chemistry in the sophomore curriculum.

Early on as chair, Pauling also worked to keep graduate students connected to the research of the division's rapidly growing staff which, bolstered by Rockefeller support, had increased by 15 people in his first year. The total number of graduate students had also increased from 25 to 45, each of them receiving annual stipends of $600 to $860. When he first became division head, Pauling was only a decade or so older than most of these graduate students and he made a point of inviting them to his home or to desert camping trips to learn more about their work, ambitions and outside interests. Pauling wanted this closeness to permeate the division and, in

September 1938, proposed that faculty participate in regular seminars where they would present their research internally. Pauling gave the first talk in this series, providing an update on his hemoglobin studies, and made it clear that he expected others to follow his lead.

***

During the final phase of Pauling's ascendence into leadership, the construction of the Crellin Laboratory lurked ever-present in the background with several adjustments to the building plans needing to be made. In May 1937, recognizing that the project was over budget, the Division Council began looking for ways to save money on equipment like tabletops and hoods. When September arrived and the project was still short, lead donor Edward Crellin agreed to make an additional gift of $5,000 specifically for floor coverings.

(Pauling was so pleased by Crellin's contributions that he named his third son, born June 4, 1937, Edward Crellin Pauling. Even though the two never got to know each other well — Edward died when Crellin was only 11 — the benefactor was still flattered by Pauling's gesture and left $5,000 in his will for the young boy.)

That fall Pauling was in residence at Cornell University as George Fischer Baker Lecturer and he took the opportunity to investigate the floors that had been installed at the Baker Laboratory. Once done, Pauling wrote to Arnold Beckman, who was overseeing the furnishing of the Crellin facility, and told him that battleship linoleum had performed well in its 14 years of covering the halls and offices at Cornell. Resolite, on the other hand, had not endured quite so ably.

Beckman followed up by testing Resolite against textile, which Caltech's contractor had recommended as a possible alternative. Beckman reported that both materials "softened" when they came into contact with organic solvents, but Resolite would be a more economical purchase. As a result, Beckman decided to use Resolite in the laboratories, despite Pauling's misgivings, and linoleum in the hallways and offices.

Meanwhile, Pauling was continuing to pursue stable funding for operations within the facility and once again reached out to the Rockefeller Foundation for support. In May, Warren Weaver responded with a four-page letter that detailed the ways in which Caltech could improve its grant application by including the anticipated costs of equipment along with more specifics on how the Division of Biology would use their allotment. Weaver also warned Pauling that his request for an increase from $10,000 to $15,000 per year for his own research in structural chemistry was a "retrograde step" that was best avoided. Further suggestions from Weaver laid out an ideal path for distributing money from a potential $60,000 annual award, with $10,000 going to Biology, $10,000 to structural chemistry, $35,000 to organic chemistry research, and another $5,000 earmarked for organic chemistry equipment.

These ideas, which Weaver also conveyed to T.H. Morgan in Biology, were incorporated into a revised application that was submitted by the two divisions in August. At that same time, Executive Council chair Robert Millikan told Weaver that if Caltech received the grant, they would prefer that it begin the following July, when the new Crellin Laboratory would be ready.

While Pauling ironed out the details of the Rockefeller request, he also took steps to safeguard support for his own projects. In his correspondence with the Executive Council, Weaver had asked for assurance that the Institute would continue to fund biochemical work even after the Rockefeller grant had been exhausted. Pauling too was seeking a guarantee that the council would continue to back his structural chemistry program, since the Rockefeller Foundation would not increase its contribution. Wearing both his administrative and his researcher hats, Pauling eventually asked that the Executive Council provide $50,000 a year to satisfy Weaver's request and $5,000 a year for his own. Millikan and Richard Tolman advanced this appeal to Caltech's Board of Trustees the following month, and the board agreed to move forward as proposed.

Parallel to these conversations, Weaver also pushed Pauling on personnel decisions, characterizing his reactions to Pauling's ideas thus far as "not entirely enthusiastic." For Weaver, Pauling's suggestion of a dual hire, brothers R.R. Williams and Roger Williams, to help run the Crellin Laboratory represented "a somewhat unsatisfactory compromise between the ideal of a young, well-trained and exceedingly brilliant man, such as [Alexander] Todd or [Carl] Niemann, and a thoroughly experienced and broadly interested world leader, such as we should like to find but cannot." In response, Pauling proposed that they shift their focus to another hire in the vein of Niemann, who would be permanently relocating to Caltech the following year. Pauling and Weaver alike assumed that Todd, a Scotsman and future Nobel Prize winner based in London, was probably not available, but both did their best to recruit him anyway.

***

As summer 1937 neared its close, Pauling was becoming increasingly certain that the Rockefeller Foundation would award a mammoth $300,000 grant to both the Division of Chemistry and Chemical Engineering and the Division of Biology. In fact, Pauling had become so confident that he requested Executive Council permission to leverage the forthcoming funds to immediately purchase a combined recording microphotometer, densitometer, and comparator. Once this permission was granted, Pauling secured the services of Fred Henson, who had constructed an apparatus of this sort for J.W. McBain at Stanford. Henson agreed to deliver a similar device to Caltech within one year for $2,600 — a discount of a thousand dollars from what he normally charged. Pauling also used $850 from funding provided by the British hydrocarbon company M.W. Kellogg to purchase a different instrument.

These new purchases would be put to immediate use in the nearly finished Crellin Laboratory, the staffing and outfitting of which remained an issue of pressing concern. While Pauling and Weaver continued to hold out a sliver of hope that Alexander Todd would relocate to Pasadena to head the new laboratory, Carl Niemann was in the process of equipping and stocking the facility. Niemann's progress, however, had been slowed by an unexpected hospitalization. As he later wrote to Pauling, Niemann had gone to see a doctor because he had a chunk of rust embedded in the cornea of his left eye, "and the first attempt to remove it was not particularly successful." He was then hospitalized and had to "have the disturbing element removed and the seat of the injury cauterized."

Once returned to work, Niemann proposed that a "central analytical laboratory for the entire department," with one person in charge, be identified to save space and minimize redundant equipment purchases. This shared laboratory would cost about $3,500 to $4,500 to outfit, whereas the cost of consumables at comparable four-person laboratories could require up to

$8,000. Niemann planned to finalize these arrangements in the fall before leaving for his scheduled year-long trip to Europe.

In December 1937, Pauling's feelings of confidence were validated when the Rockefeller Foundation's Board of Trustees finally approved the Caltech biochemistry grant. Funds would be dispersed beginning in July 1938 and would coincide with the division's move into Crellin. The proposed budget for the first year largely followed Weaver's earlier suggestion of $60,000, though Pauling pushed for an extra $10,000 to augment organic chemistry research salaries and equipment. When Alexander Todd, following a May visit, formally decided against coming to Caltech, the spending plan was set at $60,000, with an additional $10,000 earmarked for more hires and equipment in the next fiscal year.

In February 1938, with plans beginning to settle, Pauling wrote a letter of thanks to the Rockefeller board for approving the grant application, and to Weaver for his ideas on developing organic chemistry at Caltech, an initiative that was going smoothly. Pauling likewise acknowledged his personal indebtedness to Weaver for helping fund his own research. Colleagues of significant consequence including Robert Corey and Max Delbrück had come aboard under the grant and were doing good work. Pauling himself was also "getting more interested in biological problems every day" and was "anxious to see our new program in effect."

***

Amidst all this progress, a test of Pauling's commitment to his new position arose less than a year after his official appointment. In March 1938, Arthur Hill, a chemist at Yale University, offered Pauling a Sterling Professorship, the highest professorial rank awarded by the prestigious university. Pauling's proposed salary at Yale would be $10,000 and he would also benefit from the services of a private assistant.

Pauling thought hard about the offer, at first replying that he would need a week to think about it. That week stretched into five and ended with a decision to stay at Caltech. In turning down the opportunity, Pauling explained to Hill that he had only recently become chair of a growing division, a circumstance that presented him with "an attractive opportunity for contributing effectively to science" by shaping the development of the unit and opening new areas of research for himself. Hill was disappointed but admitted that the research capacity available at Yale could not compete with what had been built at Caltech.

Feeling more established in his leadership position, Pauling wrote to Executive Council chair Millikan in April 1938 that "[t]he outlook for the Division during the next few years is very attractive, especially for the field of the organic chemistry of biological substances." Nonetheless, Pauling was looking for further backing from upper administration. Though he already had a promise in hand that $40,000 would be made available as a complement to the Rockefeller funds, Pauling wanted additional reassurance that the Institute would support the division's nascent biochemistry ambitions well into the future. Buoyed by the momentum of recent months, Millikan was glad to reply in the affirmative.

***

On May 16, 1938, the Crellin Laboratory was formally dedicated at a ceremony featuring addresses from Pauling and the building's patron and namesake, Edward Crellin. In his remarks, Crellin told the audience,

> "It is pleasing to note the physical union of this building with the new unit of the great William G. Kerckhoff Laboratories of the Biological Sciences, thus enabling Dr. Thomas Hunt Morgan and Dr. Linus Pauling and their associates literally to join hands in the search for, if not the elixir of life, a better understanding of vital processes, leading to better health and longer and happier lives."

With Crellin now online and the large Rockefeller grant deployed, Weaver and Pauling continued their brainstorming about who best to head research in the new laboratory. In the summer of 1938, Weaver mentioned that Laszlo Zechmeister, a biochemist from Hungary who had developed chromatographic methods to separate enzymes, was visiting the United States and suggested that Pauling invite him to Pasadena. Pauling obliged, arranging for Zechmeister to give three lectures on chromatography, carotenoids, and polysaccharides, which he delivered in November.

Pauling was impressed by the quality of these presentations and began to think more seriously about hiring the Hungarian following a weekend trip to Mexico that included Ava Helen, Zechmeister and his wife. Pauling's estimation of the candidate further grew when he learned more about the extent to which he had managed to produce as a scientist despite working in a poorly equipped Budapest lab. Though Zechmeister would depart from Pasadena without a formal job offer, he soon became an important figure at Caltech's new research facility.

***

With its pristine building and innovative, well-funded research program, Pauling's Division of Chemistry and Chemical Engineering was increasingly coming to be recognized as a model for other universities around the country. This boost in stature is evident in Pauling's correspondence from the period. In one instance, John C. Bailar, Jr., Assistant Professor and Secretary of the Department of Chemistry at University of Illinois, Urbana, wrote to ask about ways in which they could follow Caltech's example to improve their own research and instruction. Likewise, Joseph George Cohen, Director of the Division of Graduate Studies at Brooklyn College, noting his ambition to "pattern [their] administrative practices on those schools which occupy a position of leadership in American Chemistry," asked how Caltech ran its laboratories and what responsibilities they gave to their Ph.D.'s and graduate assistants. Pauling was always willing to give his advice.

Pauling's personnel recommendations were also commonly sought by other programs looking to hire new staff. And though his picks were not always a perfect match, Pauling increasingly found himself in a position to influence the shape of research around the country.

During the fall of 1938 for instance, Pauling recommended a colleague, Gilbert King, to fill a position at Duke University. King ended up taking a job at Yale, after which Pauling's contact at Duke, Paul Gross, reached out for another name. This second time around, Pauling leaned more heavily on his title in advocating for an up-and-comer within the division, Lindsay Helmholz, whom Pauling described as "one of our best men."

Though he was reluctant to lose Helmholz's skillset, Pauling worked hard in selling him to Gross, highlighting his research and teaching skills while also emphasizing the range of material that Helmholz was familiar with, including topics in physical, structural, and general chemistry. Pauling further noted that, if offered a permanent position, Helmholz would likely accept. He confided from there that "Dr. Helmholtz is married to a very pleasant and attractive young woman, and I am sure you would consider the Helmolzes an addition to your community." As it turned out, Helmholz ended up not going to Duke, and instead wound up at Washington University after working on the Manhattan Project during World War II.

Pauling saved his second-tier recommendations for less prestigious institutions, including the Department of Chemistry at his alma mater, Oregon State College (OSC). In one instance, Francois Gilfillan, Dean of the School of Science at OSC, was looking for teaching fellows and was eager to receive any ideas that Pauling might have in mind. In response, Pauling recommended Robert J. Dery, who had taught at Caltech for three years but had not completed his doctorate since he seemed "for some reason to lack the ability to carry on experimental research." Dery, according to Pauling, was also "slow spoken, so that in conversation with him one may become impatient." This habit did not impede Dery as a teacher, however, as he spoke very well in front of an audience and, in Pauling's estimation, was a good freshman instructor.

***

By April 1939, with the Rockefeller grant approaching its second year, the division was still looking for someone to head its biochemical research, leading Pauling to ask Weaver if he might further delay the hiring as no clear-cut match for the position had been identified. That June, the Foundation's Board of Trustees, while discussing the following year's budget, noticed that no one had been hired and questioned the need to approve the full $70,000 request for the following year. Ultimately the board agreed to the original ask, but their hesitation to do so caused Pauling to accelerate his search.

In the meantime, Pauling was also obligated to deal with stray administrivia that landed on his desk. One specific issue was the need to make sure that graduate students could access specific areas of the new Crellin facility — most pressingly the student shop — that were normally open during business hours only. To solve the problem, Pauling requested that a lock be made for the shop that matched the locks on Crellin's entry doors so that the students could enter as needed.

**

An immediate and exponentially larger problem came to pass on August 10, 1939, when two Caltech researchers, Leo Brewer and Thurston Skei, were conducting an experiment in Crellin room 351. As they went about their work the bottom of a container fell off, spilling six liters of liquid ether all over the floor. Brewer and Skei quickly cleaned the spill up and checked to make sure the room was safe, which it appeared to be.

At that point, Skei exited the room to attend to matters elsewhere, leaving Brewer alone. Five minutes later, a spark from a motor running in the building's ventilation ducts ignited ether fumes, which had been sucked into the exhaust system. The air in the room quickly ignited, severely burning Brewer who immediately, and fortunately, ran out the door. Three seconds later, Crellin 351 exploded, destroying all the windows on that half of the floor and blowing apart the room's main entry as well as part of a wall.

At the same time, the laboratory's ventilation fans sucked the flames upward into its fume hoods, which ignited another set of drums containing ten gallons of liquid ether and started a massive blaze that spread to two adjoining rooms. The force of this explosion shattered almost every adjacent piece of glass and knocked over numerous storage shelves. As a result, various chemicals began to mix, and the entire third floor was inundated with toxic fumes.

In quick response, graduate students and staff alike grabbed gas masks and fire extinguishers and charged up to the third floor. Amazingly, they succeeded in containing the fire and prevented it from spreading into other areas, including the building's library. They also managed to extinguish the burning walls in the main hallway. Not long after, the Pasadena Fire Department arrived. Racing to room 351, the firemen used their axes to dislodge the fume hoods from the wall and eventually put out the last of the fire.

In the aftermath, Pauling passed along word of the explosion to several of his colleagues but waited seven days to contact Warren Weaver in New York and did his best to downplay the news.

"Perhaps you read in the papers that we had a fire in the Crellin Laboratory. Fortunately no one was injured and the damage was restricted almost entirely to the undergraduate organic laboratory, with very little research lost. We had complete insurance coverage and shall have the laboratory in shape for the students when the Institute opens next month."

In reality, the blast and fire had destroyed almost $3,300 worth of equipment, and by the time the rather extensive repairs were done, the accident had cost about $14,000 (equivalent to over $300,000 today). Luckily, nobody was killed — Brewer's was the only injury, and he made a full recovery. It is worth noting as well that lab fires were common enough at the time that the emergency procedures for the facility only required personnel to call the fire department if the staff and graduate students on hand couldn't contain the blaze themselves.

***

In calmer moments, Pauling was feeling increased pressure to find someone to head biochemistry research within the division, and the more he thought about it, the more he was convinced that Laszlo Zechmeister was the man for the job. While current staffer Carl Niemann had shown promise, Pauling thought him too young to lead the full program. Zechmeister, who was in his fifties, was arguably too old, but Pauling had come to see his age as an advantage since he was certain that a strong replacement would emerge from within once the Hungarian had retired.

In the months following his visit, Zechmeister had stayed in touch with Pauling, keeping him informed in particular of Hungary's increasingly close relationship with Nazi Germany. Zechmeister also told Pauling that, if there was a position available for him elsewhere, he would take it immediately. Pauling finally offered Zechmeister a professorship in organic chemistry in October 1939, which was quickly accepted. The only sticking point was that the Caltech appointment would have to be limited to a single year, as that was the longest period that Zechmeister was legally allowed to reside outside of Hungary. Sadly, Zechmeister's wife became sick before they came to Pasadena in 1940 and died the following year. But ultimately Zechmeister was able to remain in the United States and he continued on at Caltech until his retirement in 1959.

***

Just two and a half years into his tenure as division chair, Pauling saw to completion a variety of initiatives originally set in motion by his predecessor A.A. Noyes, including the construction of a new building. With the onset of the Second World War, an influx of federal funding would create fresh opportunities as more researchers and projects filled out the division's spaces. This period would also set the stage for Pauling to further build on Noyes's foundation while implementing his own ideas on the shape of the rapidly evolving division.

Chapter 18

Leading the Division During World War II

Pauling's National Defense Research Committee authorization papers, 1944.

The entry of the United States into the Second World War brought a shift in focus for the Division of Chemistry and Chemical Engineering. As its chair, Linus Pauling was tasked with staying on top of continual staff changes, a student population that had moved away from pure science in favor of engineering, and an "abnormally low budget" of $2,500 for supplies.

The war also prompted a reorientation of Pauling's own research towards wartime imperatives like explosives, propellants, and medicine, a recalibration that was reflected across the division. Some of Pauling's government contracts also required him to coordinate with researchers elsewhere in the country. One such individual was Villiers W. Meloche at the University of Wisconsin, Madison, with whom Pauling worked to determine the age and stability of diphenylamine compounds used in explosives. At one point, Meloche sent a researcher from his group to Pasadena to review some of the Caltech work in person, as their collaboration involved confidential details that could not be shared through the mail.

***

One of Pauling's main administrative priorities during this time was to track and, frequently, influence the draft status of his division's researchers. Guidance for this was provided by W. V. Houston, who was Acting Dean of the Graduate School at Caltech. Houston informed the Institute's division chairs that they would need to inquire into the draft status of each graduate assistant and teaching fellow working in their area, and that they should ask for deferments "for everybody who is to be depended upon for next year's teaching."

To do so, Pauling began by creating a form letter that was meant to influence the perspective of local draft boards by outlining the reasons why a given researcher should be granted deferred status. In addition to the boilerplate contained in each letter, Pauling provided his own tailored thoughts on the work that the individual was conducting, once again for the benefit of the draft board's review.

Most of Pauling's comments strongly recommended deferment and argued that the person in question was already working in support of national defense through their chemical research. But in at least one case, Pauling's argument hinged more on potential. This specific student, Werner Baumgarten, was nearing the end of his Ph.D. in organic chemistry, an area where workers were in short supply. Pauling told the draft board that, were he deferred, Baumgarten would likely stay on at Caltech after graduating and recommended that he be given a "short period" to see "whether or not his services in scientific work are to be important to the national defense."

As he compiled comments of this sort, Pauling leaned heavily on the idea that "chemistry is one of the fields vital not only to the national interest but also to the national defense." He often used this line to close his letters, including one written for Andrew Alm Benson, whom Pauling also described as "easily one of the more able of the men" on track to receive his Ph.D. within a year. Six months later, Pauling requested an extension for Benson's deferment to ensure that he could continue teaching, as others were being diverted more and more toward research in support of the war. Benson would go on to become a leading plant biochemist, spending his career at the Scripps Marine Biology Research Division at the University of California, San Diego.

While Pauling did his best to provide support to draft-eligible personnel, it is important to note that he was not simply acting as a rubber stamp in his evaluations. In the case of one master's degree-seeking student, Pauling judged the individual's work to be "satisfactory" and projected that he would "probably be a competent chemist." But unlike most of the other letters, Pauling made no explicit call for deferment. Nonetheless, this student remained at Caltech throughout the war years and later built a career working to combat air pollution in southern California.

***

In addition to retaining the talent that he had on site, Pauling also regularly needed to bring in new employees to replace departures or fill new salary lines. As chair, Pauling received many inquiries concerning possible appointments within the division. These inquiries were not always directly addressed to Pauling, but to other faculty who then passed them along. Pauling also commonly sought input from other faculty concerning aspiring applicants who might potentially work in their research area.

Some of the hires that Pauling brought on board were former graduate students. Since many were a bit rusty on matters related to current chemical research, Pauling arranged for them to audit organic chemistry class lectures led by Laszlo Zechmeister and others. Pauling requested that the Caltech Executive Council not charge these new hires for attending the lectures, since there was no need from them to earn academic credit through their attendance.

One especially notable job applicant was Margaret Sinay. In a February 1944 letter written to Pauling, Sinay noted that her husband had recently been transferred to Los Angeles for work, and that she too was looking for a position in the area. Sinay described herself as an "analytical chemist with about 15 years' experience in medical research and routine biochemistry" who had been a senior chemist at the Vick Chemical Company's vitamin research laboratory. Pauling replied that the division might have an opening for a "war research job, in which analytical training would be useful," and asked that Sinay come in for an interview once she and her husband had finished their move.

What became of Sinay is unknown, but it is significant that Pauling expressed an interest in bringing her aboard. While there were no formal policies in place at Caltech forbidding the hiring of women into technical positions, they still were not allowed to take courses at the Institute. World War II made a dent in these guidelines as Caltech began permitting women to attend no-credit night classes related to war work. None of these offerings were in chemistry, but the division began to receive a growing number of inquiries from women expressing a desire to study there. To each of these requests, Pauling was forced to reply that it was against Institute policy to admit women at any level, except for those involved in war work.

In October 1944, about a year before the war came to an end, Caltech's Graduate Committee on Post-War Policies discussed the possibility of admitting women as advanced degree candidates, but these talks ended up going nowhere. The idea was brought up anew in 1948, and again it was set aside. Clearly, while the war ushered in many adjustments for the Division of Chemistry and Chemical Engineering, it did not change everything.

***

In the early 1940s, a $300,000 biochemistry grant provided by the Rockefeller Foundation set the tone for research in the division, but it was not the only source of funds that the Foundation was providing. In addition to the large biochemistry budget, the Rockefeller board also approved smaller supplementary awards to support several promising immunological projects being pursued by Caltech faculty. This secondary line of funding would ultimately make a significant impact.

In 1940, geneticist A.H. Sturtevant received the first of the immunology grants, a three-year, $36,000 award. A year later, Pauling was provided with his own three-year, $33,000 grant to support a separate track of immunological research being housed in his division. Prior to finalizing the allocation, Rockefeller administrator Warren Weaver suggested that Pauling ask for an additional $20,000 for the second year alone, a request that was quickly approved. As time passed and Caltech's immunochemistry profile rose, a growing number of undergraduate and graduate students came to Pasadena through the support of these Rockefeller funds. Well aware of its heightening status, Pauling pushed for immunology to be institutionalized with its own administrative apparatus and advocated that Dan Campbell, a faculty member who arrived in 1942, be placed in charge.

In 1943, with Sturtevant's grant set to expire, he and Pauling decided to collaborate on a joint proposal that would combine the work being pursued by the biology and chemistry divisions. This new bid asked for an $18,000 supplement to an $11,000 fund balance that remained from the last year of Pauling's 1941 award. Immunology at the Institute was also receiving material support from the military, and the Office of Scientific Research and Development expressed its hope that the joint project would continue after the war. Encouraged by the direction that these efforts were taking, the Rockefeller Foundation approved the request, and Pauling and Sturtevant began their collaboration.

**

The division's advancements in immunology also piqued the interest of the private sector, which saw ample potential for commercializing this research. One company, Lederle Laboratories, offered to collaborate by providing large amounts of antisera and toxins needed as experimental inputs. Pauling argued against this proposed joint venture, feeling that the investigations had not yet progressed to the level of "commercial exploitation." Frank Blair Hanson, who was overseeing the immunology funds for the Rockefeller Foundation, also recommended against the partnership, but for a very different reason. It was Hanson's view that breakthrough medical applications of the work were potentially imminent and that the Foundation's proprietary interest in the research needed to be protected.

Later, in the fall of 1944, Pauling took steps to clarify the division's position on accepting funds from — and working with — large companies, a conversation that would only intensify following the war. The need for this policy clarification arose from a meeting where division faculty expressed concerns that industrial interests were sometimes being considered separately from basic questions in chemistry. Communicating on their behalf, Pauling emphasized the faculty's overwhelming preference that no strings be attached to grants offered by private interests.

Towards the end of 1941, one such private interest, Shell Development Company, offered Pauling a position as its Director of Research. Pauling visited Shell in San Francisco to tour their facility, but never seriously considered accepting the job. Instead, as he had done in the past, Pauling used the opportunity as leverage with his current employer.

In November, Pauling wrote to J.F.M. Taylor at Shell, indicating that he was waiting for a counteroffer from the Institute that might convince him to stay. Ten days later, Pauling wrote to Taylor once more, saying this time that he would decline Shell's proposal. In doing so, Pauling explained that he likely would have accepted the offer were he earlier in his career, but that "I have now gone too deeply into fundamental science, including the biological applications of chemistry, to tear myself away." It appears that the promise of a pay increase may have helped firm up Pauling's decision, as Caltech's Board of Trustees agreed to raise his annual salary from $9,000 to $10,500 a little over a month later.

***

With Pauling once again solidly in place as division head, he began to focus more intently on maintaining a balance between the Rockefeller-funded biochemical and immunological programs, and the new obligations ushered in by the onset of war. In January 1942, Weaver checked in with Pauling specifically to see if those new responsibilities were interfering with the biochemical work. Hanson also wrote, asking the same question about the immunological program. In his replies to both, Pauling stressed that, despite losing two graduate students to military service, the activities funded by the grants had remained largely unaffected. There was, however, the potential that the division might lose more student assistants in the near future.

As summer approached, the division appeared to be mostly hitting its targets. In a May progress update, Pauling reported that the biochemical grant had succeeded in completing many of its projected goals for the year, despite the war. That said, personnel turnover had been larger than normal, especially in structural and physical chemistry, since those were areas where a lot of war work was being done. Other projects, however, had not been interrupted at all.

The immunology studies faced a new challenge when the War Production Board began limiting the division's supplies. Pauling contacted Frank Blair Hanson to communicate this turn of events and put forth the idea that they solicit a $1 contract from the Committee on Medical Research to keep their supply lines open. Pauling further explained that the investigations being carried out under the grant were making a consequential impact on the war effort, highlighting especially a project on the synthesis of quinine. Hanson agreed that it was a good idea to pursue the $1 contract for the purposes outlined.

Despite these and other distractions, the Rockefeller-funded research moved along briskly, so much so that it began to outpace its budget. The Foundation's base grant was originally set at $300,000 to be spread over at least five years, but for each of the first three years the chemistry and biology divisions had requested $70,000. When that request was repeated for the fourth year, Weaver warned Pauling that there would not be enough money left to support the final year of the award.

Nonetheless, the Institute's Board of Trustees approved an even larger request for year four — $75,000 — in part because Pauling provided assurance that the two divisions would not spend the complete budget due to an increased emphasis on war work. Pauling also told Weaver that the divisions would have no problems addressing his concerns.

***

Bolstered by stable funding and a string of research successes, Pauling was inspired to formulate a broad-ranging and farsighted biochemical research program for the division. In 1942, Pauling sent a draft of this vision to the Board of Trustees. Noting that no program of the sort existed on the West Coast, Pauling expressed his belief that Caltech could collaborate with the University of Southern California Medical School, the Huntington Memorial Hospital, the Good Hope Hospital and others to launch a "cooperative scientific attack" that drew on existing research in physics, chemistry, and biology.

Pauling went so far as to propose the creation of a small institute. Initially staffed by two researchers working on hypertension, Pauling believed that space for the institute could be carved out of existing facilities at Caltech and would cost around $15,000 a year. Once established, Pauling hoped that this operation would grow in stature to the point where it would require its own building on the corner of campus. While the board did not approve Pauling's idea, he continued to persist, advocating for it as a component of Caltech's post-war plan, and in 1952 the proposal became a reality at last.

***

In the early 1940s, some began to fear that the Rockefeller-driven focus on applying chemical methods to biological subjects would increasingly overshadow the chemical engineering branch of the Division of Chemistry and Chemical Engineering. But as with biochemical medical research, it was also clear that there was a lack of fundamental chemical engineering research being conducted in the western United States. Recognizing this gap, faculty member B.H. Sage decided to stand up on behalf of his chemical engineering colleagues.

In the fall of 1944, Sage wrote to Pauling, advocating for future lines of support for research in chemical engineering. In his letter, Sage reported that the chemical engineering faculty were shifting away from investigations of unit operations as the basic steps in the chemical engineering process, a topic that had dominated the previous 15 years. Instead, the group was now interested in analyzing unit operations themselves. Pauling listened to what Sage had to say and, the following year, began pushing for revised courses in fundamentals of chemical engineering. But Sage's new line of research would also require time and money, and resources were stretched in other directions.

One promising source of funds was the Texas Company, now known as Texaco. Sage had helped build a relationship with the company and was overseeing a contract that provided funding for investigations on the molecular weight of hydrocarbons in methane and other natural gases. This $20,000 annual award was up for renewal in June 1946, and through his communications Sage had come to believe that the contract could be boosted to as much as $100,000 per year. The range of techniques that the project would incorporate was also seen as an attractive foundation for exploring basic research in chemical engineering.

However, the firm's patent requirements were likely to place restrictions on both publication opportunities as well as Sage's time, and the division ultimately decided to recommend against board approval of the contract unless the company changed its views on the dissemination of research findings. This recommendation was also motivated by a secondary fear that the corporate funding could cause an imbalance in chemical engineering research within the division, privileging Texaco's interests at the expense of the unit operations analyses that Sage wanted to pursue. Without the private money, chemical engineering at Caltech would clearly remain relatively underfunded, but its researchers would remain free to follow their own interests more closely. In the division's view, this was a trade-off worth making.

***

In addition to steering operations at the macro level, Pauling was also tasked with overseeing smaller matters related to protecting confidential research. One early incident that required Pauling's input came about in fall 1942 when Foster Strong, a physics instructor, observed an undergraduate student using a master key to enter unauthorized rooms in the Crellin Laboratory. After confronting him, Strong broke the key while giving the student, according to Pauling's subsequent report, a "severe lecture about the seriousness of the offense." As it turned out, the lecture did not take.

The following February, Elizabeth Swingle, the division's stockroom keeper, saw this same student using a key to enter her area. The first time she approached him, the student claimed that he was only able to get in because Swingle had left the door open. When Swingle later walked in on the student — this time accompanied by others — in the stockroom after hours, she went to Pauling. Describing this second encounter in a formal report, Swingle noted that "a look of surprise and a flush spread over [the student's] face when he saw me."

Once summoned by Pauling, the student said that he had copied another student's key to obtain entry. Later he admitted that this was a lie, and that he had originally had two keys made. Being in possession of an unauthorized master key was, in Pauling's judgement, a "serious offence" because of the "confidential nature of some of the work being done" at the labs. Pauling directed that the student no longer be permitted to work in the Crellin facility at all.

On a Monday morning just over a month later, Swingle found the Crellin stockroom unlocked. Upon inspection, she found that 60 liters of an anesthetic, absolute diethyl ether, and 150 grams of vanillin, an extract of vanilla, were missing. A few weeks later, Swingle found two keys left on a table by the stockroom, one of which permitted entry into the room.

At the next division council meeting, it was decided that this series of events merited the hiring of a security guard to keep an eye on both Gates and Crellin during nights and weekends. The council also agreed that identification cards would be issued to those with clearance to access the building after hours. Master keys would be restricted to faculty members only, and the provision of additional master keys to others would be up to Pauling. Elizabeth Swingle, along with two other women, had their master key privileges revoked.

At the beginning of June, the Institute also reached a verdict on a just punishment for the student whose activities had caused initial concern. It was found that he had violated Caltech's honor system and that he would be placed on disciplinary probation for the remainder of his tenure as an undergraduate. This meant that he could not hold elected office nor work on campus.

Three days after the decision was rendered, Swingle reported that someone had left the stockroom a mess. "Some chemical had been spilled on my desk leaving the finish injured," she wrote. In addition, "there was a yellow-colored chemical on the floor, in the wastepaper basket, and on two towels." Further examination revealed that the yellow chemical was quinone and that 100 grams of the substance were missing from the stockroom's inventory. Pauling asked around, inquiring if any research groups had been using quinone and if any lab workers may have removed it without filling out a charge slip. None of the colleagues with whom he inquired were using the chemical at the time.

By July the ID badge system was in place, restricting access to Gates and Crellin between 6:00 p.m. and 7:45 a.m. on weekdays, and between 1:00 p.m. Saturday and 7:45 a.m. Monday. A year passed without incident. Then, in

1944, biochemistry professor Arie Haagen-Smit saw a research assistant for the NDRC-Chem-13 project enter a laboratory in the Kerckhoff basement, where secret war work was being carried out. Haagen-Smit approached the researcher and asked how he got in. He replied that there were "a number of keys around, which open practically all the doors on campus." A friend who was a graduate student in mathematics had given him the key to see if there was any equipment in the restricted lab that he could use for his research in NDRC-Chem-13. In response, Pauling again directed that limitations be placed on master keys, and the division encountered little trouble thereafter.

***

One very difficult moment that Pauling was forced to reckon with as division chair was the tragic death of Elizabeth Swingle. As Pauling wrote in a subsequent letter,

"On September 23, 1943, Mrs. Elizabeth M. Swingle, a young woman, twenty-nine years old, who had received a Master's Degree in Bacteriology and had worked as Stock Room Keeper of the chemical stockroom in the Crellin Laboratory for about a year, went to the chemical vault in the sub-basement of the Laboratory and removed from the shelf a one kilogram bottle of ethylchlorocarbonate (Pract.), stabilized with calcium carbonate, Eastman P591. She walked to the foot of the elevator shaft and, while she was standing there, the liquid sprayed out of the bottle and over her head and shoulders, the cap apparently having been blown off."

Subsequent reports would suggest that the container had shattered due to pressure caused by the decomposition of the substances that it held.

Swingle was immediately rushed to a nearby laboratory shower by Bill Lipscomb — then a Caltech graduate student and later a Nobel Prize winner in Chemistry — who had been standing just down the hall. A few minutes later, emergency services personnel arrived and transported Swingle to the Huntington Memorial Hospital, where she was given oxygen therapy to combat the accumulation of hydrochloric acid in her lungs. Unfortunately, the young woman had inhaled too much of the gas to recover and died eight hours later.

Four days after the accident, Pauling spoke at Swingle's memorial service. In addition to a recitation of her work at Caltech, Pauling offered a more personal recollection:

> "…during her first year in Pasadena she came almost daily to our house on the hill, to care for the rabbits which were being immunized; and during these visits she made a most intimate friend and great admirer of our boy Crellin, then five years old, who looked forward to each day's conference, and who has ever since numbered his talks with Elizabeth among the greatest experiences of life in this marvelous world."

Pauling concluded his remarks with a meditation of sorts that reflected his fundamentally scientific perspective on nearly everything, including questions of meaning and existence.

> "My friends, while we are thinking about Elizabeth, about her beautiful, friendly spirit, we may well ask why? Why was she taken away from us suddenly, without warning, while going happily about her work? I think that we can find the answer to this question by contemplating the wonders of the world in which we live — the wonderful order which underlies all natural phenomena.
>
> "No one can study deeply the physical world without experiencing again and again a feeling of amazement, of transcendent exaltation, at the beautiful intricacy of the structures which constitute the physical world and the beautiful order of the laws which determine its course; and this feeling becomes ever stronger as we turn our attention to life, to man himself, and begin to understand, even though dimly, the almost unbelievably complex mechanisms of the physiological processes upon which life depends. Our faith in the future rests upon our faith in these laws of nature.
>
> "And so we can understand that these laws cannot be broken, but must pursue their inexorable course even when, because of an accidental, unavoidable concatenation of circumstances, this course is such as to take from us our friend Elizabeth early in her life, while her spirit was still growing, her large circle of friends still expanding, her due cycle of life's rich experiences only to the midpoint traversed."

Shortly after the service, Pauling contacted Eastman Kodak, the company that had supplied Caltech with the ethyl chlorocarbonate, and demanded that they investigate the accident as well as their manufacturing and shipping techniques, and provide written warnings to other customers. He received updates from the company on their internal review through the spring of 1944. Ultimately, the accident helped force both Caltech and Eastman Kodak to seriously examine safety precautions and make changes accordingly.

***

As World War II neared its end, the division recognized a pressing need to reassess its graduate offerings, which had been rapidly updated amidst the pressure to meet wartime demands. One area that had suffered because of this update was organic chemistry, a point that was emphasized to Pauling by his colleague Edwin Buchman. Noting that there was a lack of instruction and organization when it came to graduate training in organic chemistry, Buchman requested that Pauling form a committee to define policies around research and teaching in the division.

Pauling also worked beyond the division level to update graduate programs across the Institute. A member and, on occasion, acting chair of the Graduate Committee on Post-War Policies, Pauling also served on the Faculty Board and Curriculum Committee. During one meeting, Pauling became especially intrigued by physicist E.C. Watson's idea that Caltech accelerate its graduate work through the implementation of new teaching methods. By doing so, Watson saw the potential for Caltech to repeat its successes from the 1920s, the decade during which they had become an "excellent graduate school" frequently "copied by other technical schools." On a practical level, this would mean dropping applied courses like industrial design, for which there had been an urgent demand during the war. Though expressed with the intention that they be applied across the Institute, Watson's ideas lined up well with what Pauling had in mind for the division as well.

Caltech's master's degree offerings were also in need of scrutiny. Differing from many other technical institutes, Caltech's Master of Science degree was essentially a continuation of its undergraduate curriculum, with the degree awarded following the completion of a fifth year. To attract outside students to the program, Pauling proposed creating a scholarship similar to one offered by the Massachusetts Institute of Technology, which covered tuition and provided a stipend of between $700 to $1,000 for eight months of study. Much later, in the spring of 1953, Pauling suggested further expanding the master's program by treating undergraduate seniors as first-year graduate students, thus allowing them to focus more on research and gain entry to laboratory space.

As part of their fifth year, master's degree-seeking students at Caltech were required to take one course in the humanities: an introductory survey of English literature, history, philosophy, or economics. During a December 1944 meeting of the Graduate Committee on Post-War Policies, which he was overseeing as acting chair, Pauling pointed out that the addition of a compulsory humanities course had been made in 1928 as a result of faculty action and was "not a part of the general policies of the Institute as expressed by the Trustees." Pauling's comments came on the heels of a previous committee recommendation that the Board of Trustees "abolish" the humanities stipulation for the master's degree.

The humanities requirement was brought up again two weeks later, but no decision could be reached. At a following meeting, Pauling put forth an alternative notion — that the Institute consider adding courses in the history and philosophy of science. The committee liked this idea and recommended that the Division of Humanities investigate hiring someone in the field. A few years later, Caltech brought aboard Rodman W. Paul, whose research interests were in the histories of mining and agriculture. Student enthusiasm for these offerings was such that Caltech eventually created an entire program in the history and philosophy of science.

***

By the fall of 1944, Pauling had been chair of the division for seven years and was taking time to formally reflect on both its past and its future. In doing so, Pauling acknowledged the role played by his predecessor, A.A. Noyes, in developing a robust and highly esteemed research staff in physical and inorganic chemistry. Organic chemistry, which had evolved largely under Pauling's watch, was now approaching a similar position and, Pauling thought, would continue to gain in prestige. In his view, each of these three areas — physical chemistry, inorganic chemistry and organic chemistry — had arrived at an optimum point in their development and would not be in need of expansion anytime soon.

That said, Caltech's greatest contributions to chemistry were in Pauling's primary field, structural chemistry. Comparing Caltech with peers like London's Royal Institution and Cambridge's Cavendish Laboratory, Pauling opined that "the California Institute of Technology may well have made more contributions to the field than any other single laboratory in the world."

Maintaining and building upon the Institute's stellar reputation would require new financial support to make up for the eventual expiration of the Rockefeller grants currently in hand. In his report, Pauling mentioned that the Foundation was ready to fund "an intensive attack on the problem of the structure of proteins and related substances." But in the meantime, Pauling suggested that Caltech increase its annual support for structural chemistry research from $5,000 to $7,500.

One area of obvious need was applied chemistry, which had been without a professor since the onset of the war. Pauling made a goal of addressing this, expressing a desire to hire an instructor, expand facilities for graduate research, and tailor undergraduate and master's degree programs in metallurgy.

As part of his broader vision, Pauling also revived his 1942 proposal that Caltech start a major new research track focusing on the fundamentals of medicine. Pauling saw the application of chemistry to physiological processes as a field that was coming into its own and foresaw rapid advancement in the discipline over the next handful of years. The research portfolio that Pauling proposed centered around the structural chemistry of drugs, hormones, enzymes, and poisons, and included investigations into their physiological properties and explorations of their genetic and pharmacological applications. In putting forth these ideas, Pauling wrote,

"I believe that it can be predicted safely that work along these lines will in a few years lead to great advances in the fields of physiology, bacteriology, immunology, and even medicine."

Pauling very much wanted Caltech to be an important player in this future but felt that the Institute was far too understaffed to arrive at those heights. As a corrective, he suggested they hire a bacteriologist, physiologist, enzyme chemist, and virologist, ideally possessing backgrounds in biology, chemistry, and physics. As with the 1942 proposal, Pauling once more suggested that the Institute raise funds for a new building that would house these researchers, and that the group collaborate with local hospitals.

***

The war had demonstrated to Pauling that scientific research could be conducted efficiently if personnel were organized in a hierarchy, a perspective that had concerned A.A. Noyes back when Pauling was originally considered for the division chairmanship. Emboldened by the successes of the scientific war effort, Pauling expressed his desire to apply a similar approach to studying protein structures, envisioning a large research team of about 20. Pauling feared that, absent such an intensive effort, the structures of most proteins would not be solved in his lifetime.

To strengthen his proposal, Pauling hinted that the Rockefeller Foundation might be inclined to fund work of this sort, as they had already provided $433,000 to the division to support research in structural, organic, and immunochemistry over the previous 12 years, and had also set up a $1 million organic chemistry endowment. As 1944 neared its conclusion, Pauling felt ready to bring his post-war plan to the Foundation for consideration.

Pauling's first conversations were with Frank Blair Hanson, who oversaw immunological funding, and then Warren Weaver, who remained in charge of natural science research. Weaver found Pauling's ideas intriguing but unrealistic, particularly given projected shortfalls in the number of qualified researchers who would be available after the war. And while he largely agreed with this assessment, Pauling continued to push his case, drawing Weaver's attention to a new technique for measuring the absorption of isotherms in water vapor that could be applied to the structure of proteins. Pauling wanted to pursue this technique and asked for six to eight years of funding at $25,000 to $40,000 a year, but the Foundation demurred.

***

In addition to keeping the money flowing in, Pauling also focused on taking care of his fellow chemistry faculty, making sure that their pay stayed commensurate with their accomplishments. Pauling's ability to do so was enhanced in 1945, when he began a three-year term as a member of Caltech's Executive Committee. Service on this group provided Pauling with sway over the shape of his own division and the Institute as a whole, and from this position of strength, Pauling pushed for salary increases for his colleagues in chemistry. Among his successes, Pauling negotiated a raise for Don Yost, whose annual pay was boosted substantially from $5,000 to $8,000. Pauling gave additional wage bumps to Verner Schomaker and Carl Niemann by readjusting spending from the Molecule Structure Fund and the Rockefeller grant.

Pauling also oversaw the appointment and retention of the division's staff. As part of this, in the summer of 1945, the division extended assistant professorship appointments to Richard Dodson and Charles Coryell, but both turned down the offers down, Coryell going to the Massachusetts Institute of Technology and Dodson eventually leaving two years later for Columbia University.

Having learned from this experience, Pauling went to the Executive Committee and implored that they act "without delay" to make sure that Cornelius Rhodes of the Cancer Memorial Hospital in New York — and soon to be paying a visit to Pasadena — not "entice" immunologist Dan Campbell to head a laboratory group on the East Coast. Specifically, Pauling wanted Campbell to receive an assurance that his position at Caltech was permanent as he was central to Pauling's plans for the division's future, especially regarding the Rockefeller funding. Campbell ultimately did stay, securing a $3,000 grant from Wescar Investment Company within a year. Pauling moved to offer him a full professorship in 1950.

Occasionally, Pauling had to make tough decisions about which staff would receive external funding. One such instance came about in December 1945 when Joseph Barker, Chairman of the Executive Committee and Acting President of the Research Corporation in New York City, asked for Pauling's opinion on proposals submitted by three different researchers within the division. Don Yost had requested funds to support work on nuclear chemistry and the chemistry of metals; E.R. Buchman was interested in thiamine; and James Bonner wished to explore flowering plant hormones.

Barker told Pauling that the corporation preferred to award only one grant to applicants from a given institution and asked Pauling to provide direction on a worthy recipient. While he was initially hesitant to favor one of the three, Pauling ultimately leaned towards Yost, since biochemical work was already well-funded. Having done so, Pauling also put forth the alternative possibility that Barker might prefer Buchman, since he had already established a relationship with the Research Corporation. Pauling's deflection left the final decision up to Barker, which may have been Pauling's intention all along. As an administrator who was constantly on the lookout for ways to fund and support his colleagues, Pauling likely would have most preferred that grant monies go to all three.

Chapter 19

Chairing the Division After the War

Linus Pauling, 1946.

As the U.S. government's pitched demand for research and development projects began to wane following the end of the Second World War, Caltech's Division of Chemistry and Chemical Engineering began a period of reorganization. One early shift was a reversion back to a pre-war policy for graduate studies stipulating that only those able to attend full time be admitted. But amidst these and other changes, a very important theme continued to persist: the division's involvement with military research.

One such activity that ultimately involved Linus Pauling was led by two division members from the Chemical Engineering group, B.H. Sage and Dean Lacey, who were responsible for a Bureau of Ordnance contract to improve double-base propellant processing. This classified work to develop "smokeless powder" was more of an engineering task than a chemical one, and the contract was also in potential conflict with the division's policy on accepting funds that restricted publication.

Noting this, Sage went to Pauling for his approval as division chair. After giving it some thought, Pauling judged the research to be worthy of an exception, telling Sage, "I do not see how you can avoid doing work of this sort; it seems to me to be clearly your duty, in view of your experience."

***

Typical of the administrator's multifaceted burden, Pauling then turned his attention to mediating a dispute between Sage and chemist J. Holmes Sturdivant over shared responsibilities at the Chemistry Shop. The problem came to Pauling's attention in November 1946 when Sturdivant informed Pauling that the Chemical Engineering group had "contributed essentially nothing toward the maintenance of the Chemistry Shop facilities" despite spending 10% to 17% of their time using the shop. Sturdivant subsequently requested that Chemical Engineering be required to contribute at least 5% of their time over the course of each year towards maintenance. To make up for past neglect, Sturdivant also suggested that the engineers allocate 100 hours to shop maintenance over the next month.

Sage, speaking on behalf of the Chemical Engineering group, expressed annoyance with Sturdivant's requests and claimed that the engineers had in fact attended to routine housekeeping while also taking "complete responsibility for the maintenance of the small so-called Chemical Engineering Shop." (Pauling, apparently puzzled by Sage's reference to a Chemical Engineering Shop, underlined this line of text with a question mark next to it.) Sage also pointed out that the extra hours that Sturdivant wanted the engineers to spend on maintenance would accrue to an additional 17% of their time spent in the shop. Sage concluded with a request that the group receive $50 a month more to help cover shop-related overhead costs paid to the division, a charge that they had not been asked to pay in the past.

In response, Sturdivant told Pauling and Sage that the increase in overhead charges had emerged from changes made during the war, as research shifted to other areas. Specifically, the Committee on Institute Shops had recommended that all campus shops charge overhead to the groups that used them. Sturdivant, seeking to appease Sage somewhat, then recommended lowering the Chemical Engineering group's overhead charge to $20 per month for the next 18 months. Pauling agreed that this was a reasonable compromise and the dispute was settled.

Anticipating future conflicts of this sort, Pauling subsequently inquired with upper administration about the possibility of devoting new space to chemical engineering. As it turned out, nothing major was to happen for another decade — not until 1956 did Caltech break ground for a new building dedicated to the division's engineers.

***

As a member of the Institute's Executive Committee, Pauling was well-positioned to work with an incoming figure of major importance, Lee DuBridge, the first person to officially hold the title of President of Caltech. Before starting in Pasadena, DuBridge had spent six years as the first director of the radiation laboratory at the Massachusetts Institute of Technology. And though he officially became Caltech's leader in the fall of 1946, he began working with Pauling and others months before then to lay the groundwork for future strategic actions.

DuBridge and Pauling shared a similar point of view on faculty pay: to recruit and retain the best, the Institute had to offer high salaries. With this idea in mind, they first collaborated to hire John G. Kirkwood as Professor of Chemistry, meeting in Washington, D.C. (DuBridge hadn't yet moved to California) in April 1946 to discuss the best way to attract him. At the end of their meeting, they decided that $10,000 per year would do it.

Once Pauling was back in Pasadena, he wrote to DuBridge that the Executive Committee thought the amount too high, as only a "few people" —

Pauling was one of them — made that much at the Institute. DuBridge was disappointed by this news, writing that

"My first reaction is to say a salary of this amount has got to come as a fairly common figure in the near future if we are to get and keep good men — and therefore let's go ahead in this case."

Thus emboldened, Pauling reached out to Kirkwood, who replied that he would need to think about the offer. DuBridge had not expected this response and was also a bit perturbed at Pauling, since he had meant for him to merely inquire with Kirkwood about his potential interest at that salary level. In the end Kirkwood accepted the position but only remained at Caltech until 1951, leaving to take up the Sterling Professorship at Yale.

For Pauling, membership on the Executive Committee also obliged him to assist with filling positions outside of his own division, including an attempt to secure employment at Caltech for Robert Oppenheimer. In the fall of 1945, the Institute offered Oppenheimer $10,000 to return to Pasadena now that his wartime service at the Los Alamos Laboratory had concluded. The Institute then tried to sweeten the deal by offering him the chairmanship of the Division of Physics, Mathematics, and Electrical Engineering. Neither offer proved convincing and Oppenheimer ended up as Director of the Institute for Advanced Study in Princeton.

The next year, at John Kirkwood's urging, Pauling tried to persuade Cornell physicist Hans Bethe to accept the position. Like Oppenheimer, Bethe — who would receive the Nobel Prize in Physics in 1967 — was not sure that Pasadena would make for a good fit and suggested that Pauling make an inquiry with Robert F. Christy, who had worked at Los Alamos and had also been involved with the Manhattan Project at the University of Chicago. Pauling heeded this advice, and Christy went on to spend the rest of his career at Caltech.

***

In January 1946, Pauling presented the latest iteration of his plan for a joint research program to be shared between the Division of Chemistry and

Chemical Engineering and the Division of Biology. Delivered for the third time to the Institute's Board of Trustees, Pauling's vision called for

> "an expansion of the work of these Divisions during the next fifteen or twenty years, in order that a very promising field of investigation intermediate between chemistry and biology may be cultivated; this field of investigation is also very closely related to medicine."

In putting forth these ideas, Pauling sought to build and expand upon previous research successes that had emerged from support provided by the Rockefeller Foundation.

In his talk, Pauling noted that the past two decades had brought about the development of immunochemistry, chemical genetics, and the use of radioactive tracers. These breakthroughs had made more feasible the potential determination of the "structure and nature" of substances smaller than the cell — enzymes, proteins, genes, and viruses — that are not visible under a microscope. But determining these structures, Pauling told the board, would require

> "a considerable expansion in chemistry and biology, with the addition to the staff of specialists in fields such as enzyme chemistry, nucleic acid chemistry, microbiology, general physiology, and virology."

To bolster his argument, Pauling brought Warren Weaver back into the mix by sharing "that in his opinion there is no place in the world so well suited for this work as the California Institute of Technology." If the trustees agreed to go along, Pauling believed that the program could potentially bring in millions of dollars' worth of Rockefeller support to split between divisions and spur the construction of two new buildings.

While he had faith that the Rockefeller Foundation would provide significant external funding for the plan, Pauling also had his eye on other sources. One useful contact was E.K. Wickman of the Commonwealth Fund, whom Pauling queried about granting capacity at the National Foundation for Infantile Paralysis (NFIP). Wickman reviewed the NFIP's documentation and reported that they likely had $10 million in their national reserves at the start of the year and had since established a goal of raising another $25 million through their annual March of Dimes. Wickman added that his was a conservative estimate, and further suggested that

> "Considering that the National is now pricked by criticism for large accumulations, that it has just had fresh increases, and that as a relative newcomer in the philanthropic field it may want to establish a reputation in competition with the old foundations, you may well be coming to them at the right moment for a substantial grant."

Thus encouraged, Pauling, along with Caltech biologists George Beadle and Alfred Sturtevant, drew up "A Proposed Program of Research on the Fundamental Problems of Biology and Medicine." The proposal asked for $6 million over the next 15–20 years and was submitted to both the Rockefeller Foundation and the NFIP. The overarching goal of the proposed program was to "uncover basic principles" in the biochemistry of medicine including the structure and mechanism of genes, a general understanding of viruses and antibodies, and the physiological basis of drugs. The authors also expected that plenty of practical discoveries would be made along the way.

The proposal placed special emphasis on the need to attract graduate students and post-doctoral researchers trained in biology and medicine. It pointed out that the number of graduate students working in the two divisions had dropped by more than 20 since the end of the war, a trend that would need to be reversed were the Institute to achieve new heights. Fortunately, at least in the authors' views, Caltech was well-positioned to support an ambitious program, one that would usher in "a period of great and fundamental progress, similar to that through which physics and chemistry have passed during the last thirty-five years."

Once they had evaluated the proposal, the Rockefeller Foundation, as was their custom, asked for assurance that Caltech would continue to support biochemistry and biophysics with its own institutional resources. The Foundation was also not prepared to finance the construction of new buildings (after which Pauling and Beadle began to press President DuBridge for internal funding). The Rockefeller trustees did ultimately agree to provide a measure of support, though it fell far short of the proposal's ambitious ask. A semiannual grant of $50,000 was allocated, to be paid out over seven years for a grand total of $700,000. The NFIP also agreed to a partial measure: $300,000 over five years.

Pauling, Beadle and Sturtevant were glad to have these commitments in hand and immediately began probing other routes to the original $6 million ask, including a $2.3 million private bequest that had recently been made to the Institute. With funding momentum gathering, Pauling also decided that he would shorten his forthcoming Eastman residency at Oxford University so that he could devote more time to creating action items and managing budgets.

Once implemented, it did not take long for the new plan to bear fruit. By 1947, Institute researchers had launched into multiple lines of novel work that included studies of the structure, composition and molecular weight of amino acids, peptides, proteins, and viruses; the chemistry of enzymes and nucleic acids; serological genetics and embryology; chemical genetics; virology; and intermediary metabolism in plants and animals.

And yet, despite the new money, concerns about adequate funding reemerged as news of a $240,000 budget shortfall began to circulate. As a corrective, the divisions started to look at other pots of money to cover the gap, including another large grant that had been promised by the Rockefeller Foundation, as well as smaller sources like a $3,300 award that George Beadle had received from the Eli Lilly Company to work on the biosynthesis of vitamins. Certain funding lines, however, remained out of bounds, including a five-year $75,000 allocation that Pauling had secured in 1945 from Union Carbide to support fundamental research on the structure of metals and alloys.

**

The demands of the war years had steadily pushed the Institute toward contract work that was funded by the government and private entities. These contracts were especially attractive to faculty, as the deals often served as a source of extra income that could augment their Caltech salaries.

Increasingly, more money for Pasadena households was becoming a necessity. According to a 1947 report commissioned by President DuBridge, the cost of living in the area had increased "well over 40 per cent" since the start of the decade, but Caltech had only boosted its salaries by about half that amount. To keep pace with Harvard, Berkeley and MIT, the report argued that Caltech would need to raise its salaries by 50% above 1940 levels, followed by an additional 75% increase over the next three years. One solution that DuBridge found to address this problem was to change the compensation structure for faculty such that they were paid a 12-month salary at the same monthly rate as their nine-month salary. In instituting this change, DuBridge effectively gave his faculty a raise that was equal to three months of pay.

Amidst this rapidly changing landscape, Pauling continued to recruit new researchers into the Institute. While overseas, Pauling assisted E.C. Watson, Caltech's Dean of Faculty, in searching for both a mathematician and solid-state physicist. One name that Pauling put forth was Mary Cartwright of Cambridge, who had recently been named the first female fellow of the Royal Society and who came recommended as the "most outstanding younger mathematician in England."

Pauling had a harder time finding good physicists, but did recommend Clarence Zener of the University of Chicago's Institute of Metals. The following month, Pauling suggested that Paul Dirac, then of the Institute for Advanced Study, be invited to Caltech. Ultimately none of these suggestions worked out, but Pauling's grander vision for post-war science at Caltech was inarguably moving forward.

***

At the same time, Pauling's Caltech colleagues began to notice that he was pulling back from some of his administrative responsibilities. One such individual was Rodman Paul, a member of the Caltech History department who arrived at the Institute in 1947 and soon realized that Pauling had delegated oversight of day-to-day operations to Carl Niemann. According to Paul, this observation was widely shared across the Institute.

On December 26, 1947, Pauling and his family set sail for England to begin a nine-month residency at Oxford University, a move that further convinced some of Pauling's colleagues that his main priorities no longer centered on chairing the division. That year, John Kirkwood, George Beadle, and J. Holmes Sturdivant authored a ten-year plan that they mailed to England for Pauling's feedback. After reading over the document, Pauling's main suggestion was that the authors add the possibility of creating a nuclear chemistry program within the division. This note was in keeping with Pauling's attitude toward administrative work by that point: while his engagement with routine affairs appeared to be fading, he remained passionate about a bold and expansive future for the division, one that would rely on dependable sources of funding.

***

In 1949 Pauling served a one-year term as President of the American Chemical Society, and in this capacity, he penned an editorial for *Chemical and Engineering News* that bemoaned the lack of industry support for "pure chemistry" and other areas of "pure science." This dearth of engagement had been felt in his own division and had slowed progress toward fulfilling his vision for fundamental research in biochemical medicine. Traditional sources of funding like endowments were not doing well at the time because of high rates of inflation. And while other avenues of support — including UNESCO fellowships and the newly formed National Science Foundation — had come online, their resources were not substantial enough to make up the difference, since they only provided the minimum needed to keep research going.

Pauling believed that government entities, corporations, private endowments and foundations were all capable of doing more, but he targeted industry most of all, writing that

> "The industrial corporations, the chemical industries that depend upon fundamental science for their success, are, I believe, failing to do their part in the support of pure chemistry."

Pauling believed that the benefit these companies received from basic research was such that they should fund up to 30% of the work being done — a number far higher than the 1% they did support — and that they should do so without placing restrictions on patenting or publication.

The editorial made an impact. In one instance, University of Chicago administrator Theodore Switz reported using the piece in his fundraising outreach to chemical companies like Sinclair Refining. Another apparent response to Pauling's call came from the DuPont Company. Over the previous three years, the corporation had funded a lone postgraduate fellowship for a Ph.D. in chemistry at Caltech, providing $1,200 for a single man or $1,800 if he was married, with an additional $1,000 going directly to the Institute. After Pauling's article was published, DuPont revised the annual donation to $10,000, made generally to the Division for distribution as it saw fit.

Caltech's internal response to the evolving funding climate was to create a group called the Industrial Associates of the Institute. The first meeting of this collective, held in November 1950, brought in research executives and staff from companies including DuPont, Douglas Aircraft, Lockheed Aircraft, North American Aviation, Shell Development, Socoy-Vacuum, Magnolia Petroleum, General Petroleum, Union Oil, and Standard Oil. In the meeting, Caltech put forth a great many research areas where the interests of the Institute and industry might overlap. These points of intersection included heat transfer, fluid mechanics, the influence of shock loading on design, soil mechanics and engineering structures, radioactive tracer usage, X-ray usage, and the influence of molecular structure on material properties.

Pauling never assumed a leading role in the group, but he did take part in affiliated conferences, visited industry labs, and hosted private sector guests in his own lab spaces. By 1957, President DuBridge was applauding the success of the program and emphasizing how it had opened up communications between Caltech and industry.

***

As his focus turned increasingly toward the big picture, Pauling relied more and more on division colleagues to keep unit operations running smoothly. In the spring of 1949, Pauling asked Carl Niemann, Howard Lucas, and Laszlo Zechmeister to form a committee to recommend "likely young organic chemists" for permanent staff appointments, a big shift from past practice where Pauling had typically comprised a search committee of one. Even more broadly, two years later, Pauling asked the entire faculty in Chemistry to recommend a "young man" who could be appointed "in some field of work that is not strongly represented" in the division.

Sometimes Pauling would reach outside his administrative unit to look for recommendations, as was the case in the winter of 1949. The previous fall, Pauling had made an internal announcement of the need to fill the physical chemistry vacancy created by the death of Roscoe Dickinson, his former Ph.D. adviser. Four months later, Pauling asked John Kirkwood, a physics professor, to chair a committee on behalf of the division and to make the appointment in physical chemistry. Pauling had initially hoped that Michael Szwarc at the University of Manchester would accept the position and later he suggested Wilhelm Jost from Germany, but ultimately both remained in Europe.

As he had always done, Pauling continued to look out for the existing staff whom he had helped to shape and worked to ensure the retention of those whom he especially valued. In June 1946, Pauling offered an instructor position in chemistry to one such colleague, Norman Davidson, whose prior research had focused on molecular biology. It was Pauling's hope that the instructorship would turn into a tenured position at some point.

That moment came closer in 1949, when Pauling recommended to his dean that Davidson be appointed Assistant Professor for two years as part of a general round of raises being considered for five chemistry professors. Pauling further asked that Davidson be offered a permanent position within the division, and this soon came to pass. Davidson received tenure in 1952, spent the rest of his career at Caltech and received the National Medal of Science in 1996.

***

Not all of Pauling's staffing decisions revolved around faculty and research; he was also interested in building up staff to support the work of others. One instance involved the possibility of hiring a chemistry librarian, an idea that Pauling first floated in the fall of 1949. After some discussion, it was decided that this position should be put on hold until a new facility was built — the chemistry library was split between two buildings at the time. The following year, Pauling signed on to library director Roger Stanton's suggestion that chemistry and biology share a librarian, but this did not come to pass. Indeed, it would be nearly 20 years until a new library facility was built at Caltech.

A new chemistry and biology building, however, would come much sooner. At the beginning of 1949, Pauling reported to the division that fund-raising efforts had already generated $700,000 toward this goal. Potential locations had also been identified, with options including a space near the north end of Crellin or possibly an addition to the Kerckhoff Laboratory. Pauling favored the latter, since an extension of this sort had always been envisioned for Kerckhoff.

As progress on this initiative moved forward, the new truth about Pauling's chairmanship continued to solidify. While it rightly appeared to some that he was pulling back from certain administrative tasks, he clearly retained an active interest in many of the details and, above all else, remained strongly focused on the division's long-term objectives.

Chapter 20

A Cold War Division Chair

Linus Pauling, 1950.

As the 1940s neared their conclusion, Linus Pauling's growing inclination to delegate some of his administrative duties became more pronounced, an evolution that was prompted in part by the need to spend time advocating for and securing funds to support biochemical medical research within the division. But without doubt, another significant factor propelling the need to delegate was Pauling's increasing involvement in peace activism and, particularly, his schedule of public speaking on issues related to nuclear weapons. These activities ultimately brought Pauling before Caltech's Board of Trustees, who contemplated his dismissal.

In 1949 the board communicated to Caltech President Lee DuBridge that Pauling's public statements on peace topics were "damaging" to Caltech's reputation. In response, Pauling is said to have "pledged" to DuBridge that he would cut back, since he did not want his political views to interfere with his scientific work. However, President Harry Truman's decision to develop a hydrogen bomb the next year changed Pauling's mind, and he was again brought to the attention of Caltech's leadership. This time, Pauling told DuBridge that he wished to speak with the trustees directly.

Meanwhile, the board had formed a committee made up of five trustees and five faculty members who were asked to determine whether Pauling should be dismissed from Caltech. In a statement dated July 14, 1950, Pauling expressed shock at the very idea, writing that his tenure rank and hugely successful 28-year career at Caltech had not prepared him for such an extreme possibility.

Three days later, Pauling was given the chance to speak to the board. At that meeting, as in his earlier conversations with DuBridge, he reiterated a desire to cut back on his political activities, but by now the ambition was somewhat couched. "I still propose to do this, at a rate determined by the world situation," he offered. "However, I remain unwilling to pledge myself to cease all political activities." Pauling made it clear to the trustees that he did not want to harm Caltech and would do "anything compatible with my conscience and my principles" to protect its reputation.

But in actual fact, Pauling did not believe that he was harming Caltech's reputation at all. Rather, after surveying several colleagues and students who told him that his activities had caused no "appreciable damage" to them, Pauling concluded that he was *helping* the Institute's standing.

President DuBridge harbored a decidedly different point of view and informed the board that "many staff members" had told him that Pauling's actions had "damaged them greatly." These sentiments focused especially on Pauling's support for Sidney Weinbaum, a Russian émigré who became a United States citizen in 1927, completed his Caltech physics doctorate in 1933, spent the next decade as a research assistant in Pauling's lab, and then served two years in prison.

***

Sidney Weinbaum, 1950.

After completing his doctorate, Israel Sidney Weinbaum (1898–1991) received a research fellowship in chemistry and spent much of it solving complex mathematical calculations in support of Pauling's work on crystal structure determinations. In 1934 he co-published two papers with Pauling — one on the crystal structure of enargite and another on calcium boride — and his handwriting can be found in four of Pauling's research notebooks from the 1930s. He also delivered a number of lectures in Pauling's quantum mechanics course and was one of several students who helped put together a published textbook based on the course's content. In addition, Weinbaum and his wife Lina enjoyed a relatively close relationship with Linus and Ava Helen Pauling, meeting socially several times a year over the course of their time in Pasadena.

Outside of his academic responsibilities, and distinct of his enthusiasms for the piano and chess, Weinbaum was very active in politics. Though he later maintained that he never spoke with Pauling on topics of the sort,

Weinbaum was himself engaged in political discussion groups throughout the 1930s and 1940s. He distributed petitions and was well acquainted with a variety of activists, some of them communists, with whom he would discuss the issues of the day.

In 1943 Weinbaum left Pauling's lab for a position in the aviation industry. He returned to Caltech in 1946, this time working at the school's Jet Propulsion Laboratory. As part of his new job, which involved classified projects, Weinbaum was required to hold a security clearance, which he obtained and held for three years. His circumstances abruptly changed on July 7, 1949, when both the Jet Propulsion Laboratory and Weinbaum were sent a letter from the 6th Army Headquarters in San Francisco, denying Weinbaum continued access to classified materials. His work assignment was immediately shifted, and his position was terminated soon thereafter.

According to the authorities who revoked his security clearance, in his completion of a 1949 Army security questionnaire, Weinbaum had "willfully omitted" previous involvement with any communist organization. After he was dismissed from his Caltech position, several hearings concerning the matter were held before the Industrial Employment Review Board, during which Pauling himself submitted an affidavit that attested to Weinbaum's loyalty. Weinbaum then appeared under oath before a military review board, where he denied charges of communist membership.

Though he continued to maintain his innocence, Weinbaum was eventually arrested on charges of perjury (but not disloyalty), with his attendance of "communist club" meetings and his association with known communists during his days as a student held against him as proof. In response, Pauling and several other "Friends of Sidney Weinbaum" began raising money to help defray Weinbaum's legal expenses, and the Pauling family offered Lina a room in their home.

President DuBridge did not like the optics of the situation and suggested that Pauling raise money by word of mouth, and not through the mail. An undated form letter authored by Pauling and five others, and appealing for money to support Weinbaum's case, suggests that Pauling either ignored DuBridge's advice or that Caltech's chief executive was trying to reel Pauling back in.

In September 1950, at the Federal District Court in Los Angeles, Sidney Weinbaum was convicted of perjury and sentenced to four years' imprisonment. During his incarceration, and despite the suspicion that he knew it would engender, Pauling wrote to Weinbaum with some frequency. Both men remained cordial in their correspondence, but it is evident that each was resentful for what had happened. At one point Weinbaum blamed Pauling and his associates for not helping more. Pauling responded that his reputation had been significantly damaged by the affair and that he had lost a $4,800-per-year consulting job as a direct result.

Indeed, both before and after Weinbaum's trial, Pauling himself was under investigation by several authorities for any ties that he might have developed with communist actors. Though his support for Weinbaum did not help his cause, no convincing evidence could ever be found to suggest that Pauling himself was a communist. The investigations did not end any time soon, however, as the FBI continued to monitor his activities, and even the social time that Ava Helen spent with Lina was marked by federal agents. After the perjury trial, Pauling's relationship with Weinbaum was frequently held against him by investigatory committees, and though he maintained his composure throughout the process, the strain of the moment deeply affected him. Pauling was especially hurt by the distrust and distancing that occurred between him and many of his colleagues at Caltech.

After spending over two years in prison, Weinbaum was released on parole in 1953, after which he faced difficulties finding employment and was left debilitated by illness. The stress and misfortune of their circumstances led the Weinbaums to divorce soon after Sidney was paroled. And though Lina continued to send the Paulings letters up through the early 1990s, Linus received his final letter from Sidney in 1953, shortly after the conclusion of his sentence.

Despite having suffered through great difficulties, Weinbaum's story ended on a sunnier note. A man he knew through chess associations eventually found him a job at a factory and, shortly afterward, he met the woman who would become his second wife. Though his scientific pursuits were ended, Weinbaum lived a relatively happy life thereafter and passed away well into his 90s.

***

In October 1950, not long after Weinbaum's sentencing, Pauling came under further scrutiny after being named a communist by Senator Joseph McCarthy. The Republican from Wisconsin fanned the flames of this allegation by further defining Pauling as an atomic scientist who had received classified information from the Atomic Energy Commission as a result of his connections with the Guggenheim Foundation. The Senator was quick to add that the Foundation was rumored to have "a flagrant record of giving fellowships to Communists."

Responding internally, Pauling explained to Charles Newton, President DuBridge's assistant, that he was only on the Committee of Selection for the Guggenheim Foundation and could in no way have been involved with the organization in the manner that McCarthy had alleged. Pauling added that McCarthy was likely targeting him for his peace work.

In the midst of it all, the Caltech select review committee continued to deliberate, though they never called on Pauling to testify. In fact, another 12 years would pass before DuBridge finally informed Pauling that the committee had recommended, in May 1952, that nothing be done to punish Pauling. Instead, the group suggested that Pauling be continually pressed to end his political activities and, as time went on, the internal pressure on Pauling most certainly rose.

***

While Pauling's political activism was increasingly taking calendar space away from his daily responsibilities, he continued to take pride in heading the division, which was racking up successes. In 1950 Pauling reported to division staff that the Committee on Professional Training had given the chemistry program an overall grade of A, as well as top marks in physical chemistry, inorganic chemistry, and analytic chemistry, coupled with a lone B for organic chemistry. Pauling delighted in these accomplishments and made a point of budgeting time to actively work with new crops of incoming students.

One such student, Fernando Carraro of Brazil, first wrote to Pauling in 1950, expressing an interest in Pauling's research on antibodies. Over subsequent exchanges, it became clear that Carraro wanted to study with him at Caltech. In 1953, Pauling suggested that Carraro apply to the Institute and that he also seek funding from the Guggenheim Fellowship for Latin

American students. In the meantime, Carraro wanted to know what he should study, to which Pauling offered the idea of mathematics.

In providing guidance related to the Guggenheim application, Pauling further advised that Carraro focus on the structure of proteins, the application of quantum mechanics to molecular structure, or the analysis of gas molecules by electron diffraction during his stint in Pasadena. But Pauling also warned Carraro that he would likely not be given a graduate assistantship since he had not attended an American university.

Carraro's fellowship was ultimately approved, and he studied under Pauling during the 1954–1955 academic year. Though a small story in the grand scheme of Pauling's life, his interactions with this student from Brazil serve as evidence that, despite everything else that was vying for his attention, he continued to set aside time for those wishing to learn.

***

As division chair, Pauling was obligated to deliver an annual report each year to President DuBridge. As he compiled these reports, Pauling solicited comments from members of the division that focused primarily on their research progress over the previous year. In 1952, using the Division of Biology's report as a model, Pauling added some specifics to his usual request for comment. This time around, he needed 1) information about funding sources; 2) a short (100 to 500 words) description of work completed that would be accessible to the general reader; and 3) a list of awards received and publications authored. These extra details painted a generally positive portrait of the division, though they did not always reflect the budgetary strains being felt across the unit.

While Pauling oversaw an influx of funding that allowed the division to expand, the money on hand never seemed to be quite enough. As a result, Pauling was forced to keep a close watch on the bottom line, tracking staff salaries, fellowships, supplies, and special funds. In doing so, Pauling sometimes uncovered what appeared to be frivolous spending in unexpected places. In 1951 for example, Pauling asked division staff to be careful about publishing in journals that charged page fees for article reprints. Pauling had found that, in just that year, the unit had spent $2,500 on reprints, including $700 for covers alone. Going forward, Pauling asked that, unless absolutely necessary, reprints be ordered without covers; a small sacrifice to conserve money. Unfortunately for the division, Pauling's personal activities were beginning to impact operations to a far greater degree.

***

By 1951, Pauling's work as an anti-nuclear activist had been targeted for suspicion by multiple external forces as well as Caltech's own administration. One tangible outcome of this suspicion hit the division hard when the United States Public Health Service denied Pauling a $40,000 grant on the basis of his alleged communist ties.

Around this time, Caltech decided that all individuals serving on the Institute's Contracts Committee be required to pass a low-level security clearance, to be administered by the Industrial Employment Review Board (IERB). In part because it was meant to be a routine, low-stakes review, this directive was something that should have posed no complications for Pauling. However, problems did indeed arise after Pauling's name was erroneously (and accidentally) included on a list of upper-level administrators connected to Project Vista, a top-secret hydrogen bomb research program with which Pauling, in actual fact, had no affiliation.

As a scientist, Pauling's IERB case was to be evaluated by a military panel. At the beginning of August 1951, Pauling received notice from Lieutenant Colonel W.J. King that he had been denied a security clearance due to his being a "member" and "close associate" of the Communist Party since 1943. As backing for this claim, King cited Pauling's support of "known Communists," a likely reference to the fundraising effort that Pauling had helped lead for Sidney Weinbaum's defense the previous year. Pauling adamantly denied the charges, calling anyone who accused him of being a communist a "liar." Pauling did concede that he may have defended communists in the past but also maintained that he had the right to defend those who "deserve to be defended."

King informed Pauling that he could submit evidence in his own defense before a final decision was made, at which point Pauling turned to his scientific colleagues, asking them to vouch for his character through letters of reference. In making this appeal, Pauling sent out a form letter describing how he had signed the Espionage Act several times over the previous 11 years, a period during which he had carried out war work that made use of large amounts of classified information. The letter also stressed that Pauling wanted to continue working on similar projects as a "service to the Government," but that he would not be able to do so without approval from the IERB. The letter concluded with a statement of Pauling's strong belief that his own political actions to "help improve our national politics and to prevent and rectify injustices to individuals" should not be held against him.

One ally, Frank Aydelotte of the Rhodes Scholarship Trust, told the IERB that Pauling was definitely not a communist. In his letter of support, Aydelotte wrote,

> "Professor Pauling is a liberal; he is a man of great personal courage who would not hesitate to defend anyone whom he believed to be the victim of injustice, but he is at the same time a man of complete integrity and proven loyalty to the United States Government."

Not all of the colleagues solicited by Pauling gave their unconditional support. One of them, Karl Compton, the Chairman of the Corporation at the Massachusetts Institute of Technology, asserted that Pauling's scientific and personal character were admirable, but hedged that he had never spoken with Pauling about communism and could not offer an opinion on whether Pauling was a communist or not.

Lee DuBridge also agreed to write a letter for Pauling but added that he was not surprised by the decision that the IERB had reached, an admission that shocked Pauling. For the Caltech president, the issue was mostly a headache that he wanted settled "one way or another." As Pauling's activities had come under increasing scrutiny, some of the Institute's trustees had threatened to leave the board, and DuBridge himself was in danger of being replaced if he could not find a way to keep Pauling in line.

By the end of September, it appeared that the letters of support had fallen on deaf ears as Pauling's clearance was once again denied. Following an immediate appeal, another month passed until the Project Vista clerical error came to light and the baseline clearance was restored. Tensions were lessened for the moment, but problems related to Pauling's public persona would only continue to interfere with his duties at Caltech and within his division.

***

Somehow, despite the mounting pressures, Pauling was largely able to maintain effectiveness in his position as division chair. In particular, Pauling was able to keep momentum going for the new biochemical laboratory that he had championed. By 1952, following a three-year fundraising effort, logistics for the new facility were coming together, architects were working out some of the design specifics, and research groups within the division were jostling for pieces of the new space.

Chief among the claimants was Pauling himself. With Dan Campbell, Pauling put in for 15,000 feet to be used by the immunochemistry group. A separate request for 12,000 square feet was also put forth by Pauling, joined this time by Robert Corey and the X-ray diffraction group. Both requests were double the allocation afforded to these activities in the Crellin Lab. In addition, Pauling wanted an increase in his personal office and laboratory space, bumping up from 1,000 to 1,500 square feet. By comparison, Carl Niemann's enzyme group was to receive 4,500 square feet and the analytical chemistry group would get 3,000.

Near the end of August 1952, fundraising for the new building received a big boost when Caltech accepted $750,000 from a local businessman, Norman W. Church. The gift was large enough to pay for the shell of a 70,000 square foot biochemical laboratory, though not enough to connect the new laboratory to Crellin, which was the ambition. (Caltech thought that it could potentially cover the cost of doing so itself.) Estimates to complete the entire building ran to $1.4 million, with another $270,000 needed for furnishings. By the next May, as the Institute began accepting construction bids, the estimate had risen to around $2 million to complete the main wing.

As the project advanced, it became clear that the funds on hand were not sufficient. In August 1953, J. Holmes Sturdivant, who was supervising the Biology component, conveyed the bad news that the full initiative was short to the tune of $500,000. To make amends, Sturdivant felt it likely that the design would have to scrap two planned basements and not furnish the building as originally conceived.

Pauling appointed a committee to address this matter, and they in turn concluded that the two basements were critically important because the new facility would be too small to function effectively otherwise. The group further suggested that it was fine to keep portions of the building unfurnished at the outset — the basement was the priority. Pauling agreed, telling President DuBridge that they could always raise more money to fully furnish the space later on.

Not merely interested in its internal layout, Pauling also hoped to make a stylistic imprint on the building's exterior. During his visits to England and France four years prior, Pauling toured many 11th- and 12th-century Norman churches and was impressed by the elaborate receding arched door-ways — some of them decorated with depictions of spirits — that adorned these houses of worship.

Since Caltech's new building would be called the Norman W. Church Laboratory, Pauling thought it appropriate to similarly decorate its main entrance using depictions of bacterial cells, viruses, animals, crystal struc-tures, and scientific instruments. In the end though, this idea did not come to pass — be it through lack of interest or lack of funds, the final building boasted a much less ornate entryway.

***

In 1952, Warren Weaver at the Rockefeller Foundation informed Caltech's Biology head George Beadle that the Foundation was planning to reduce its support for biological studies due to the presence of an increasing number of alternative funding sources that were becoming available. Despite this, Caltech would still be at the top of the Foundation's list for research funding requests of "a more general nature."

Knowing full well that the Institute was nearing the end of a seven-year $700,000 Rockefeller grant covering biochemical research, Weaver told Beadle of the potential for a new $1 million allocation that could fund 10–20 years of biochemistry research, with additional money allocated for equipment. In his communications with DuBridge and Pauling, Beadle emphasized that "Weaver made it quite clear without saying so directly that we would not be left high and dry."

That Weaver had communicated this possibility to Beadle rather than Pauling was perhaps a sign of Pauling's waning influence with upper administration at Caltech. But while he no longer appeared to take the lead, Pauling did play an important role in securing funding as head of his division. And as he worked to generate financial support to broaden the division's horizons, he continued to cling to the original post-war vision for biochemical medical research that he had first formulated eight years earlier, a vision that had thus far received insufficient financial support.

In making their formal appeal for further backing to the Rockefeller Foundation, Pauling and Beadle first laid out the extant funding sources that were currently supporting their two divisions. They began by noting that the Institute itself provided $325,000 per year for chemistry and $250,000 for biology. Another $600,000 in soft money was also on hand, including $100,000 per budget cycle from the current Rockefeller grant. Finally, $300,000 more was channeled annually to the two units from endowment funds.

Some of those endowments came from individuals who had willed their assets to the Institute. In May 1952, Pauling wrote to Caltech trustee George Farrand to express his gratitude to a Mrs. Robinson who had left "the bulk of her estate" to Caltech for cancer research. In his note, Pauling touted the Institute as being the perfect place to conduct fundamental research that might supplement clinical trials going on elsewhere but hastened to add that these kinds of donations were not enough to sustain the levels of medical research that he and his colleagues thought possible.

In their pitch to the Rockefeller Foundation, Pauling and Beadle stressed that future support needed to be long-term — at least 15–20 years — if they were going to attract tenured faculty capable of conducting world-class research. Covering a time span of this length would require around $3 million and the duo requested that the Foundation commit to providing half this total, with the other half to be raised by the Institute from other sources. Were funding of this magnitude to be acquired, and with endowment interest accruing, the two divisions would be able to spend about $150,000 a year for a minimum of 14 years.

Happily, the Foundation agreed to Pauling and Beadle's proposal, with the proviso that the Institute generate the matching funds within the next three years. This major commitment from a key external partner appeared to finally secure Pauling's post-war plan for long-term biochemical research at Caltech.

Chapter 21

Final Years as Division Chair

Dr. Pauling Steps Down to Teach

Independent June '55

Dr. Linus Pauling, Nobel Prize-winning c h e m i s t, resigned yesterday as chairman of Caltech's Division of Chemistry and Chemical Engineering to devote his time to teaching and research.

Succeeding him as division chairman is Dr. Ernest H. Swift, formerly professor of analytical chemistry at Caltech and an internationally-renowned scientist.

Dr. Pauling will remain at Caltech as professor of chemistry.

President Lee A. DuBridge of Caltech commented on the change saying:

"Dr. Pauling requested a year ago that he be relieved of administrative duties in order to devote full time to teaching and research. He has served as division chairman for 21 years, and his desire for relief is understandable. Caltech is fortunate at this time that it can name to [Ph.D. degrees at Caltech in

LINUS PAULING
. . . resigns

ERNEST H. SWIFT
. . . new chairman

As the 1950s moved forward, Linus Pauling's increasingly public stances on nuclear weapons and peace issues emerged as a public relations problem for the California Institute of Technology's administration, which was repeatedly forced to respond to charges that Pauling was a communist. As this issue became more pronounced, three of the Institute's trustees made good on threats to resign because of perceived damage caused by Pauling's image.

Pauling's reputation took a big hit when accusations were levied again against him by one Louis F. Budenz. Before starting what was effectively a career as a professional informant, Budenz was managing director of the *Daily Worker*, a nationally distributed socialist news outlet. Budenz began consulting with the FBI after submitting to an inquiry by the House Un-American Activities Committee. He then became a staff member at Fordham University, where he wrote multiple articles and books about his former association with the Communist Party. Making his living as a lecturer, writer, and testifier, Budenz claimed in 1953 to have earned $70,000 as a witness.

In a November 1951 article published in *American Legion Magazine* and titled, "Do Colleges Have to Hire Red Professors?", Budenz described a speaking trip that Pauling took to the University of Hawaii as being, at least in part, motivated by a desire "to spread Stalin's views of 'peace' among the students of that institution." The next year, at an appearance before the House Select Committee on Foundations, Budenz claimed that,

"In connection with Dr. Pauling's many memberships on Communist fronts, I was officially advised a number of times in the late-, that is, in the middle-Forties, that he was a member of the Communist Party under discipline. The Communist leaders expressed the highest admiration and confidence in Dr. Pauling."

Budenz' incendiary accusations drew the attention of Pauling's colleagues around the country, many of whom wrote to seek clarification. In one exchange with W.H. Eberhardt, a Georgia Institute of Technology chemist who had worked at Caltech during the war, Pauling explained that the assertions actually meant that there was no tangible evidence showing Pauling to be a member of the Communist Party. Pauling added that he was "pretty irritated" by Budenz's testimony, but as division chair he was obliged to set aside his exasperations as he dealt with a wide array of professional responsibilities.

***

For the entirety of his tenure in leadership, Pauling had, by default, been forced to deny entry to any women applicants who wished to pursue graduate study within the division. Caltech's formal policy was that women not be admitted at any level, though they could be employed as post-doctoral research assistants. In practice, even a post-doctoral appointment rarely happened.

One potential post-doc, I.E. Keszler, was considered in the spring of 1953, but ultimately turned away by Lazlo Zechmeister because her interest in developing tests to determine the dyes used in wines was considered to be too tangential a research question. Other women seeking to become graduate students did not receive even that level of consideration, but this would change the same year that Keszler was denied.

The woman who broke through Caltech's glass ceiling was Dorothy Semenow, a graduate student in chemistry working with John D. Roberts at the Massachusetts Institute of Technology. When Roberts accepted a position at Caltech, Semenow expressed a desire to relocate as well, so that she could continue working with him. After receiving an "informal application" from Semenow in February, Pauling brought her case to William N. Lacey, Caltech's Dean of Graduate Studies. Lacey replied that he would have to take this request to the Graduate Committee, who would then need to make a recommendation to the Faculty Board and also gain the approval of the Board of Trustees.

The following week, division faculty deliberated on the precise wording of the request, which would ask that the Committee on Graduate Study be empowered to admit women as graduate students. The language emphasized in the document, which would be echoed as the process moved along, was that women be admitted

"when, in the opinion of the sponsoring Division and of the Committee on Graduate Study, the applicant possesses exceptional qualifications for admission to graduate study and gives unusual promise of continuing scientific productivity."

Since the Committee on Graduate Study would be weighing each candidate separately, Pauling requested that they also communicate their opinions to the Faculty Board. Carl Niemann seconded Pauling's request and it was passed without dissent, though some in attendance abstained. A month later, Pauling forwarded the division's recommendation to the Faculty Board who quickly approved it. By the end of the spring, the Board of Trustees had agreed to admit women as graduate students on a case-by-case basis.

Semenow's next hurdle was for the faculty of the Division of Chemistry and Chemical Engineering to vote to admit her as a Ph.D. student. She cleared this hurdle by a vote of 17 to 3 in favor, with Donald Yost, Dan Campbell, and J. Holmes Sturdivant voting against. William Corcoran and B.H. Sage voted in the affirmative, but with the stipulation that Semenow be permitted entry only in the absence of applications from comparable male candidates. An additional faculty member, William Lacey, prefaced his vote with a statement that he was generally against the admission of women to Caltech, but thought that "since the Faculty voted to permit it, I believe that Miss Semenow's case is a suitable one to try out on the Committee of Graduate Study."

To get the committee to approve Semenow, the division next had to prove that she was a promising student. Carl Niemann was asked to assist with this process and, in due course, met with Mary Sherrill, chair of the chemistry department at Semenow's undergraduate alma mater, Mount Holyoke College. Following this conversation, Niemann reported to Dean Lacey that "Miss Semenow has a genuine interest in chemistry as a profession and has every intent to continue in this field after receiving her Ph.D. degree." This affirmation was enough to secure her spot at Caltech.

***

With Semenow's case decided, the Institute felt freer to admit other women as graduate students. In April 1954, the Committee on Graduate Study recommended that Estelle Maxine Fowler be admitted as a Ph.D. student in mathematics, and the following spring they approved Elizabeth Rosenthal in biochemistry and Caroline Teichemann in aeronautics. Seventeen years after Semenow, Caltech took similar action on the undergraduate level, welcoming women into the freshman class of 1970.

In 1955, just two years after being accepted, Dorothy Semenow completed her Ph.D. in chemistry and biology, becoming the first woman at Caltech to do so. Following that, she opted to remain in Pasadena and attempted to continue her research but was denied access to laboratory space.

As it turned out, this denial caused problems. In July 1957, Carl Niemann, writing as acting division chair, informed Semenow that she had been seen entering the division's laboratories twice after hours with a key. Niemann asked that she return her key and warned her that she could be arrested if she was caught again without the accompaniment of an authorized staff member. She was also informed that she was not to use the laboratory facilities under any conditions. Thus admonished, Semenow continued to pursue her research while staying out of trouble with the division.

In the years that followed, Semenow went on to earn another Ph.D., this one in psychology. She also remained engaged with chemical research, focusing on the molecular components of potential neurotransmitters of acetylcholine and norepinephrine, and often collaborating with her spouse, Donald Garwood, whom she met while at Caltech.

In 1968 the couple reached out to Pauling in hopes that he might help them secure a research position at the University of California, San Diego (UCSD), where he was working at the time. After looking over their research statements, Pauling's colleague Robert Livingston judged that Semenow and Garwood's work did not overlap sufficiently with the interests of UCSD's Neuroscience department and thus ended the inquiry. This appears to have been Linus Pauling's last documented contact with Dorothy Semenow, a pioneering figure in the history of the California Institute of Technology.

***

The beginning of the end of Pauling's years in charge was punctuated by at least two major events. One was the construction of the Norman W. Church Laboratory for Chemical Biology, which commenced in August 1954. Earlier budget concerns, which threatened to reduce the size of the building and leave much of it unfurnished, had been partially overcome when an extra $250,000 was allocated for the project. But even this fresh infusion of cash was not enough to get the building over the hump. Full funding was not in hand until the next spring, when a $368,000 construction grant was awarded to provide for the laboratory's furnishings.

All told, the Church facility cost more than $2 million to complete. It was dedicated in November 1955 as part of the National Academy of

Sciences' annual meeting, which was held at Caltech at Pauling's urging. Faculty and staff did not fully occupy the building until the following summer, but everybody was pleased with the result and excited by the opportunities for collaboration between chemistry and biology that the shared space helped facilitate.

Another extremely significant highpoint for 1954 was Pauling's receipt of the Nobel Chemistry Prize. Pauling learned that he was to receive the award on November 3, 1954, just 45 minutes before giving a lecture at Cornell University. He recalled later that he "had a little trouble with the seminar." Soon after finding out that he had won the Nobel, congratulations began to come from his nominators, who were colleagues and friends from around the world. Hundreds of letters and telegrams followed from there.

The tenth American to win the Chemistry Prize, Pauling was honored by the Nobel Committee for his study of the structure of matter and of the seemingly invisible forces that hold its building blocks together. When asked for his thoughts on this work, Pauling first explained that it was the support and environment that fellow scientists and collaborators had fostered at Caltech that helped him to win the prize. He likewise noted that he had been able to develop his theories as a result of many years' worth of work — by him and others — on X-ray crystallography and the behavior of electronically irradiated chemicals.

On December 3, 1954, just a few days before Pauling was to depart for Sweden, the Institute hosted an enormous dinner celebration attended by 350 faculty members and guests. Held at the Caltech Athenaeum, the event featured quips from molecular biologist Norman Davidson, the night's master of ceremonies, a harp solo by a toga-clad faculty member, and a series of hilarious parodies put on by Pauling's colleagues.

The performances, collectively titled "The Road to Stockholm," included a number of songs, skits, and speeches. Much of the night's entertainment was masterminded by a young humanities professor, Kent Clark, and British post-doctoral fellow Ted Harold. The two men were responsible for such inventive songs as "Pauling's Courses" ("They will teach you all the facts you need to know / And maybe some that are not so") and "The Gates and Crellin Laboratory" which declared,

"If you have an intuition that is clear and keen, and you love to pound your fingers on the desk machine / If you are fond of polyhedra and the way they pack, and for first approximations you have got the knack / Then the only place in the world to be, is the Gates and Crellin Laboratories of Chemistry."

The students and faculty spent the evening lampooning Pauling's discoveries, loudly expounding on his achievements, and gently poking fun at the controversies surrounding his peace work. At the close of the event, Pauling took his place on stage and briefly lectured on the academic environment and his newest research. He then heartily thanked the performers and the audience, declaring the event to be "the high point of my life."

***

The division's annual report for 1956 reflected the optimism of the moment. Echoing the tone of earlier years, the report recognized its head for developing a division that used chemical methods to advance critical questions that cut across disciplines. Decades before, A.A. Noyes had attained similar heights in building a unit that was capable of applying physical methods to chemical questions. Now Caltech could boast of a collective that was "strong" in chemistry and chemical engineering and "especially outstanding" in research on molecular structure and the application of chemistry to biology and medicine.

The division's strengths had a lot to do with the funding that it had attracted and, despite some setbacks, Pauling remained fundamental to making this so. One major achievement during this period was success in securing matching funds for a $1,500,000 challenge grant that had been put forth by the Rockefeller Foundation earlier in the decade. Pauling had played a key role in solidifying the Rockefeller money in many ways, including his successful authorship of a $450,000 Ford Foundation grant to support his research on the molecular chemistry of mental disease. In addition to being well-funded, the intellectual heft of this line of inquiry greatly impressed many of the unit's younger faculty in particular.

***

While the view inside the division was in many respects rosy, Pauling continued to come under scrutiny elsewhere. Time and again, his loyalty to the United States was being questioned by government and media sources alike, a circumstance that led Caltech's administration to apply more pressure on him to reduce his profile as an activist.

One noteworthy instance came about in March 1958 when Fulton Lewis Jr., a conservative newspaper and radio commentator, attacked Pauling's recent circulation of a petition that called for an international agreement to end nuclear testing. Lewis accused Pauling of making money from the petition, estimating earnings of $10 per signature. Lewis also charged Pauling with damaging national security by focusing on the cessation of U.S. testing efforts and overlooking tests being conducted by other nuclear-capable countries.

After reading Lewis' column, T.C. Coleman, President of the Engineering Company of Los Angeles, wrote to Caltech President DuBridge that this was just another reminder of how Pauling was tarnishing the Institute's image. Coleman then threatened to withhold any future financial support from the Institute

> "...unless I again become convinced that a truly loyal attitude prevails, and that prominent staff members such as Dr. Pauling will be required to show cause why their political activities are not detrimental to the college and the country which deserves this loyalty."

In issuing this warning, Coleman claimed that he was not trying to suppress Pauling's ability to express his opinions, nor was he concerned that Pauling would influence Caltech's students, since "as they mature they will grow more conservative." Rather, he was mostly concerned that Pauling was not taking seriously the "heavy responsibility" that came with representing Caltech to a public who might be more easily persuaded by his views because of his scientific credentials. Coleman was also sure that Caltech would have already dismissed Pauling had he not been such a well-respected scientist internationally and leader within the Institute.

DuBridge forwarded Coleman's letter to both Pauling and Albert Ruddock, the chair of the Institute's Board of Trustees. Ruddock responded directly to Coleman and in vigorous defense of Pauling, noting that his "unconventional opinions" were not evidence of disloyalty — a trait that Caltech would not tolerate. Ruddock further pointed out that accusing Pauling of being disloyal was absurd since his opinions on banning nuclear tests had been adopted by the nation's president, Dwight Eisenhower. From there, the trustee suggested that Pauling's political activities had not interfered at all with his science, and that he was still "supreme in his field." In fact,

> "The very independence of thought that leads Dr. Pauling into certain attitudes and opinions to which you and many others object is that which lies at the very basis of investigational research."

The board, Ruddock explained, would be hypocritical if they punished Pauling for exercising independent thought in one area, and encouraged it in another. Furthermore, disciplining Pauling would open a "Pandora's Box of difficulties" that would "explode" as other faculty members rushed to defend Pauling, even if they did not agree with his activities.

Pauling's only response was to DuBridge. Having read Lewis' article, Coleman's letter, and Ruddock's reply, Pauling explained that the inciting column was misleading in more ways than one. For starters, his petition called for both the United States and the Soviet Union to stop their nuclear tests. Pauling also corrected Lewis' calculation of how much each signature cost, putting it at three cents each. He added as well that he had borne most of this cost as he had hired a secretary to help him with the circulation effort, which was global in scope.

Pauling then confided his intent to file a libel lawsuit against Lewis using, with DuBridge's permission, Coleman's letter as evidence. Pauling concluded as follows:

"Let me say that I feel that the United States of America is in great danger from the group of powerful but misguided men, among them T.C. Coleman, who attempt to misuse their power in the way illustrated by Mr. Coleman's letter."

Unfortunately for Pauling, letters like Coleman's would continue to come across DuBridge's desk, and the pressure on Pauling would continue to mount.

***

The final phases of Pauling's tenure in administration were marked by steady change within the division and the Institute. One shift, first discussed in 1956, was a revision of the freshman chemistry curriculum to incorporate newer and more physics-based theoretical approaches to the discipline. By covering these topics in the first year, Caltech's undergraduates would be well-equipped to take organic chemistry as sophomores. Other curricular edits focused on removing a handful of junior- and senior-year requirements, thus freeing up room for electives.

While the undergraduate population was robust enough to merit a rethinking of the coursework being offered, graduate studies at the Institute were suffering by comparison. The difficulties that Caltech was facing with attracting ideal numbers of graduate students were in turn leading to a lack of people available to teach lower-level undergraduate courses. Seeking to address this problem, President DuBridge suggested that the division chairs consider increasing the stipends offered to graduate instructors.

But amidst these conversations, a steady buzz of concern about Pauling's worsening public image, and its potential impact on Caltech's well-being, could not be ignored. The tremendous strain that this generated ultimately undermined Pauling's position of leadership at Caltech and resulted in his stepping down from his administrative post.

***

On May 11, 1958, just two months after Fulton Lewis published his critical newspaper column, Pauling made an infamous appearance on the long-running NBC news television program, *Meet the Press*.

At first, the show's four-man panel, led by moderator Lawrence Spivak, cordially questioned Pauling about his petition to ban nuclear testing. But as the interview went on, Spivak and others began to ask more leading questions about Pauling being a communist. The set became especially heated once the panel brought up Fulton Lewis' allegations that the petition had been a money-maker for Pauling to the tune of $10 per signature. Pauling attempted to respond but was cut off before he could complete his rebuttal. In 1984 Pauling recalled that,

> "When that program came to an end, Spivak took off down the hall, running as fast as he could go, with my wife after him, waving her fists. I guess she had a hard time restraining herself during the program. But he managed to escape."

Following the episode, Pauling received scores of letters thanking him for his work and bemoaning the treatment that he had received from the interviewers. Some correspondents even sent him money in hopes that their donation might assist with the petition initiative. Activist and Unitarian minister Stephen Fritchman spoke for many in his assessment, writing,

> "You held the camera, you gave no quarter, you hit every target raised… and gave millions tremendous moral support. This was your finest hour as a public figure for peace and humanity."

***

A decidedly different cache of letters came into President DuBridge's office. Among them were expressions of outrage at Pauling's activities accompanied by threats to withhold future donations to Caltech. DuBridge forwarded a representative example ("one of many") to Pauling, authored by someone who simply signed off as "Disturbed." Describing themself as being "truly frightened" by Pauling, a "psychopathic" character who was in "consistent alignment with Communist activities," Disturbed judged Pauling's appearance on *Meet the Press* as having had no redeeming value for the Institute. The author also pointed out that they had no children and threatened to remove Caltech from consideration in their will. For Disturbed, it was beyond the pale to have "men like this fellow in top faculty positions — men who may, for all I know, be in a position to share the secrets of defense production that go on at Cal-Tech."

This latest controversy was the last straw for Lee DuBridge. After forwarding Disturbed's letter, the Caltech president called Pauling into his office to tell him — according to Pauling's notes — that he was the "laughing stock of people everywhere" and to ask that he cease his peace work. DuBridge also stressed that Pauling's activities were directly interfering with Caltech's ability to raise money, noting specifically a development fund of $16 million that had gone to Harvey Mudd College instead. DuBridge then confided that the Institute's trustees had seriously considered dismissing Pauling but did not do so because of concerns related to academic freedom.

Pauling replied that he would continue to follow his conscience but also put forth a scenario that would give DuBridge a way out of his political bind. Pauling's proposal was that he be appointed a Research Professor with no teaching duties, a circumstance in which his salary would be solely charged against research funds. Importantly, Pauling also suggested that he step down as Chairman of the Division of Chemistry and Chemical Engineering. In Pauling's mind, this revision of duties would help reduce potential damage to fundraising efforts caused by his public appearances.

A month later, Pauling again suggested to DuBridge that his duties be shifted, but this time asked that he not be appointed a Research Professor — after thinking about it for a while, he realized that he enjoyed teaching too much and did not want to give it up. He was, however, prepared to cede his administrative position immediately, telling DuBridge that he would stay on as chair for longer if needed, but would prefer to exit as soon as he could.

Extra time was not necessary, and Ernest Swift took over the position on June 30, 1958. The annual report issued by the division marked this major transition by avoiding the politics of the situation. In the document,

DuBridge gave his own spin on why Pauling was stepping down, writing that "He has served as Division Chairman for twenty-one years, and his desire for relief is entirely understandable."

***

Once empowered, Swift took the division in a different direction, granting more sway to the sub-disciplines within the unit than had Pauling. Lab assignments were also reorganized, with Pauling and his research teams afforded less space. His administrative duties removed, Pauling's salary was likewise reduced from $18,000 to $15,000.

And despite the hope that Pauling might be able to express his views more freely without harming Caltech's finances, his public persona remained a problem. Even after Pauling had resigned from the chairmanship, DuBridge continued to forward letters that he had received from those who were upset with his activism and his alleged communism. Pauling's response to one such letter was that the author "…is a hopeless case — probably a junior member of the John Birch Society."

Sadly, Pauling's second Nobel Prize in 1963 only increased the tension and, in stark contrast to nine years earlier, neither the Institute nor his own division made any formal acknowledgement of the honor. Rather, DuBridge went so far as to disparage Pauling's methods for seeking peace. The snub proved to be too much for Pauling, and not long after receiving notice of the prize he announced that he would be leaving Caltech in favor of the Center for the Study of Democratic Institutions.

Although he had already cleared out his office, in early December colleagues in the Biology Division invited Pauling back to campus for a small gathering over coffee to celebrate his Nobel Peace Prize. This event proved to be the only recognition of Pauling's achievement hosted at the Institute. It also marked the close of a hugely productive association that had spanned 41 years. From that point on, Pauling's affiliation with Caltech was reduced to that of Research Associate and was defined by a series of unsalaried contracts that were renewed annually into the late 1960s.

Part III
Period of Wandering

Chapter 22

The Center for the Study of Democratic Institutions

AWARD—Robert M. Hutchins, right, president of the Center for Study of Democratic Institutions, Santa Barbara, accepts bronze likeness of scientist Linus Pauling, left, contributed in recognition of center's peace efforts. Pauling, winner of two Nobel Prizes, is a staff member of the center.

In October 1963, just after it was announced that Linus Pauling would receive the Nobel Peace Prize, a press conference was held in the living room of his Pasadena home. Pauling naturally spoke of his happiness with the Nobel committee's decision, but he also caught the media's attention with some news of his own: he intended to take a leave of absence from the California Institute of Technology and would resume his work in science, peace and medicine at the Center for the Study of Democratic Institutions (CSDI) in Santa Barbara, California.

This disclosure came as a shock to most of his colleagues, who were largely unaware that Pauling was mulling his departure after more than four decades of historic achievement at Caltech. By the time of his announcement, however, Pauling had already arranged for the research currently under his supervision to continue to completion while he was in absentia. And though he didn't come right out and say it, Pauling's leave of absence effectively marked his resignation from Caltech.

Pauling's decision was radical not only because he was leaving the institution where he had worked for 41 years, but also because the CSDI was a dramatically different place in nearly every way. Where some, like Pauling biographer Thomas Hager, could describe Caltech as "a monastery devoted to science," the CSDI was an educational enterprise created by a $15 million grant from the Fund for the Republic (a subsidiary of the Ford Foundation). Its mission, in the words of director Robert Hutchins, was "to promote the principles of individual liberty expressed in the Declaration of Independence and the Constitution of the United States." At its core, the CSDI was a think tank that sought to study the effects of democracy and constitutional rights on modern institutions such as corporations and labor unions. Suffice it to say, this was a far cry from the hard and applied sciences that Pauling was used to at Caltech.

***

In his departure communications, Pauling described his move to the CSDI as being motivated by a desire to more fully incorporate world affairs into his studies of science and medicine. Among other work as an activist, Pauling had become one of the world's most prominent advocates for the partial nuclear test ban treaty that was agreed to by the United States, Great Britain and the Soviet Union in 1963. Though the treaty limited its scope to banning nuclear explosions in the atmosphere, outer space and underwater, it nonetheless marked a major turning point in the short history of the nuclear age.

While Pauling was honored by the Nobel committee and others for this work, he was also routinely criticized, with many framing his efforts as hindering the strength and effectiveness of the U.S. military. Print and television journalists frequently accused Pauling of harboring un-American sentiments, and these increasingly public allegations led to an erosion of trust within the Caltech power structure.

By contrast, the CSDI seemed to be a welcoming and potentially supportive environment. In a letter to Pauling dated October 17, 1963, Hutchins, who was also a past president of the University of Chicago, gave Pauling his full support and noted how closely aligned their views and goals were, so much so that their most recent publications had similar titles. Hutchins also saw Pauling as a potential candidate to take over the center's program of research in medicine, which was still in its early days. The director was likewise clear on the impact that Pauling's mere presence would immediately make upon the center's profile.

In Pauling's October statement to the press, he expressed excitement at the idea of working in an environment where he would already be familiar with many of his new colleagues and could more freely choose areas of study. Indeed, "increased freedom of action" was one of the main points of emphasis in his remarks. But despite the potential benefits of the move, Pauling also knew that his scientific work would, by necessity, become increasingly theoretical. Back at Caltech, Pauling's many years of research, experimentation and funding had allowed him to amass vital, and expensive, instruments for his laboratory. The CSDI, on the other hand, had hardly engaged with the natural sciences and its research in medicine had not yet included any experimental work. These were uncharted waters for Pauling, but in 1964 he ventured forth.

***

His first objective as a newcomer to the CSDI was to continue exploring and pursuing the ideals put forth in his Nobel lecture, "Science and Peace," delivered in Oslo on December 10, 1963. In that address, Pauling expressed his feeling that scientists, through their technical contributions, were partly responsible for the horrors of nuclear warfare. He also strongly affirmed his long-standing belief that the scientific community should act as a forceful leader in bringing the world to a period "where no greater nor more destructive weapons can be discovered, leading countries to realize that matters should not be solved by war or force but by a world law."

Even before the first nuclear weapons were detonated over Hiroshima and Nagasaki, Pauling noted, a committee of atomic scientists had been urging the U.S. Secretary of War to refrain from using atomic bombs in surprise attacks. Actions of this sort were in line with Pauling's model of how scientists should view their duties as public citizens. For Pauling, scientists bore a particular obligation to help educate and guide others in the real-world application of scientific developments, particularly when it came to technologies of warfare. The social and political activism that helped define Pauling's career from the mid-1940s on is clear evidence of his striving to live by this ideal. While the CSDI would require him to leave his nicely outfitted laboratory in Pasadena, it would also give him the chance to further his work as a "complete scientist" empowered to inform the public about science's role in social issues.

As he spent more time thinking and researching, Pauling cultivated additional ideas about how humanity might move toward an era of peace. One position that he pushed was that peace could be achieved through implementation of a world democratic law. Pauling had been interested in this concept since at least the early 1940s, when he and Ava Helen participated in activities sponsored by the Pasadena Chapter of the Federal Unionist Club. He returned to the idea with some frequency throughout the 1950s, incorporating it into talks like "Our Choice: Atomic Death or World Law," which he gave at Hiroshima University in August 1959. His new affiliation with the CSDI afforded him an opportunity to spend more time on the concept and to help champion it in concert with like-minded thinkers.

By "world law," Pauling and other proponents were envisioning institutions and legal systems invested with the authority to moderate between countries during times of conflict. Advocates of the concept based their thinking on an initial assumption that nations would insist on maintaining their sovereignty. That given, what was needed was a toolkit that could insure against catastrophic conflict in this new nuclear age. This toolkit would be comprised of a set of international agreements that would then be enforced by a global governing institution. The system would also institutionalize representative voices charged with separately expressing the views of both national governments and their citizens. An arrangement of this sort, Pauling believed, would allow for the peaceful resolution of disputes both between nations and within nations as well. Failure to do so would inevitably lead to increased human suffering, be it through violent revolution or iron-fisted dictatorship.

In effect, what Pauling was proposing was that the United Nations, or an entity like it, become a far more powerful institution, one capable of serving as the world's foremost democratic authority. Though he recognized that it may take many years to achieve, for Pauling this was a "path of reason" rather than a "path of insensate militarism." His vision was seconded by peace advocates around the world and formed an important component of his rhetoric while at the CSDI.

***

On April 11, 1963, in the wake of the Cuban missile crisis, Pope John XXIII (now a canonized saint of the Catholic Church) issued a papal encyclical, *Pacem in Terris*. The statement exhorted members of the Catholic faith and "all men of good will" to seek peace, which the Pope viewed to be the divinely ordained nature of the world. "The world's Creator," Pope John wrote, "has stamped man's inmost being with an order revealed to man by his conscience and led the consciousness of all societies to prefer peace over war."

The encyclical was based on ideals of human dignity and political justice, both of which John XXIII defined in the document. In his view, human dignity was founded on the principle that each individual "is truly a person" with inalienable rights to intelligence and the exertion of free will, to be expressed in a manner that would not infringe upon these same rights enjoyed by others. In discussing political justice, the Pope similarly offered that a state must satisfy its needs and the needs of its people but not hinder the rights of other nations in pursuing these same aims.

John XXIII further put forth that international conflicts must "be settled in a truly human way, not by armed force or by deceit or trickery." He also stated that, to resolve conflicts while maintaining political justice and human dignity, "there must be a mutual assessment of the arguments and feelings on both sides of an argument, a mature and objective investigation of the situation, and an equitable reconciliation of opposing views."

In February 1965, nearly two years after the encyclical was issued, religious, political and social leaders from the world over were called to New York City by the CSDI for a meeting titled "The Convocation on the Requirements of Peace." The Center's main goal for this gathering was to establish a series of concrete steps by which the ideas set forth in *Pacem in Terris* could be applied throughout society. The convocation was led by CSDI president Hutchins, who issued the central question that participants were asked to consider: "If the principles in *Pacem in Terris* are sound, how can they be carried out in the world as it is?" Not surprisingly, Pauling was sent to the conference as a representative of the CSDI.

Pauling's fame as a recent recipient of the Nobel Peace Prize garnered him a prime slot among the main speakers at the meeting. And as with the Pope's message from two years earlier, Pauling's contributions to the gathering also focused on the ideals of political justice and human dignity. Appropriately titled "Peace on Earth," Pauling's discourse explained his assessment of John XXIII's encyclical as it applied to the secular world.

Pauling and the Pope agreed on many fronts — each believed in an end to warfare, and that countries should be guaranteed equal rights and advantages during a dispute. Pauling and John XXIII also shared a belief that conflicts could be ameliorated by the action of an international authority, but the two held critically different views on the shape that this authority might assume. Where Pauling had proposed that a world government could and should fill this position, John XXIII believed that submission to the authority of God would lead individuals to discover a life in which their more violent or aggressive tendencies would be calmed and controlled. While appropriate for his position as Pope, John XXIII's views on the authority of God were more difficult to accept for a non-Christian mindset. To this end, Pauling, an atheist since the age of 11, viewed the New York convocation as an opportunity to situate his own ideas on authority and politics alongside other perspectives on the pursuit of world peace.

In so doing, Pauling elaborated on a variety of additional issues not addressed by John XIII, including the problems of suffering, evil and economic inequality. He began by expressing his oft-repeated acceptance, "as one of the basic ethical principles, the principle of the minimization of the amount of suffering in the world." He then spoke of the various causes of human suffering in the world, listing accidents, natural disasters and disease, as well as "... man's inhumanity to man, as expressed in economic exploitation, the maldistribution of the world's wealth, and especially the evil institution of war."

Noting that "we now spend more on war in South Vietnam than the total annual income of the miserable people of South Vietnam," Pauling emphasized that,

> "I believe that it is a violation of natural law for half of the people of the world to live in misery, in abject poverty, without hope for the future, while the affluent nations spend on militarism a sum of money equal to the entire income of this miserable half of the world's people."

Reflecting on the atomic bombing of Hiroshima, Pauling then offered that

> "War has become increasingly unjust and immoral both in the magnitude and in the distribution of the suffering that it causes. Great nations claim the right to sacrifice human lives and to take human lives. ... It is chance that determines whether or not the soldier will be killed and also whether the civilian will be killed."

Chillingly, the World Wars of the first half of the 20th century would be "far transcended by a great war in the nuclear age...instead of tens of millions, hundreds or even thousands of millions of human beings might be killed."

But unfailingly, Pauling returned to his core belief that human intelligence and the scientific method could ease many forms of suffering, referencing breakthroughs in the control of infectious disease and the potential for scientific and technological solutions that could end world hunger. He then extended these feelings of optimism to the subject of nuclear war. By 1965, Pauling was becoming more confident that the specter of nuclear holocaust would not remain an issue for much longer, as the certainty of human extinction should such an event come about had rendered nuclear war an irrational option for the world's governments. Pauling also optimistically believed that humankind was nearing a point where it would be ready to develop a rational alternative to war of any kind, "to be replaced by a system of world law based on principles of justice and morality."

Correspondence from numerous observers who wrote in after the New York meeting indicates that Pauling and the other conference participants succeeded in satisfying Hutchins' mandate. In the view of these authors, the conference had developed sensible mechanisms for carrying out the principles of *Pacem in Terris* in the world of 1965. Many of the letters further point out that print commentators from around the country had taken note of the discussions held at the convocation and that Pauling's discourse was particularly impactful. Though just a small part of a much larger conversation, the Convocation on the Requirements of Peace helped to propel deeper understanding of what the idea of "peace" actually meant some two decades into the atomic age.

***

While at the CSDI, Pauling also served on an ad hoc committee that drafted a memorandum addressed to President Lyndon B. Johnson and titled "The Triple Revolution." Once finalized, the document was circulated outside of the committee and ultimately signed by 35 academics, journalists and leftwing activists. Noteworthy among these signatories were James Boggs, an auto worker and author of *Pages from a Negro Worker's Notebook*; Todd Gitlin, President of Students for a Democratic Society; retired Brigadier General Hugh B. Hester; Gerard Piel, publisher of *Scientific American*; Bayard Rustin of the War Resisters League; and socialist leader Norman Thomas. Also included were Pauling and W.H. "Ping" Ferry, vice-president of the CSDI. Submitted in March 1964, the memorandum was later published as a pamphlet.

Reflective of its title, "The Triple Revolution" posits that three socioeconomic revolutions were simultaneously occurring during the 1960s: the Weaponry Revolution, the Cybernation Revolution, and a Human Rights Revolution that included the many civil rights movements then gaining momentum around the world. The authors also argued that all three revolutions were the result of technological development and changes in the economy.

The section on the Weaponry Revolution discusses nuclear warfare and disarmament and speaks to one of Pauling's greatest hopes: the end of war as a means of conflict resolution. Echoing Pauling's rhetoric elsewhere, the pamphlet expresses the view that the specter of nuclear holocaust would incentivize an end to all wars, for fear that escalation could result in the destruction of modern civilization. While acknowledging the supreme difficulty of this undertaking, "The Triple Revolution" nevertheless holds fast to the idea of a "warless world," stating that it is a need acknowledged by most people.

The memorandum also proposed that a Cybernation Revolution was underway, one in which the use of machines promised to radically change the roles traditionally assumed by people in the economy and society. For the authors, this revolution was concerning in part because those lacking the resources to purchase or develop machines would be left without the opportunity to earn a living, should the economy switch to a more heavily automated means of production.

The members of the committee also feared that the Cybernation Revolution would lead to an increasingly unequal distribution of wealth and, in turn, a more unsustainable national economy. Indeed, one of the proposals put forth by "The Triple Revolution" is that a redistribution of wealth be deployed as a necessary action to prevent future economic instability. In defining what it termed "the right to an income," the memorandum encouraged the development of an economic system that would compensate those who were left behind by this changing industrial age. One group specifically mentioned was the African American community, whom the authors viewed as being largely excluded from the ownership of machines and thus limited in their capacity for new economic development.

The memorandum likewise stressed that societies around the world must recognize the dignity of each individual. In discussing this imperative, the authors framed the U.S. Civil Rights Movement as a local manifestation of a worldwide push to reform political systems such that individuals could not be excluded on account of their race. It was in this sense that "The Triple Revolution" connected civil rights activities in America to its broader idea of a Human Rights Revolution.

Having stated its view of the situation, the document then suggests that the U.S. government could ably lead American society through this period of change by decreasing the quantity of resources allocated to military endeavors and increasing the attention given to those who were at a social disadvantage. The pamphlet closed with a warning that failure to seek solutions to the issues that arise when human labor is replaced by machines would exacerbate social inequality and lead to "misery and chaos." But so too did it remain hopeful that, given proper leadership, societies could overcome the challenges presented by the turbulence of the 1960s.

In its presentation of "The Triple Revolution," the ad hoc committee identified itself as a group of concerned citizens. And though the text made sure to include government action in its proposals for a brighter future, the document was considered by many to be anti-government, and by some to be anti-American. Certain critics worried about the pamphlet's apparent lack of appreciation for the military, as indicated by its recommendation that resources used for military efforts be curtailed. Media pundits were also quick to disparage the idea of a guaranteed income and its implications of creeping socialism.

Two weeks after receiving the document, the White House issued a short reply to the authors of "The Triple Revolution," stating that the President had already taken measures to address the problems that it identified. The letter of response was signed by Lee C. White, Assistant Special Counsel to the President, who is remembered today for having advised Presidents Kennedy and Johnson on civil rights strategies. For the most part, the response uses general statements and stresses that President Johnson's "War on Poverty" was tackling the issues brought up by the memorandum. It is unclear whether Johnson ever received, read or was briefed on the document.

Despite its quick dismissal by the administration, other peace activists shared and built upon the views expressed in "The Triple Revolution." One group in particular, Women Strike for Peace — an organization in which Ava Helen was especially active — voiced several analogous points when it organized a demonstration outside the Peace Palace at The Hague, Netherlands in 1964. Through their activism and intellectual product, the ad hoc committee and Women Strike for Peace alike sought to navigate changing times by making parallel demands that social inequalities be resolved both for the benefit of individuals and as a step toward world peace.

***

One especially provocative detail from Pauling's tenure at the CSDI was his consideration of a study of unidentified flying objects (UFOs), which he outlined in a July 1966 note to self. In discussing this moment in his career, it is important to point out that Pauling viewed the study as a "possibility," and that the investigation was not conducted. The proposal document that he crafted, however, is evidence of the approach that Pauling would have taken to engage with the phenomenon, an approach that was notably consistent with his studies of more mainstream research topics.

The circumstances amidst which it was authored are also important to consider. Though he had joined the CSDI with feelings of optimism for a new beginning, after two years in Santa Barbara, Pauling found himself growing more distant from the scientific community. During this period, he had published a few papers on nuclear physics and medical chemistry and had engaged in stimulating discussions on world peace, but increasingly he could not ignore an emerging truth: he was growing dissatisfied with the CSDI.

The center's lack of capacity to support scientific research meant that the work he could do on his own was very limited. Similarly, in terms of politics and peace, the center was mostly engaged in theoretical discussion; there were no specific causes for which it was fighting and many of the meetings that it sponsored yielded few concrete action items. And so it was that Pauling found himself at his ranch in Big Sur brainstorming ideas on what to do with his time; researching UFOs was perhaps the most unorthodox notion to emerge from this period of reflection.

Pauling was quite familiar with reports of UFOs. Scattered among the many letters that he received were a handful asking questions about strange noises in the night, rumors of alien contact, and other observed phenomena that were difficult to explain. Being a long-running fan of science fiction himself and feeling a duty to communicate scientific knowledge to the public, it's not difficult to see how Pauling could find value in addressing these matters in a formal way.

Pauling's 1966 research outline, titled "A Study of Unidentified Flying Objects," was divided into 12 main points. In each, Pauling stated his intent to examine everything from the veracity of observer reports to the possibility of an extraterrestrial origin of human life. Viewed as a whole, the proposal's 12 bullets are suggestive of an ambitious project that would have required Pauling to tap into the fields of evolutionary biology, psychology, and history, among others. An abbreviated itemization of his proposed agenda is as follows:

1. Aggregation of evidence provided by present-day observers.
2. A study of historical reports.
3. Analysis of all reports with respect to the possibility of interpretation as physical phenomena.
4. Analysis of all reports with respect to their explanation as involving man-made vehicles, apparatus or structures of known character.
5. A discussion of possible explanations of reliable reports not accounted for under 3 or 4.
6. Analysis of available information about experimental vehicles.
7. A study of the possibility that extraterrestrial sentient beings have been in contact with the earth over a long period of time.
8. Analysis of the evidence about the possibility of the extraterrestrial origin of man and of some or all other organisms on earth.
9. Analysis of the possibility of existence of sentient beings on other planets of the solar system.
10. A discussion of the existence of sentient beings on planets adjacent to the nearest stars, and of the possibility of their transportation to earth.
11. Analysis of the possible existence of sentients beings on planets of distant stars, and the possibility of their transportation to earth.
12. Conclusions about the interpretation of the evidence about observation of UFOs.

Pauling's engagement with the idea of UFOs is reinforced by a look at his personal library. While the "flying saucer" section is very small, it's clear he was interested in the big picture, pulling from both scholarly and popular sources. One title, *The Scientific Study of Unidentified Flying Objects*, is billed as "The complete report commissioned by the U.S. Air Force." This volume, though lacking Pauling's typical marginalia, sports heavy wear indicative of having been read.

For more sensational details, Pauling turned to Brinsley Trench's *The Flying Saucer Story* and John G. Fuller's *Incident at Exeter*. Pauling's copy of Trench's work is speckled with hastily scribbled questions, the word "check" next to underlined passages, and notes to contact a variety of scientists and officials. His comments belie a heavy skepticism suggesting that, despite his willingness to explore the unorthodox, he maintained a strict logical outlook. As can be expected, claims that defied conventional science readily drew criticism. At one point, Trench claims "It [a UFO] could easily withstand temperatures at 15,000 degrees Fahrenheit, without showing any traces of melting." A large question mark sits in the margin next to it, perhaps a testament to Pauling's disbelief.

Evidence of Pauling's connections with the subject does not end with the 1966 note to self. As late as 1968, Pauling wrote to Stirling Colgate, a physicist who was then President of the New Mexico Institute of Mining and Technology, inquiring into an alleged sighting of a flying saucer on the school's campus. Colgate's reply confirms that the sighting was a hoax perpetrated by a New Mexico Tech student, but Pauling's letter is indication that UFOs remained on his mind for at least a couple of years after he drafted his proposal document.

All of that said, it can also be argued that Pauling's formal, and seemingly abrupt, interest in the topic was mostly an outgrowth of his restlessness with the CSDI. For much of his time at the Center, he was engaged in a quest to use his talents in new ways for the benefit of the public. To his dissatisfaction, he was unable to find exactly what he was looking for in Santa Barbara, and after three-and-a-half years there, he began to actively seek out a new institutional home.

***

Though he didn't know it right away, a critical moment in Pauling's life came about just a couple of months prior to writing his UFO research proposal. In March 1966, Pauling received a medal from the Carl Neuberg Society for International Relations, awarded for his work to integrate new medical and biological knowledge. In his acceptance speech, "Science and World Problems," Pauling — who was 65 years old at the time — mentioned that he hoped to live for another 15 years so that he might see several advances in medical science that he believed to be right around the corner.

A man named Irwin Stone was in the audience at this lecture and, on April 4, 1966, he wrote Pauling a fateful letter in which he noted,

"You expressed the desire, during the talk, that you would like to survive for the next 15 or so years. … I am taking the liberty of sending you my High Level Ascorbic Acid Regimen, because I would like to see you remain in good health for the next 50 years."

Born in 1907, Stone attended the College of the City of New York and then worked at the Pease Laboratories, a well-known biological and chemical consulting lab, first as a bacteriologist and later as chief chemist. In 1934 the Wallerstein Company, a large manufacturer of industrial enzymes, recruited Stone to set up and direct an enzyme and fermentation research laboratory. Ascorbic acid, or vitamin C, had just been identified and synthesized by a Hungarian research team led by Albert Szent-Györgyi, who would soon win the Nobel Prize in Medicine for his work. While at Wallerstein, Stone pioneered processes for implementing the antioxidant properties of ascorbic acid in industrial settings, including its use as a preservative for food, an innovation that landed him three patents.

Once triggered, Stone's keen interest in vitamin C lasted for the remainder of his life. He began to study scurvy intensely and by the late 1950s had formulated a hypothesis that the disease was not merely a dietary issue, but in fact a flaw in human genetics. (He called it "a universal, potentially fatal human birth defect for the liver enzyme GLO.") As an outgrowth of this thinking, Stone became convinced that the amount of vitamin C that government nutritionists recommended in a healthy diet — the Recommended Daily Allowance (RDA) — to be far from sufficient. In 1968 that recommendation was 55 mg for women and 60 mg for men, which was the baseline intake needed to avoid scurvy. Current standards today are slightly increased, but neither set of figures are anywhere near what Stone was pushing.

Stone believed that humans suffer from "hypoascorbemia," a severe deficiency of vitamin C caused by our inability to internally synthesize the substance. This stands in stark contrast to most mammals, who are able to synthesize vitamin C in large quantities relative to body weight; proportionately, Stone argued, humans should be taking between 10–20 grams per day. Stone hypothesized that, about 25 million years ago, most primates lived in regions where they were able to consume relatively massive amounts of ascorbic acid compared with what we get from our diets today. These material circumstances perpetuated an environment in which a genetic mutation occurred that allowed these human ancestors to stop synthesizing the substance but continue on living. Because of changes in modern diets, Stone noted, humans were typically consuming only 1–2% of what they truly needed.

This hypothesis initially led Stone to propose a vitamin C intake of 3 grams for optimal health — 50 times the RDA — and as he further researched ascorbic acid, he recommended increasingly higher doses. He was convinced that taking less than this amount would cause "chronic subclinical scurvy," a state of lowered immunity that increased susceptibility to a variety of illnesses. He felt that large doses of ascorbic acid could be used to prevent and treat a host of maladies including infectious and cardiovascular diseases, and health problems normally associated with aging. Practicing what he preached, Stone and his wife began taking megadoses of vitamin C and found that it greatly improved their overall health. When the couple both incurred injuries from a serious car accident, they treated themselves in part with large doses of vitamin C and reported a swift recovery.

Pauling was initially skeptical of Stone's claims, but he had recently learned about other uses of megavitamin therapy and their successes, so he and Ava Helen decided to give the regimen a try, beginning with 3 grams a day. In July Pauling wrote back to Stone to report this change in habits. "I have enjoyed reading your paper and manuscript about hypoascorbemia," he replied, "[and] have decided to try your high level ascorbic acid regimen, to see if it helps me to keep from catching colds."

As it turned out, the results proved impressive. For most of his adult life, Pauling had suffered from severe colds several times a year and, as a preventative, had taken regular doses of penicillin from 1948 to the early 1960s. Pauling thought of the antibiotic as his primary defense against colds but, in all likelihood, he was actually killing off good bacteria in his system and making himself more susceptible to colds through overuse of the medication. Once the Paulings started taking vitamin C, they reported a noticeable uptick in their energy and seemed to suffer from fewer routine illnesses. Thus began a famous fascination that would dominate much of Pauling's life for the better part of three decades.

***

One of Pauling's hopes for his tenure at the CSDI was that he might find opportunities to work with science faculty at neighboring institutions. Optimistic of the support and independence that he would enjoy at the Center, Pauling believed that he could continue to pursue his scientific ambitions through off-Center collaborations. Upon their arrival in Santa Barbara, however, Ava Helen expressed a fear that the CSDI would prove "too superficial" for her husband's needs. This prediction, as it turned out, was fundamentally correct, and increasingly Linus found himself disappointed by his inability to progress his scientific research both at his home base and in partnership with others.

His options for joining nearby institutions to perform scientific work were also quite limited. Importantly, the University of California (UC) rejected Pauling's application for an adjunct position at UC Santa Barbara (UCSB) because of his controversial politics. Pauling fought for this appointment for nearly a year, appealing to Governor Pat Brown and expressing a willingness to work for no salary. But his proposal was met with resistance from UCSB chancellor Vernon Cheadle, who did not inform the university's chemistry faculty of Pauling's interest and ultimately refused to even file the paperwork necessary for the offer to come under preliminary review by the University of California system.

By August 1965, only two years after being hired by the CSDI and just one year after moving to Santa Barbara, Pauling was writing letters to the

Center's president, Robert Hutchins, asking to spend less time at CSDI headquarters so that he might advance his scientific work from a new location: the Pauling ranch at Big Sur. Pauling spent much of this time away developing new ideas on nuclear structure, a body of work that he called the close-packed-spheron model of atomic nuclei. This theory was unveiled in four different articles in September and October 1965, each of which addressed different implications of the idea.

The close-packed-spheron model was based on a previous idea called nuclear shell theory. Pauling took that concept a step further by attempting to explain why specific "magic numbers" of protons and neutrons cause greater nuclear stability. Pauling's theory proffered that the "magic" qualities associated with these numbers of nuclear components corresponded to the filling of nuclear "spherons," or nuclear sub-units, where protons and neutrons are arranged. From there, Pauling's model stated that lower magic numbers represent atoms in which the first or second nuclear shells are filled, and that higher magic numbers correspond to a special hybridized "mantle" shell that can form if greater quantities of nuclear components arrange into spheres.

In developing his model, Pauling was trying to explain the arrangement of nuclear components by simplifying previous theories and applying the principles of electron orbitals to the protons and neutrons that reside in the atomic nucleus. Pauling's past work had helped to establish the fundamentals of electron orbital hybridization, and he hoped that this new idea would yield similar fruit for the study of atomic nuclei. If such were the case, it would then be possible to more readily explain the stability of atoms and would also provide further insight into the geometric arrangement of protons and neutrons.

Though interesting and relatively easy to understand, Pauling's theory failed to spark much enthusiasm within the scientific community. For the next several years, Pauling continued to advocate for the idea and, in June 1974, he applied for a National Science Foundation grant to further advance the model. But the application was denied and, before long, Pauling turned his attentions elsewhere.

***

The solitary development of the close-packed-spheron theory and the lack of attention that it received from his peers were emblematic of the difficulties that Pauling experienced during his affiliation with the CSDI. The limited resources available to him during this time enabled only theoretical investigations on subjects with which he was already at least somewhat familiar. Likewise, his official connection with an institution that existed well out of the scientific mainstream stifled his ability to engage with scientific colleagues on a regular basis.

Pauling was only at the CSDI until 1967, and towards the end of his tenure there his eagerness to return to the sciences grew. Other publications from the period focused on molecular protein structure and the chemical bond. As with the structure of atomic nuclei, these topics were, again, among those that he had researched prior to moving to Santa Barbara. Once he found a new scientific home, the University of California, San Diego, Pauling began new investigations in medical chemistry which ultimately led to his deep engagement with vitamin C as a research topic.

Pauling's switch to a scientific focus could also be interpreted as a waning interest in world affairs, but his papers indicate otherwise. Social and political concerns most certainly remained central to Pauling's activities and continued to lay claim to large pieces of his time, especially as the war in Vietnam escalated throughout the late 1960s and early 1970s. Fundamentally, Pauling was interested in developing ideas that could lead the world towards peace, while the Center was primarily a think tank that seemed more inclined toward discussion at the expense of conclusion. In the end, superficial or not, the CSDI simply was not the institution for Linus Pauling and by fall 1967 he was gone.

Chapter 23

The University of California, San Diego

Linus Pauling, 1967.

In spring 1966, Linus Pauling received a letter from a colleague, Fred Wall, querying his interest in potentially joining the chemistry faculty at the University of California, San Diego (UCSD). An experienced chemist and Vice Chancellor for Research at UCSD, Wall knew of Pauling's dissatisfactions at the Center for the Study of Democratic Institutions (CSDI) and was also familiar with the earlier rejection that Pauling had received when appealing for a faculty position at a sister school, the University of California, Santa Barbara (UCSB). Sensing an opportunity and eager to rectify these past disappointments, Wall was working to lay the groundwork to provide Pauling with a new institutional home.

Though pleased to learn of this option, Pauling was initially hesitant. He remembered all too well the hostility that informed chancellor Vernon Cheadle's refusal to consider his appointment at UCSB and was clear in his understanding that Cheadle's position was fully supported by the UC regents. This history fresh in mind, Pauling saw no reason why he would be permitted to work at UCSD; after all, his political views hadn't changed over the past two years and he'd become, if anything, even more vocal about them.

This time, however, Pauling's case was receiving far more support at the top levels of administration. Quite contrary to Vernon Cheadle's stance, UCSD's chancellor, John Galbraith, was fighting for faculty endorsement of a petition that aimed to

"urge that every effort be made not only to induce him to accept the present appointment assured for one year, but also to press with all means possible for its renewal for whatever periods Dr. Pauling and the faculty involved agree to be appropriate."

Galbraith likewise went out of his way to praise Pauling's excellent lecturing ability, which he felt would be a potential asset to students and faculty alike. Similarly, he affirmed that Pauling's appointment would prove valuable not only to the chemistry department, but to the physics and biology departments as well. In due course, faculty in all three units signed the petition and the chemistry department voted unanimously to bring Pauling aboard.

Buoyed by this strong show of support, Pauling accepted a one-year contract with the university, with on-site work to begin in fall 1967. This agreement carried with it the understanding that a tenured position might be offered in the coming years, so long as the UC regents didn't object.

A letter from Ava Helen Pauling to her son Peter, as well as a statement made by Pauling in the *Women's International League for Peace and Freedom Newsletter*, indicate that his initial take on UCSD was a positive one. Perhaps most importantly, the university offered him the means to return to scientific research, a clear source of invigoration following three years at the CSDI, which was not capable of providing him with adequate lab space. In her letter to Peter, Ava Helen confirmed this new feeling of enthusiasm, one made all the more exciting by Linus's decision to focus on fresh topics in medical chemistry. Pauling himself called UCSD a "first-rate" institution and expressed delight that he would now be able to more easily collaborate with a range of accomplished university researchers.

It didn't take long for Ava Helen to find a house to rent in La Jolla and, in September 1967, her husband arrived at his new office on the UCSD campus. In their initial meetings, Bruno Zimm, the chemistry department chair at the time, encouraged Pauling to develop customized coursework that would explore specialized subjects of his choosing. Pauling replied that he would prefer to focus mostly on research, as his salary was coming entirely from research funds. He remained active on campus though and participated enthusiastically in a lecture series tailored for first-year students.

***

Shortly after settling in, Pauling began partnering with UCSD biology professor Arthur B. Robinson, a former student of Pauling's at Caltech. Together, the duo would tackle Pauling's latest research quest: an exploration of orthomolecular medicine, defined by Pauling as, "the use of the right molecules or [...] substances that are normally present in the human body in the amounts that lead to the best of health and the greatest decrease in disease." This fruitful collaboration eventually led to their co-founding, in 1973, of the Institute for Orthomolecular Medicine, now known as the Linus Pauling Institute at Oregon State University.

Pauling's research was being supported by UCSD as well as lingering funds from the CSDI, but soon it became clear that his team would need additional resources. As he delved further into his orthomolecular program, Pauling estimated that the work he had in mind would take at least five years to conduct, a length of time that was extended, in part, by the small size of his research team. In addition to Pauling and Robinson, the UCSD group consisted of two lab technicians (Sue Oxley and Maida Bergeson), a post-graduate resident (Ian Keaveny), and two graduate students (John and Margaret Blethen).

When applying for grants, Pauling described his research as focused on discovering better diagnostic and treatment methods for mental illness. His applications asked mainly for equipment funds, and he usually received what he wanted. Pretty quickly, the Pauling team found that vapor-phase chromatography — a process that had been suggested by Robinson at the outset of the project — was the most effective technique for engaging in quantitative analysis, and the grant applications that followed sought to enhance these capabilities in the laboratory.

Pauling's goal during these first years was to uncover and establish a link between mental illness and deficiencies of various vitamins. At the outset, the team specifically planned to look at the correlation between fluctuations in mental health and variations in intake of ascorbic acid (Vitamin C), nicotinic acid (B_3), cyanocobalamin (B_{12}), and pyridoxine (B_6). Pauling believed that the brain and nervous system were especially sensitive to molecular composition and structure, and that certain mental illnesses were actually a function of localized cerebral deficiency. This concept was, in essence, the guiding principle behind much of the team's work.

Pauling also felt that schizophrenia had not received adequate scientific study, and accordingly the group decided to focus their primary research on schizophrenics. If all went as planned, the following three years would be devoted to developing diagnostic tools to identify deficiencies as well as effective therapies for correcting them. The researchers would also use this time to explore the impact and consequences of other vitamin imbalances. Though enthusiastic about this burgeoning program, Pauling took pains to present orthomolecular therapy as being an adjunct to, and not a replacement for, traditional methods of mental health care including psychoanalysis, antipsychotics, and antidepressants.

***

During the CSDI years, Pauling's grant funding from the National Science Foundation (NSF) had been continuously delayed, mostly because he didn't have a lab in which to conduct the work. Once he was established at UCSD, however, the NSF was quick to award him the budget that he'd long ago requested. Pauling also received funding from the Department of Health, Education, and Welfare, and additional monies from the CSDI were set aside too, should he need them.

His new group began working in earnest in late 1967, focusing on measurements of vitamin absorption, and by April 1968, Pauling had published his introductory paper, "Orthomolecular Psychiatry," in the journal *Science*. The article, which proved influential, drew from the existing literature and focused especially on a study by Abram Hoffer and Humphry Osmond, who had reported improvement in mentally ill patients treated with a regimen of nicotinic acid and nicotinamide.

From there, Pauling outlined his thesis that mental disease could be caused by combinations of abnormal reaction rates and abnormal molecular concentrations of substances essential to the human body. The paper noted that most chemical reactions that take place in humans employ an enzyme as a catalyst. As such, if a certain enzyme fails to function properly, the rate of the reaction that utilizes the enzyme will be severely reduced. Pauling argued that this rate reduction could be addressed by adding large amounts of the enzyme's substrate or by somehow increasing the amount of the enzyme that is synthesized. Pauling then discussed a variety of vitamins and other essential substances and outlined the consequences that would result from deficiencies of each. It was Pauling's belief that simply introducing more of a deficient substance would generally alleviate the effects of each deficiency.

One specific disorder that Pauling studied was phenylketonuria, which is caused by poor functioning of the enzyme that processes phenylalanine and which can result in a whole host of psychological problems. According to Pauling, this disease could be treated in two ways. As noted, one option is the introduction of mass amounts of the enzyme's substrate to counter-balance the effects of the dysfunction. A second treatment option is strict adherence to a low-protein diet, which significantly reduces the amount of phenylalanine that a patient consumes. (This approach was also an important component of the successful treatment protocol that Pauling himself followed when battling a commonly fatal renal disease, glomerular nephritis, in the 1940s.) Both treatment possibilities are examples of orthomolecular therapy.

"Orthomolecular Psychiatry" was subsequently republished multiple times, including in the proceedings of a 1977 Congressional hearing and in a book, *Orthomolecular Psychiatry: Treatment of Schizophrenia*, for which Pauling served as co-editor. As a result of the paper, Pauling began to receive a growing volume of letters from community members who had been directly or indirectly affected by mental illness. He took care in replying to these correspondents, often pointing them toward additional resources for more information and encouraging them to write again if they had further questions. Meanwhile, the reaction to Pauling's paper from physicians and medical researchers was mixed, with most remaining largely unimpressed by his conclusions. This divergence between public response and institutional skepticism would only grow over time.

***

As his program on orthomolecular psychiatry began to take off, Pauling's work as an activist moved forward with as much zeal as ever. Despite criticism that his association with the CSDI and his protests against the Vietnam War made no sense in the context of his scientific career, Pauling had stopped viewing his interests as an activist and scientist as being separate branches of a single life.

Pauling happened to be visiting the University of Massachusetts five days after Martin Luther King Jr. was assassinated. Invited to deliver a series of lectures as the university's first Distinguished Professor, Pauling had fashioned his remarks around a series of thoughts on the human aspect of scientific discoveries. He used this platform to publicly reflect on the tragedy of the previous week. Telling his audience that it was not enough to mourn the fallen civil rights leader, Pauling exhorted individuals of good conscience to carry King's legacy forward by continuing the work that he had begun. "Military might, police might, the power of the assassin are being used by our country to protect an evil economic and social system, based on inequality and injustice," he exclaimed. "Now, let us pledge ourselves to follow the path of righteousness, the path shown to us by Dr. King."

In keeping with this theme over the course of his lectures, Pauling emphasized the scientist's responsibility to ensure that discoveries be used for the good of all humanity and society, rather than support of war and human suffering. Pauling also felt that scientific inquiry should prioritize solutions to current issues and pointed to lack of equal access to medical care in the United States as one such concern. Pauling saw his work in orthomolecular medicine as, potentially, part of the solution, since vitamins were inexpensive, accessible, and could, he believed, significantly improve one's mental and physical well-being.

***

Pauling made similar connections to his work on sickle cell anemia, and in so doing ventured into deeply controversial territory.

Though he was no longer involved in the daily operations of the CSDI, Pauling continued to participate from time to time in a public lecture series that the center sponsored. In one contribution to an event titled "The Revolutionary Age: The Challenge to Man," Pauling put forth his ideas on a potential solution to sickle cell disease. As science had succeeded in identifying the gene mutation responsible for the painful disorder, Pauling believed that forms of social control could be used to prevent carriers of the mutation from marrying and procreating. Once done, Pauling reasoned, the mutation would eventually disappear.

Pauling specifically called for the drafting of laws that would require genetic testing prior to marriage. Should tests of this sort reveal that two heterozygotes (individuals carrying one normal chromosome and one mutation) intended to marry, their application for a license would be denied. Pauling put forth similar ideas about restricting the number of children that a couple could have if one parent was shown to be a carrier for sickle cell trait. In a letter to Mrs. S. Leonard Wadler of the Marriage Museum in New York, dated August 15, 1966, Pauling further detailed how systems of social control might be deployed:

> "I have suggested that the time might come in the future when information about heterozygosity in such serious genes as the sickle cell anemia gene would be tattooed on the forehead of the carriers, so that young men and women would at once be warned not to fall in love with each other."

In proposing these ideas, which he called "negative eugenics," Pauling aimed to ensure that his discovery of the molecular basis of sickle cell disease be used to decrease human suffering. He also felt that whatever hardships his proposed laws might cause in the short run, the future benefits accrued from the gradual elimination of the disease would prove worthwhile.

Pauling came into harsh criticism for his point of view, which remains controversial. But in putting it forward, Pauling took pains to clarify that he was not at all in favor of "positive eugenics" — the manipulation of genetic combinations as a means for developing a "superior" human — and that he was staunchly opposed to the concept of genetic purity. Rather, his intent was to minimize human suffering by using social controls to remove harmful diseases from the gene pool. Nevertheless, by the early 1970s Pauling had stopped circulating these ideas and was turning instead to other means of improving the human condition.

***

On February 28, 1968, Pauling turned 67 years old, a milestone that the University of California regents used to hold up discussions about his obtaining a permanent appointment in San Diego. Sixty-seven, the board argued, was the typical retirement age within the UC system. Moreover, the UC regents were empowered to veto any age-related retirement exceptions. Given his radical political views, Pauling was unlikely to receive any support at all from the group, much less an exception.

One of the stated reasons why the regents harbored concerns about Pauling's politics was his increasingly strident rhetoric. Pauling frequently commended student strikes and demonstrations, and although he emphasized non-violence as the most effective means to foster social change, he counseled students to recognize that authorities may incite violence through tactics of their own. In these cases, he felt that retaliation could sometimes be justified, even necessary.

Pauling also believed that the regents and their trustees wielded too much power, and that they were part of a system that largely inhibited social progress and took agency away from students. For their part, the regents saw Pauling in a similar light: a dangerously influential radical who was constraining the university's capacity to grow.

Realizing that, in all likelihood, Pauling was soon to be forced out, UCSD colleagues Fred Wall and Bruno Zimm began searching for a way to shift the governing authority for his reappointment to the UC system's president, Charles Hitch, with whom Pauling had a positive relationship. After months of negotiations, Zimm succeeded in winning a second year-long appointment.

Though Pauling was grateful to Zimm for his efforts, the increasingly faint possibility of a permanent position at UCSD was emerging as a source of dismay. Looking for a longer-term academic home, Pauling began considering other universities that might also provide better support for his research.

Over time, Ava Helen had also found herself frustrated with both UCSD and La Jolla. In particular, she disliked their rental house and missed their previous home in Santa Barbara, where she had been able to tend a beautiful garden. As 1968 moved forward, the couple began spending more and more time at their Big Sur ranch, with Ava Helen hinting that she would like to make it their permanent home in the coming years.

***

In early 1969, Pauling announced that he had accepted an appointment at Stanford University and that he would be leaving UCSD, where he had been on faculty for the past two academic years. In making this announcement, Pauling explained that Stanford would be a better fit for his orthomolecular research, in part because of the school's well-established department of psychiatry. Palo Alto was also significantly closer to the couple's home at Deer Flat Ranch, which pleased Ava Helen immensely.

By now, Pauling and his team felt confident that they had uncovered evidence of abnormal patterns of ascorbic acid elimination in individuals suffering from acute and chronic schizophrenia. The team planned to continue their analyses of these abnormalities as they moved toward the identification of genetic defects, the creation of diagnostic tools, and the promotion of effective therapies for sufferers of mental disease.

And though they had made significant progress on their psychiatric studies at UCSD, one problem that they had yet to solve was the ability to control for other variables — especially those introduced by diet — that could contribute to variations in the levels of nutrients observed in test subjects' bodies. Because of this, the group was not able to accurately track what Pauling called "individual gene defects." Moving the project to Stanford meant that the researchers could work with mental health patients at Sonoma State Hospital, all of whom were consuming the same elemental diet, which was provided by the Vivonex Corporation. Intrigued, Pauling coordinated with Vivonex to obtain containers of Vivonex-100, a powder that one would mix into water and take every few hours instead of solid food. Pauling's idea was that his control group could follow this diet, and on at least two occasions, he and Ava Helen tested it on themselves.

***

Pauling Speaks Out

Professor Linus Pauling was joined at the
Capitol by Mel Posey after the Nobel-winning
professor addressed the thousands massed in
Sacramento to protest the presence of the National Guard in Berkeley. Pauling called the
Berkeley situation 'immoral, tragic . . .' Posey,
a student at the University of California at
Davis, was one of a delegation which met later
with Governor Ronald Reagan. For the story,
see Page 1.

Pauling's final act at UCSD was appropriately radical. Shortly after the student occupation of People's Park at UC Berkeley and the subsequent death of James Rector, a Berkeley student who was shot by Alameda County sheriffs in May 1969, UCSD students and faculty gathered to decide about an appropriate response to the tragedy at their fellow UC system school.

Many of the faculty in attendance expressed a desire to mourn the death and voice their solidarity with Berkeley, but not to disrupt daily operations. Pauling, on the other hand, stood in front of the hundreds of students who had gathered and encouraged them to actively protest this and other actions taken by the National Guard, the police, and Governor Ronald Reagan. Indeed, for Pauling the violence at Berkeley was

"part of a pattern — the pattern of the war in Vietnam, the increasing militarism of the United States, the growth of the military-industrial complex, the suppression of the human rights of young men and others."

He further explained that those who held power would do whatever they could to protect and move forward with "the continued economic exploitation of human beings" and to further their plan to "make the rich richer and the poor poorer." Pauling then made it clear where he stood on the question of the next appropriate action:

> "Everyone in the whole University of California, all the students, the faculties, the employees, should strike against the immorality and injustice of the act at Berkeley."

Less than a week later, Pauling participated in a march and rally at the California State Capitol in Sacramento, where he gave an impromptu speech that echoed his remarks in San Diego. "The university is not the property of Governor Reagan and the other regents," he urged. "We must protest until the police and the National Guard are removed from the campus of the University of California…the university belongs to us, the students, the faculty, and the people." So concluded Pauling's final remarks on the UC system and its regents while a member of the UC faculty.

***

Although Pauling never again worked for a University of California school, his short time at UCSD was undeniably productive and invigorating. While his two years in La Jolla marked a reemergence into the scientific realm following his frustrating tenure at the CSDI, UCSD also provided the opportunity for Pauling to incubate his partnership with Arthur Robinson. This collaboration provided a strong foundation from which Pauling worked doggedly to expand his research on all manner of topics related to orthomolecular medicine. Though the work ultimately proved to be very controversial, as he finally left southern California for good, Pauling had every reason to be optimistic about the bold new direction that his interests were taking.

Chapter 24

Stanford University

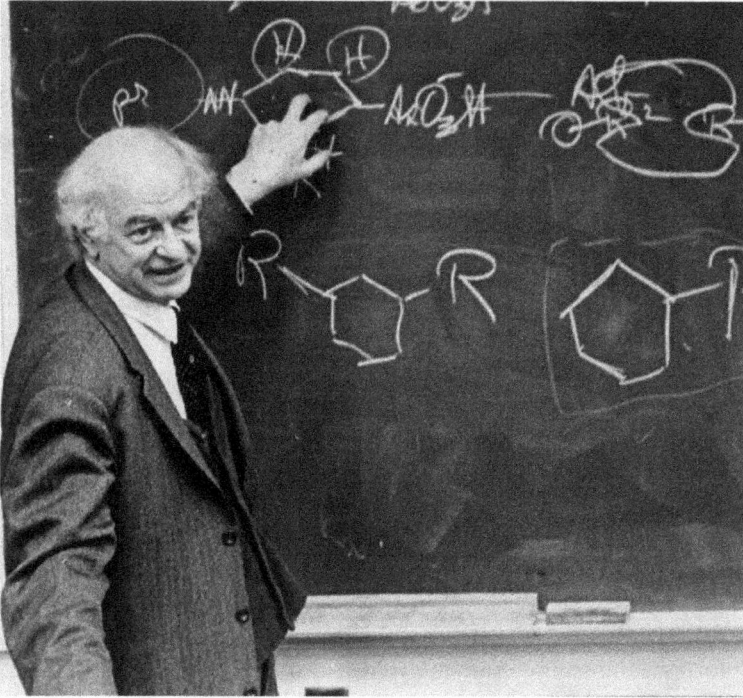

Linus Pauling in lecture at Stanford University, 1969.

Long before arriving at Stanford University as a professor, Linus Pauling had built a working relationship with the Stanford Research Institute through its branch office in Los Angeles. In February 1950, Pauling agreed to join the branch's advisory panel on atmospheric pollution. His role on the panel, according to J.E. Hobson, the director of the Stanford Research Institute, was to give "scientific and technical assistance in connection with our air pollution activities and, particularly, assistance regarding the solution of the Los Angeles smog problem." The panel was to meet monthly at the University Club in Los Angeles over a period of six to eight months and Pauling would be paid a $100 consulting fee for each meeting.

The group's gatherings typically centered around a specific topic like ozone, the chemistry of hydrocarbons, or the future of research. One other meeting consisted of a tour of the laboratory at the Pasadena Field Office and, after making the visit, Pauling offered his ideas for improving the air cleaning technologies that were under development there. Specifically, Pauling suggested to lab director A.M. Zarem that "an effort be made to fractionate the oxidant in smog by the use of a variant of chromatographic adsorption."

Unable to recall the names of those who had previously done similar research, Pauling provided his own suggestions on the best way to clean smog-filled air. Pauling's method first stipulated that water vapor be removed from a tube containing activated alumina and liquid air. From there, Pauling advised that the temperature of the system be increased, allowing a small amount of smog-free air or nitrogen to pass through the tube, in the process collecting the pollutants. Zarem liked Pauling's idea and expressed an interest in testing it.

***

A decade later, in early 1961, the smog in the Los Angeles area had gotten so bad that Pauling was seriously considering moving away, possibly to Stanford University. In addition to its sterling academic reputation, Stanford was also attractive due to its relative proximity to Pauling's ranch at Big Sur. The end of his academic career was likewise on Pauling's mind, as he would be reaching Caltech's mandatory retirement age in eight years. At Stanford he would at least have an extra two years available to him before hitting their age limit of 70.

As he thought more about it, Pauling decided that he would most like to join Stanford's Hopkins Marine Station as their Professor of Molecular Biology, ideally working under a five-year appointment. Learning of this interest, Lawrence Blinks, who worked at Hopkins, offered Pauling an office in the station's library and a possible laboratory space on Canary Row.

Before he went up to visit Blinks and Stanford president J.E. Wallace Sterling, Pauling sent a letter assuring Sterling that he would not impose any financial burden on the university since he was able to secure most of his own funding. Pauling's recent grants had been used to support an eclectic program of work, including his development of a molecular theory of general anesthesia and new inquiries into the potential biochemical basis of mental illness. During his visit, however, Pauling discovered Blinks had over-promised: a laboratory space would not be available at all and he would not have access to the library office during the summer months.

After the visit had been completed, President Sterling followed up, reiterating that the ideal arrangement that Pauling had put forth was impractical and would not work. Undaunted, Pauling replied that, even if he did not have access to laboratory space, he would still view working at Stanford as a step in the right direction. In his letter to Sterling, Pauling made his case:

"I have thought about the nature of my contributions to science, and have recognized that the important ones are the result of my theoretical work rather than of my experimental work, although the theoretical ideas have sometimes been verified in a valuable way by the experimental work… Moreover, I have got rather tired of supervising experimental work, and have decided that I want to devote my time instead to theoretical work. In particular, I do not want to administer a laboratory."

Despite this concession on laboratory space, the ability to support himself financially, and his evident usefulness to Hopkins as the study of biology shifted more and more towards a molecular focus, Pauling's request for a five-year professorship was still too much for Sterling to accept. Thus rejected, Pauling would have to wait until the end of the decade before his desire to be at Stanford was fulfilled.

***

In late 1968, now five years removed from Caltech, Pauling made contact with Harden McConnell, a professor in Stanford's chemistry department, and renewed the conversation about his potential move to Palo Alto. McConnell replied that "everyone is enthusiastic" about the possibility that Pauling might join the department.

Despite this, Pauling soon found that he was facing hurdles similar to those encountered in 1961, and once again he went out of his way to emphasize that he would not present any financial burden on the university and that he could pay much of his own salary through grants that he had received. At the end of January 1969, McConnell wrote to Pauling with an update, "I have now put the Administration here in a position where they must make a decision soon on your appointment." Hand-annotating the letter in red ink, McConnell added: "The decision had better be the right one." Within a few weeks a verdict was rendered: Pauling was in.

But even after he had been accepted, Pauling was made to understand that his future at Stanford was not fully assured and that he would have to follow through on his claims of self-support. For his part, McConnell could only promise that the chemistry department would cover half of Pauling's salary for the first year. Beyond that, there was no certainty about what future years might look like. In relaying these details, McConnell lamented that, "The Chemistry Department is unanimously in favor of your coming here, and we are all greatly disappointed that the material aspects of the arrangements are so meager." All the same, by March Pauling had been approved for a one-year appointment that would officially begin on July 1, 1969.

***

Unlike his previous attempt to join the Stanford faculty, Pauling was this time given his own laboratory. Located in the Chemical Engineering building, the space was offered for up to three years, were Pauling to stick around that long. Jumping at this opportunity, Pauling began organizing the relocation of his equipment from San Diego to Stanford, enlisting close colleague Art Robinson to steer the operation. In addition to Robinson, post-doc Ian Keaveny and lab technician Sue Oxley also followed Pauling to northern California. James McKerrow, who had sought out Pauling while he was at UCSD, rounded out the group as a research assistant.

Shortly after Pauling had completed the move to Palo Alto, he began making himself a part of the Stanford community by donating many of his scientific journals to the university. Colleagues also reached out to Pauling, beginning with faculty in the sciences who invited him to participate in various department-sponsored functions. Physics professor Alexander

Fetter, for one, asked Pauling to join a panel at an upcoming conference on the science of superconductivity. Chemist Carl Djerassi was also solicitous of Pauling's time, requesting his involvement in a symposium sponsored by the department's Industrial Affiliates Program.

Ultimately Pauling was forced to turn both opportunities down because he was already scheduled to attend a Nobel conference and give a talk at the Symposium on Sulfide Minerals in New Jersey. But many other chances to participate in all manner of Stanford faculty functions would arise over the coming years and Pauling was often eager to contribute.

***

During his years on The Farm, Pauling's experimental work partly focused on developing and refining urine and breath analyses for use in diagnosing various diseases and genetic conditions ranging from schizophrenia to cancer to skin diseases, heart disease, and Huntington's chorea. In addition to funding from the National Institutes of Health (NIH) and the National Science Foundation (NSF), Pauling and his laboratory were supported by a collection of smaller awards including a 1971 grant from the American Schizophrenia Association.

Pauling also continued to promote his research and peace work through a hectic writing and travel schedule. In January 1970, he served as Visiting Professor at the Technical University of Chile, where he also received the country's Medal of the Senate. That same year, Pauling published an influential article, "Evolution and the Need for Ascorbic Acid" as well as his book *Vitamin C and the Common Cold*. The latter captured the public's attention and sold exceptionally well.

In 1971 Pauling published six articles, one on nuclear weapons and others covering various topics in chemistry. He also completed revisions for, and saw published, the third edition of his hugely successful textbook, *General Chemistry*. In April 1971, he received the Lenin International Peace Prize at the Soviet Embassy in Washington, D.C. The next year, he partnered with Paul Wolf in the Department of Pathology to study sickle cell anemia. And in early 1973, *Orthomolecular Psychiatry* was published, which Pauling co-edited with David Hawkins. Though now in his early 70s, it was clear that Pauling had no intention of slowing down.

***

Not long after his arrival, Pauling was made aware of a need to attend to a variety of administrative tasks. One of the first items on this to-do list was to update his consent forms and put them on Stanford letterhead. Since he was now associated with the university, doing so would help should any legal problems arise with his research.

As part of this process, Pauling also had to make sure that his experimental designs were in compliance with Stanford's Committee on the Use of Human Subjects in Research. This protocol included, for one, clarifying whether or not the dose of Vitamin B_6 used in a particular study "approach[ed] the 4 GM/Kg that produces convulsions and death in animals."

But Pauling's main administrative issue was workspace. Though he fully understood the modest circumstances governing his hire at Stanford, Pauling nonetheless expressed frustration with the accommodations that had been made for him. In an undated letter to Alan Grundmann, at that time an assistant to the Stanford provost, Pauling complained about his small work area, emphasizing that space adjacent to him was sitting unused. As his mood soured, Pauling demanded that Stanford do a better job of acting in accordance with the facilities guarantees that had been stipulated in his contract. He later threatened to leave if the situation didn't improve, suggesting that he might return to the University of California, San Diego, where he knew they had enough space for him.

Though his relationship with administration may not have been perfect, other faculty members at Stanford were clearly very interested in Pauling's research and teaching. Shortly after he arrived, a variety of professors began asking Pauling to address classes varying from a general chemistry course, a psychiatry research seminar, and a postgraduate survey of basic medical science. Pauling also spoke to medical and psychiatry students about vitamin C and his evolving concept of orthomolecular medicine.

Pauling's understanding of social issues also proved a draw. In one instance, he and Ava Helen jointly addressed a freshman seminar on the social responsibility of scientists. Pauling also participated in Stanford's Professional Journalism Fellowship Program series, at which he was asked to respond to the question, "What would you do if you were Secretary of State?"

Even Pauling's personal medical examinations piqued interest within the Stanford community. Specifically, Roy Maffly from the Department of Medicine conducted a renal evaluation of Pauling, a study that was possibly inspired by Pauling's successful bout with glomerulonephritis in the 1940s (a medical triumph that had been led by a Stanford physician, Thomas Addis). Maffly was also keen to learn more about the Pauling laboratory's urine studies and agreed to interpret the results of Pauling's renal evaluation using the lab's methods.

Within the chemistry department, Pauling joined the Industrial Affiliates Committee, which was chaired by his friend Carl Djerassi. This committee sought to connect private corporations to the research being conducted within the department by addressing questions like the relationship between academic chemistry and the practice of chemical engineering. Pauling was also involved in organizing different symposia for the committee, speaking at its first gathering of the sort in November 1969. He represented the group again in 1973 when he presented on his vitamin C research at an international conference.

Pauling further integrated himself into the department by taking on graduate students; by the start of his second year, he was chairing two doctoral committees and was a member of four others. His students included Robert Copland Dunbar, who was using ion cyclotron resonance to investigate interactions between ions and molecules. Margaret and John Blethen, both of whom assisted Pauling with his schizophrenia studies, and David Partridge, who worked on the chromatographic analysis of urine samples, were also mentees of Pauling's.

Having set up his operations, Pauling also took the opportunity to offer advice derived most recently from his experiences at UCSD, where graduate students rotated between different laboratories during their initial months. Pauling suggested to the leadership of his new department that, upon their arrival, graduate students in chemistry similarly move through six different laboratories, spending six-week periods in each over the course of their first year. Pauling believed this to be an effective way for new students to get to know staff and to better understand the different lines of research underway. Armed with these experiences, the students would then be well-equipped to make a considered decision when it came time to choose the path that they would follow at the start of their second year.

***

Perhaps because it was sometimes inadequate, Pauling often took a creative approach to making the best use that he could of both the laboratory and computing equipment available on campus. In one instance, less than a year after his arrival, Pauling wrote to chemistry chair Paul John Flory inquiring about the possibility of his taking charge of the department's X-ray facilities. Noting that the supervisory position overseeing the instruments had been recently vacated, Pauling suggested that he could be in charge until the university had found someone more permanent, perhaps in two or three years.

In presenting this unusual offer, Pauling referred to his need to continue work that he had initiated at UC San Diego with Art Robinson and Ian Keaveny on the structures of inorganic compounds including delta iron (III) oxyhydroxide and tri-cadmium diarsenide. Having routine access to X-ray equipment, Pauling pointed out, would greatly assist with this effort. In addition, Pauling was also seeking to develop a new technique to measure atomic distances. To do so, Pauling wanted to attach a computer to an X-ray diffraction apparatus which would convert X-ray intensity functions into radial distribution functions. In compounds containing two metal atoms, this conversion would serve as a measurement of the distance between the pair, and the addition of a computer would greatly speed up the process by which this calculation could be made.

Clearly the advanced computing infrastructure then available at Stanford was very enticing for Pauling and he did what he could to take advantage. Perhaps most importantly, in August 1970 Pauling requested access to ACME, an IBM computing network that could be used by university researchers through Stanford's Medical School. Faculty demand for the apparatus was high and Pauling needed to submit formal letters to the ACME Subcommittee on User Charges to gain access. Once he had been approved, Pauling was obligated to pay usage fees through an account that was set up for him and that contained artificially limited funds. At the start, Pauling's research group would be allocated $200 per month for "pageminutes" which cost 1 cent, and $100 per month for disk storage, which cost 10 cents per block per month. Unfortunately, the amounts allocated to this account were not always enough to cover everything that Pauling wanted to do.

Pauling's primary interest in ACME was its potential for use in analyzing urine samples collected from people suffering from schizophrenia and other mental diseases. In addition to other types of assessments, Pauling's laboratory carried out chromatographic analyses looking at about 200 different substances in the urine both before and after a given individual had been placed on a special diet and vitamin regimen. As Pauling told Trammell Lonas, who helped to coordinate use of ACME, "the analysis of these data in a reliable way can be made only with use of a computer." As such, it was critically important that Pauling retain access to the network.

***

Maintaining a well-staffed laboratory was also crucial to moving the schizophrenia research forward. Linus Pauling, Jr. — the eldest of Pauling's four children and a practicing psychologist who lived in Honolulu — even joined the laboratory as a part-time assistant for a short period beginning in October 1970. The next month, Pauling offered a research fellowship to Paul Cary, who was on leave from The Rockefeller University. Pauling specifically wanted Cary to run the chromatography tests central to the laboratory's urinalysis of schizophrenia patients.

Cary had earlier reached out to Art Robinson, asking him to provide a reference letter as he looked for positions during his leave. Learning this, Pauling decided that it would make sense for Cary to come work for him instead. In his offer, Pauling expressed optimism "...that the work has come to a very exciting stage, after a long period of difficulty in getting problems ironed out." Cary promptly accepted.

In October 1973, as part of his laboratory's development of urine analysis techniques, Pauling also requested the grade point averages of 180 students who were participating as research subjects. In doing so, Pauling explained that "one question that is of interest to us is whether there is a difference in composition of the urine for students with different academic accomplishments."

As it happened, Pauling was particularly intrigued by testing results published by two researchers, A.L. Kubala and M.M. Katz, who had run a study in 1960 that showed an improvement in students' grades after drinking orange juice for several months. Walter Findeisen, the recorder at Stanford's Office of the Registrar, told Pauling that he was not allowed to provide grade point averages, but did indicate that the investigation could make use of letter grade indicators, such as A, B, C, etc. This is the route that Pauling ultimately decided to take.

***

Pauling's nutrition research also commonly took him beyond Stanford's campus. In the spring of 1970, Pauling became a consultant for Vivonex, a company that produced a nutritional replacement for use by people who were unable to consume solid food. According to a 1969 pamphlet that Pauling saved, individuals had lived solely off Vivonex for periods of at least three years straight, the product serving as their sole source of nutrition for that whole duration of time. The company also claimed that the elemental diet could help to move people towards their "ideal weight." While Pauling was not offered a fee for his consultancy work, he did receive 500 shares of stock.

Wishing to dig deeper, Pauling began taking Vivonex himself, as did Ava Helen Pauling and Art Robinson. In fact, the trio relied on Vivonex as their sole sustenance for three two-week periods. After taking it periodically for a few months, all three began suffering headaches and lethargy to degrees that exceeded what they had previously experienced. These symptoms prompted Pauling to ask for a complete list of ingredients, including the quantity of each, and to lessen the intensity of his self-experimentation.

In February 1971, Morton-Norwich Products, Inc. purchased Vivonex, and Pauling was eventually compelled to sell his shares for $18.91 each. Though his more rigorous personal trials did not pan out, Pauling felt good about his overall experience with the product and he continued to recommend it as occasion presented. In one instance, Pauling's friend and colleague Frank Catchpool asked him for his thoughts on the nutritional replacement Ensure. Pauling replied that, compared with Vivonex, Ensure was somewhat inferior because it was "essentially a milkshake with added vitamins" and also contained molecules like caseinate that required digestion.

Towards the end of his life, in 1993, Pauling came into personal contact with Vivonex once more, this time during a stay at the hospital. Though in ill health, Pauling still had energy enough to write to John Pepper, the president of Proctor and Gamble, which by then was manufacturing Vivonex. In his letter, Pauling complained of the product's evolution, noting that its now "unpleasant taste" made it difficult to consume. Pauling further informed Pepper that he would no longer recommend Vivonex to others. While Pepper did not respond to Pauling's complaint, another representative did, telling Pauling that the bad taste was likely due to improper preparation.

***

From the outset, Pauling knew that grants and other external funding would be of paramount importance to keeping his Stanford research afloat. In September 1972, three years into his tenure, Pauling authored a memo in which he explained that his salary was now coming exclusively from grants, that he had no assigned duties at the university besides heading research, and that he was actively working to secure new sources of money. In particular, he had "negotiated" a sickle cell anemia contract with NIH in June and was estimating that $92,000 would be forthcoming from the agency.

The previous year, in spring 1971, Pauling applied for a grant from the Department of Health, Education, and Welfare to build a field ionization spectrometer for use in his urine analysis studies. This piece of equipment had only recently become available, the result of new advances in instrument design. In his application, Pauling detailed the potentially profound impact that the spectrometer would have on his work, projecting that

> "This device would make possible simultaneous quantitative analysis and identification of 500–1000 chemical substances in a human body fluid in a time period of a few minutes and with an expenditure of only a few dollars per sample."

Pauling requested $387,554 for the project and it appears from a later report that he received the grant.

While Pauling enjoyed a long track record of success in attracting funding for his work, it was not always enough. In August 1972, Perry West, an administrative officer at Stanford, wrote to Art Robinson to inform him that the Pauling laboratory's current NIH and NSF funds would be exhausted two months earlier than anticipated. As it turned out, Pauling's group had been using more computer time than they had been allocated, and had "drastically overdrawn" one account which they needed to reconcile for themselves. The team had also overdrawn a second computing account that West had been funding for them.

***

In addition to finding money, establishing institutional support for his research proved important to a much larger ambition that Pauling was fostering: the formation of a new Department of Orthomolecular Medicine and Nutrition. Pauling began developing this idea in a pre-proposal that he drafted in August 1972. A few months later, in January 1973, he brought a more finished document to Stanford provost William F. Miller. In making his pitch, Pauling emphasized the potential for orthomolecular medicine to bring in "millions of dollars" of funding and described the ways in which interest in orthomolecular research had already been taking off. As evidence, Pauling noted several talks that he had given the previous fall, details of which had made their way into the press.

By this point, Pauling was also seeing potential for vitamin C to treat a number of maladies including cancer, skin diseases, schizophrenia, the common cold and other infections. To begin actively investigating these tantalizing possibilities, he wanted to establish research centers at both Stanford and the University of Chicago. Miller replied that he would consider his proposal and discuss it with the Dean of the Medical School.

In the meantime, Pauling was receiving encouragement from others reaching out to him, particularly Ewan Cameron, a surgeon and medical researcher at the Vale of Leven Hospital in Alexandria, Scotland. In their early correspondence, Cameron shared data related to his own promising use of vitamin C to treat bladder cancer patients in the UK. Pauling sought to follow up on Cameron's successes and, in 1972, the pair attempted to publish a paper in the *Proceedings of the National Academy of Sciences (PNAS)* on ascorbic acid as a treatment for cancer and other diseases. This paper was rejected twice by *PNAS*, an action later described as "professional censorship" in an editorial published by the *Medical Tribune*.

Undaunted, Pauling continued to push his interest in growing orthomolecular medicine at Stanford and, in May 1973, proposed that the university consider building a new laboratory dedicated to the field. In addition to the direct benefit of providing support for orthomolecular research, Pauling argued that a new space would remove this work from the chemistry building, thus freeing up space that could be more closely tied to the department's more traditional points of emphasis. Pauling again approached Provost Miller, telling him that a donor had already promised to give $50,000 for construction — about half of the estimated total cost — and that other grants were expected to come in as well.

But Miller did not think it wise to pursue construction of Pauling's orthomolecular facility. In rendering this judgement, the provost explained that Pauling had only been at Stanford for a short period and that his position was subject to annual renewals. This being the case, Miller did not want to "institutionalize" Pauling's work unless he was able to convince others in the chemistry and medical departments of its importance.

In effect, Pauling was told that, if he wanted his space, he would have to win over his colleagues first and convince them to initiate their own research programs in orthomolecular medicine. Were this to come about, then Miller would be more open to considering a new capital project. If the reality fell short of that, however, then donor funds would instead be steered towards a general-purpose facility that would be made available to all the chemistry faculty.

Miller's decision was important in that it directly led to Pauling's departure from Stanford University. Motivated to develop a space to pursue what he believed to be an exciting line of research, Pauling began to look for a laboratory facility off-campus. This search led him to a building in Menlo Park near the Stanford Linear Accelerator. Not long after, the location became home to the Institute of Orthomolecular Medicine which, in 1974, was renamed the Linus Pauling Institute of Science and Medicine.

***

It should come as no surprise that, while at Stanford, Pauling kept a close watch on political activism, both on and around campus. While much of the material that Pauling saved would suggest that he was mostly an observer, a look through the *Stanford Daily* student newspaper archives shows that, in fact, he continued to speak on topics related to peace and non-violent protest.

During the years of Pauling's association with the university, both faculty and students alike were involved in demonstrations related to the Vietnam War and its expansion into Cambodia in early 1970, Pauling's first academic year in Palo Alto. Pauling collected several newspaper clippings documenting the protests and occupations that arose in response that spring. Pauling also retained a copy of a letter that Stanford President Kenneth Pitzer had sent to President Richard Nixon in which he urged Nixon to rethink any additional expansion of the United States military's presence in Southeast Asia, arguing that it would only serve to further polarize the citizens from their government.

Around this time, Pauling also received a letter from a group called The Vigilantes, who wrote,

"We are coming to Stanford to show you our form of demonstration and violence. The first one to get the bullet between the eyes will be you... We know all about you from San Diego... Your days are numbered... we'll get you."

Though unsettling, this was hardly the first time that Pauling had received a death threat. It is unclear who the group was or why they had decided to target Pauling, but fortunately nothing more came of their note.

***

It appears that Pauling avoided much of the direct action but kept close tabs on those who did participate in demonstrations, chronicling in particular the ways in which they were treated. One key incident involved a tenured English professor, H. Bruce Franklin, who had been involved in several rallies protesting the U.S. military's actions overseas. Pauling was clearly interested in Franklin's case, collecting and saving numerous press releases, newspaper articles, and other documents that told his story.

Franklin appears to have first come to Pauling's attention in early January 1971. At that time, a group of students and faculty had disrupted a speech given at the Hoover Institute by Henry Cabot Lodge, the U.S. Special Envoy to the Vatican. Lodge had previously served as ambassador to Vietnam, as appointed by President Kennedy in 1963. In this capacity, Lodge was involved in the development of both diplomatic and war strategies up through the late 1960s.

After being interrupted during his speech, Lodge moved to a smaller room to continue his talk, commenting that those who shouted over him were "afraid of the truth." Bruce Franklin was among those subsequently reprimanded by Stanford's administration for interfering with the event.

In explaining his actions to Richard Lyman, Stanford's president, Franklin argued that his "heckling" was not in any way a punishable offense. Lyman disagreed with Franklin's use of the term "heckling" and specified that he had been charged with "deliberately contributing to the disturbance which forced the cancellation of the speech." Lyman continued,

> "...the gravity of the charge cannot be lessened by giving it an amusing-sounding name, for it is an offense that strikes at the University's obligation to maintain itself as an open forum."

The Stanford president believed that Franklin's offense was severe enough to merit suspension without pay for the academic quarter following the resolution of the case. John Keilch, a 24-year-old library staff member who was alleged to have also participated in the demonstration, faced a similar suspension as did six students.

***

Nonetheless, Franklin continued to speak out. At the end of January, he took part in a demonstration with about 200 others in support of *Los Siete* — six Latino youths who had been charged with armed robbery and car theft. At the event, protesters clashed with police and Franklin was charged with felonious assault for elbowing a police officer in the ribs while the officer pushed Franklin in the back with a baton.

According to the *Stanford Daily*, *Los Siete* had previously been acquitted of murdering a police officer, and these new charges of theft had been brought forward afterward. The paper also reported that five police officers had grabbed Franklin, kicking him in the groin and striking him with clubs.

A different newspaper article clipped by Pauling was far less sympathetic towards Franklin. This piece, which also centered on the *Los Siete* demonstration, described Franklin as a "proclaimed Maoist" and a member of the "militant" *Venceremos*, and also published his home address. Pauling's reactions, written in red ink, indicate clear disagreement with the charges brought against Franklin:

> "Provocation? Marchers had permit for sidewalk. Arrested at RR [railroad] crossing where sidewalk is not well demarcated from street. Police cars + other cars blocked intersection."

Pauling also drew quotes from the article in support of his position:

> "'Line of marchers was <u>impeded</u> + some spilled into roadway.' 'Franklin elbowed a policeman in the ribs.'"

At the end of his notes, Pauling simply wrote, "!<u>FELONIOUS ASSAULT</u>!"

But none of this seemed to slow Franklin down. According to a chronology of events created by contemporaries supporting his activism, Franklin was also involved in a rally in early February. At this gathering of roughly 750 people, Franklin advocated "shutting down the most obvious machinery of war" on campus, the Stanford Computation Center. In due course, some 150 people — Franklin not included — occupied the center for three hours until it was cleared out by the police.

A second rally of roughly 350 people followed immediately afterwards. Franklin again spoke, telling those in the crowd to return home to form smaller groups and to plan actions that would avoid the attention of the police. The chronology states that, later, "beatings of both conservative and radical students occur, and a high school student is shot in the thigh."

President Lyman blamed the violence on Franklin, declaring that he "threatens harm to himself and others." In a Statement of Charges against Franklin, which expresses a point of view that is very different from the chronology, Lyman wrote,

> "During the course of the rally, Professor Franklin intentionally urged and incited students and other persons present to engage in conduct calculated to disrupt University functions and business, and which threatened injury to individuals and property. Shortly thereafter, students and other persons were assaulted by persons present at the rally, and later that evening other acts of violence occurred."

In addition to documenting Franklin's history as they had viewed it, the authors of the chronology circulated a petition. Created for submission to Stanford's Advisory Board, the document argued in favor of Franklin's activities and right to free speech. Linus Pauling's signature was among those included on this petition.

***

In the wake of these heated and, at times, violent anti-war protests, President Lyman moved to have Franklin, a tenured professor, dismissed from the Stanford faculty. In addition to accusations of inciting violence, Lyman viewed Franklin as an enduring threat to others at Stanford.

Pauling disagreed with this course of action and decided to question Lyman directly. In a handwritten note that he referred to while speaking to the university's Academic Senate, Pauling stressed that

> "The 'misbehavior' which he [Franklin] is accused was not in connection with his academic duties. It is my understanding that Professor Franklin has **not** been charged with misbehavior or neglect or malfeasance in connection with his teaching or other academic duties."

Nor did Pauling see "any credible justification" that Franklin was a threat to others. As such, Pauling concluded that Lyman's case stood as "an extraordinary and unprecedented act of violation of the principles of academic freedom and individual rights — a really dangerous introduction of authoritarianism in the University." [Pauling's emphases]

Pauling also saved a copy of a letter that Franklin wrote to Lyman at the end of February 1971. In it, Franklin accused the president of using the press — and especially the university's media apparatus — as a lever to turn Stanford's faculty against him. Instead of taking this approach, Franklin felt that Lyman should issue his accusations directly, rather than operate in innuendos such as "acting in an unlawful manner" and "playing a role in tragic events." Franklin further noted that these vague charges, as issued by Lyman, would appear in affidavits submitted for his forthcoming court appearance, thus putting Franklin in a position that he characterized as "First the sentence, then the defense, and finally the charges."

***

Franklin's judicial hearing came the next week, but he was not fighting a solitary battle. The day before, Pauling and 54 others appeared in court on his behalf in an attempt to block an injunction that had been issued against him and over 1,000 others. Pauling and his colleagues argued that the injunction would have "no effect on the underlying causes of campus unrest. If anything, it may serve to hinder the analysis and correction of Stanford problems." The group further described their action as taking inspiration from the lack of a coordinated response by academics against the Nazis in the 1930s. In tandem, over 100 members of the Stanford Academic Council issued their own warning against Lyman's actions, stating that his decisions would "create an institutional orthodoxy which makes 'heretics' out of those who disagree."

The following day, Franklin made his appearance in court. A subsequent press release described a portion of Franklin's closing argument, in which he stated,

"I would say frankly that when I read of the bombing of the [U.S.] Senate yesterday [by the Weather Underground], I thought that that was a wonderful act and I understand that according to what is left of our rights in this country, that one supposedly has the right not only to believe that, but to say what I just said. The advocates of free speech are not prepared to allow free speech to people who think those thoughts and say those things…when a peaceful sit-in or advocacy of a strike is threatened as criminal behavior, the state teaches us a lesson — that our revolutionary analysis is correct and that at some time we should advocate immediate armed struggle against the state."

When the petition in support of Franklin was delivered to the Advisory Board at the end of April, faculty members also addressed the Academic Council on the matter. A statement that Pauling saved from this meeting described how faculty were most "concerned with the intimidating effect upon all of us, in carrying out our obligations to our consciences and to the University community, if the exercise of the First Amendment rights on this campus can be penalized by loss of tenure and dismissal." They then invoked the Nuremberg trials as a precedent to question Henry Cabot Lodge's role in "criminal war policies," and cited the First Amendment in support of Franklin's protest of Lodge's appearance at Stanford.

***

About a month later, with the situation at Stanford beginning to calm down, Pauling gave the commencement speech at the University of California, Berkeley, stressing his own commitment to the peaceful resolution of conflicts. In his address, Pauling stressed a basic belief system that had guided him for decades:

"I believe that it is possible to formulate a fundamental principle of morality, acceptable by all human beings, and that this principle of morality can and should be used as a basis for making all decisions. The principle is this: that decisions among alternative courses of action should be made in such ways as to minimize the predicted amounts of human suffering."

In early June, at about the same time as Pauling's speech, Bruce Franklin was formally suspended from Stanford without pay. That September, at the beginning of the next academic year, Pauling voiced his continuing objection to Franklin's treatment by adding his name to a "Statement of Faculty Opposed to Political Firings."

The statement not only addressed the Franklin affair, but also the firing of Sam Bridges, an African American janitor at the Stanford Medical Center. In so doing, the document connected the Franklin and Bridges incidents, noting that they were both "sharp reminders of the acute problems of racism and war" and arguing that "the time and energy of the University should be directed towards the solution of the problems, not toward the punishment of protesters."

Pauling does not appear to have been as involved in the Bridges case, but he did save newspaper clippings and press releases issued by Stanford Medical Center officials surrounding the April 1971 affair. According to a *Stanford Daily* article published after the incident, Bridges had been speaking with fellow employees about racist hiring policies at the center. Specifically, Bridges told his colleagues that he had been prevented from advancing within the hospital while others from the outside had been brought in to fill vacancies for which he was qualified, vacancies that would have served as a step up the ladder. Other employees responded with similar stories, and Bridges shared them as well.

Not everyone that Bridges spoke with was sympathetic, however, and some complained. Within a week of these complaints being issued, Bridges was fired without any possibility of submitting a grievance. The Black Advisory Council at the medical center investigated the firing and found that there had been several complaints made against Bridges for not doing his work and for being verbally abusive. Some of these accusations were subsequently withdrawn, an action that precipitated an occupation of the medical center building, the participants calling for Bridges to be rehired. Once the occupation had passed its 30th hour, police intervened by breaking down the door of the office in which the occupiers had sealed themselves and clearing the space with tear gas. Afterwards, the medical center allowed Bridges to file a formal grievance but he chose not to pursue this option, believing that it would not lead anywhere productive. Instead, he devoted more of his time to coordination efforts with the medical center's Black Worker's Caucus.

While the Bridges affair resolved fairly quickly, Bruce Franklin's case dragged on. In January 1972, nearly a year to the day of his initial demonstration against Henry Cabot Lodge, the Faculty Advisory Board voted to formally dismiss Franklin, effective August 1972. Assisted by the American Civil Liberties Union, Franklin attempted to fight the decision, but to no avail.

Pauling saved a March 1972 article from *Science* which reported that Franklin "hoped" for violence in response to his dismissal, and that arson and vandalism on campus had been observed in the aftermath. The article also quoted Pauling on the decision, which he described as "A great blow, not just to academic freedom, but to freedom of speech."

***

From the outset, Pauling knew that his time at Stanford as a full professor would be short-lived. Hesitant from the beginning, Stanford had stipulated that Pauling's contract go up for renewal every year and that his reappointment hinge on his effective supervision of research. Furthermore, at the start of his second year in Palo Alto, Pauling's salary was reduced by about half.

The 1971 academic year coincided with Pauling's 70th birthday, Stanford's mandatory retirement age, but Harden McConnell, a close friend of Pauling, defended his right to remain on staff. In making his argument, McConnell wrote that no current professor "should have an automatic right to office and/or laboratory space after standard retirement age, nor should any outstanding and active scientist be denied such space merely because of age." McConnell added that he could easily prove that Pauling fell squarely into the latter category.

In fall 1972, though Pauling's research activity had remained undeniably high, Stanford provost William Miller informed the Associate Dean of the School of Humanities and Sciences, Calvin Quate, that he thought it "unwise to approve further reappointments of Dr. Pauling as a regular Professor" and recommended that Pauling be appointed Professor Emeritus beginning in September 1973. In other words, Pauling would be allowed just one more year on staff.

But as it turned out, Pauling would receive an additional year beyond Miller's recommendation. This information was formally communicated in a November 1973 letter written by chemistry chair Henry Taube. In it he wrote, "Your colleagues in this department hold you and your work in very high esteem and place great value on your continued association with this department." Taube also told Pauling that when he became Professor Emeritus in fall 1974, he could continue his research contacts with graduate students, though an active professor would need to serve as a "nominal sponsor." Taube also told Pauling that he could be called back as a full professor at any time.

***

In fairness, Pauling had begun to retreat from active participation in academic life at Stanford at least a couple years prior to Taube's letter. During winter term 1971 he taught his last course, a special topics class on the structure of atomic nuclei. By then, the only notes that he needed for his teaching were sets of equations that he had worked out step-by-step in advance.

Pauling also served in the Academic Senate while at Stanford, a stint that lasted for two years and that also came to a conclusion in 1971. In submitting his resignation from the body that spring, Pauling told H. Donald Winbigler, Stanford's Academic Secretary, that "decisions about the University should be made by younger men who can look forward to a longer period of association with the University."

As his connections lessened, the support that he received withered in kind. By fall 1972, Pauling was no longer receiving a Stanford salary, only office and laboratory space. As such, his sole form of professional funding was coming by way of external grants that he had received.

In August 1974, Pauling officially retired from Stanford and formally became Professor Emeritus of Chemistry. But even in this capacity he maintained a connection with the university and his former colleagues. Perhaps most notably, Pauling continued to sit on graduate student committees and steadfastly updated his still-growing list of publications for inclusion in department pamphlets.

After his retirement, colleagues at Stanford also regularly contacted Pauling, often inquiring about the progression of his research. *Stanford Magazine* likewise profiled Pauling in 1979, some five years after he had left, asking him to reflect on his career and his engagement with the world. In the piece he explained his approach to life, which had remained consistent over the years.

> "I have a sort of general theory of the universe. I try to fit everything I read into the general picture. If I read something that doesn't fit, I wonder about it. Or, if I think something seems to fit, I try to follow through."

He also reflected on the varying degrees of satisfaction that he had derived from his work as a scientist and peace advocate:

> "With the Chemistry Prize, I was just enjoying myself, learning about the nature of the world, having a good time and making a living, too, as a professor... The Peace Prize came for work that I was doing as a sacrifice... I was taking time away from the things I really like to do, but doing it because of a sense of duty."

Pauling maintained a residence on the Stanford campus and, by 1984, was still delivering guest lectures for courses taught at the university. One of these was for an Optimal Health and Fitness course taught by Dr. Jack Martin. Following Pauling's appearance, Martin shared some of the student reviews that had been submitted, with comments covering a range of impressions including: "smart guy but not very interesting"; "vehement and extremely knowledgeable, not to mention amusing"; "entertaining, but a grain of salt is necessary"; "it is always great to hear from someone as famous as he"; "new stuff, good presentation"; "a little weird"; and "incredible man."

As late as spring 1993, Pauling remained on call to represent Stanford should an appropriate occasion arise. In one instance, he participated in a meeting with the Swedish Minister of Education and Science, Per Unckel, who was visiting to explore a potential research collaboration on environmental problems. Associate Dean of Research Patricia Devaney had asked Pauling to meet with Unckel during his visit and Pauling, then 92 years old and in fading health, obliged.

In the years following his passing, Pauling remained of interest to the Stanford community. An undergraduate student, Kristine Yu, wrote about

Pauling for the spring 2003 issue of *The Stanford Scientific Review*, fashioning her article partly on conversations with those who had known him. One anecdote centered on a visit that Pauling had made to Henry Taube's home. The story had it that Pauling was interested in an "unusual" geode that Taube had brought back from Brazil. As Pauling looked at the specimen, Taube shared his personal theory of how it had been formed. Pauling responded, "If you feel that strongly about it, you should write a paper on it." Not long afterwards, Pauling sent Taube a long letter explaining how Taube's theory was wrong.

Part IV

Travels

Chapter 25

The Guggenheim Trip

Pauling at the Temple of Neptune, Paestum, Italy, April 1926.

Simon and Olga Guggenheim established the John Simon Guggenheim Memorial Foundation in 1925 to honor their son, who had died in 1922 just before he was to enter college. The Guggenheims' intentions for this new foundation were "to improve the quality of education and the practice of the arts and professions in the United States, to foster research, and to provide for the cause of better international understanding."

The Foundation aimed to do so by offering "promising scholars, both men and women, opportunities under the freest possible conditions to carry on advanced study and research in any field of knowledge, or opportunities for the development of an unusual talent in any of the fine arts, including music." That charge supported 15 fellows in the organization's first year and 43 in its second. Within its first decade, the Foundation had extended its purview by supporting applicants from Latin America and Canada, in addition to the United States.

Since the first awards in 1925, many major figures including numerous Nobel and Pulitzer Prize winners have received Guggenheim Fellowships. Ansel Adams, Aaron Copland, Martha Graham and Langston Hughes are among those who benefitted from the Foundation's support, as did Henry Kissinger, Paul Samuelson, Wendy Wasserstein, James Watson and Linus Pauling.

Shortly after setting up the Foundation, Simon Guggenheim stated that

"It has been my observation...that just about the time a young man...is prepared to do valuable research, he is compelled to spend his whole time in teaching. Salaries are small; so he is compelled to do this in order to live, and often he loses the impulse for creative work in his subject, which should be preserved in order to make his teaching of the utmost value, and also for the sake of the value of the researches in carrying on of civilization."

At the time of his 1926 fellowship, Pauling's circumstances closely approximated the archetype that Simon Guggenheim had in mind. And as hoped, the support that he received from the Foundation proved to be a crucial step forward for both his career and, indeed, the broader trajectory of 20th-century chemistry.

***

By the mid-1920s, scientific institutions across Europe were producing top-notch researchers in physics and chemistry. New and exciting work was being conducted across the continent and the scientific community was booming. To many, the California Institute of Technology seemed a relative backwater compared to powerhouse laboratories in cities like Göttingen, Munich, and Copenhagen. It was in this context that, in 1925, Pauling applied for a Guggenheim Fellowship seeking support for a European tour during which he would visit the continent's most esteemed research centers and learn from its scientific leaders.

Prior to his application, Pauling, who had not yet completed his Ph.D., was supported by a fellowship from the National Research Council (NRC). The NRC funding was designed such that Pauling would work at Caltech for six months and then move to the University of California, Berkeley for the second half of the year. But Caltech's chemistry head A.A. Noyes had other plans.

A prominent figure in the American scientific community, Noyes was able to provide Frank Aydelotte, the head of the Guggenheim Foundation, with a reference so strong as to guarantee Pauling a fellowship. In doing so, Noyes also proposed that Pauling be sent to Europe early so that he and Ava Helen could enjoy the sights of the continent before beginning an intensive work schedule. (Of this, biographer Thomas Hager writes, "Noyes, a romantic at heart, may have hoped that Pauling's Italian tour would bring to flower a latent aesthetic sensibility. But Pauling wasn't Noyes.") In return, Noyes suggested that Pauling forfeit his NRC award and remain at Caltech rather than serving the half-year term at Berkeley. Pauling readily accepted the proposal, which no doubt pleased Noyes greatly, given his concern that Pauling might choose to leave Pasadena once he had finished his doctoral research.

***

At this time, Linus and Ava Helen Pauling were also young parents. Their first child, Linus Jr., was born on March 10, 1925, and just six days shy of his first birthday they said goodbye to him, leaving him in the care of Ava Helen's mother Nora and sister Lillian. Historian Mina Carson notes that

> "Linus later remembered protesting in 'shock,' but he yielded as Ava Helen argued that the journey would be hard on the baby and would keep them from doing the things they wanted to do in Europe."

Ultimately the couple chose to equip the caretakers with an extensive list of instructions for Linus Jr.'s physical care and social well-being and, after a four-day period of acclimation in Portland, left their son en route to the East Coast. They would see him next in September 1927, some 18 months later.

***

After a trans-continental train trip and a brief stay in New York, the couple stepped onto the steamship *Duilio* and officially embarked on their first trip to Europe. After a week of rough seas and a "young hurricane," the Paulings and their shipmates finally set foot on dry land when they arrived at the island of Madeira off the coast of Portugal. Reflecting on this moment, 22-year-old Ava Helen wrote in her diary,

> "For two hours we sailed along the southern edge of Madeira, watching the pretty villages made of toy houses with red roofs scattered along the terraced slopes, and seeing light lovely waterfalls beneath the snow-topped hills."

Her journal entries, filled with romantic imagery and exclamations of delight, contrast sharply with her husband's letters to his mentor, A.A. Noyes, in which he deemed Naples "not spotless," the Roman ruins "disappointing" and Rome itself "terribly crowded."

The Paulings' wedding in Salem, Oregon had been a quiet affair followed by a one-day honeymoon in their college town, Corvallis. Though three years late, their stay in southern Europe evolved into the post-nuptial vacation that they had missed, and even Linus's complaints couldn't stifle the fun of the trip. Ava Helen's travel diary, given to her by Linus and inscribed "For my dear Ava Helen," included all manner of details about their activities, including a diagram of the Rock of Gibraltar. In contrast, Pauling's own diary briefly notes their transit across the U.S., a few sights in New York, and several days' weather reports before ending in a long series of blank pages.

Among the more colorful of Ava Helen's entries is her description of an assassination attempt on Benito Mussolini in which an English woman "shot him through the nose." She then recounts her impressions of Mussolini's car, containing the undoubtedly shaken leader as it passed quickly through a crowd. This was not the only encounter the couple had with Italian politics during their stay. On a train ride from Pisa to Florence, the Paulings found themselves in conversation with a leader of the Fascist movement in the region. During the trip, he regaled them with stories of his war wounds, the evils of Communism, and the successes of Mussolini. The couple, while entertained by the man's exotic tales, were "glad to return to Florence and to dinner."

At Linus's insistence, the Paulings' vacation, originally meant to continue through the end of April, concluded a week early. In a letter to Noyes, the restless young scientist wrote, "We have come to the end of a very pleasant trip, and I am glad; for even though Italy is wonderful, and everything was new to us, traveling becomes tiresome. Moreover, I am very anxious to get back to work after nearly two months of idleness."

***

The Paulings' arrival in Europe coincided with a period of seismic reform in quantum theory. At its inception, physicists and chemists had attempted to apply the classical laws of physics to atomic particles in an effort to understand the motion of and interactions between nuclei and electrons. This application proved flawed as it became clearer that the classical laws, such as those developed by Isaac Newton, mostly pertained to macroscopic systems only. Theorists soon discovered that these same principles were not necessarily relevant for atomic systems and that the microscopic world does not always consistently align with experimental observation. As Pauling would write in 1935, the work being conducted across the continent to develop quantum mechanics at that time constituted nothing less than "the most recent step in the very old search for the general laws governing the motion of matter."

A series of breakthroughs in the early- to mid-1920s accelerated the decline of the old quantum theory. In 1924 Louis de Broglie discovered the wave-particle duality of matter, and in the process introduced the theory of wave mechanics. Then in 1925, just one year before Pauling began his European adventure, Werner Heisenberg developed his uncertainty principle and began applying matrix mechanics to the quantum world.

In 1926, shortly after the Paulings arrived, Erwin Schrödinger combined de Broglie's and Heisenberg's findings, mathematically proving that the two approaches produce equivalent results. Schrödinger then proceeded to develop an equation, now named for him, that treats the electron as a wave.

The adoption of wave and matrix mechanics led to the development of a new quantum theory and the overwhelming acceptance of a burgeoning field known as quantum mechanics. The effect was revolutionary. As noted by Pauling in 1929,

"Where the old quantum theory was in disagreement with the experiment, the new mechanics ran hand-in-hand with nature and where the old quantum theory was silent, the new mechanics spoke the truth."

***

Linus and Ava Helen arrived in Munich during the last week of April and the top item on Pauling's agenda was a meeting with Arnold Sommerfeld at the Institute for Theoretical Physics. Collaborating with Niels Bohr, Sommerfeld was responsible for a contemporary model of the atom — the Bohr-Sommerfeld model — that deeply informed current quantum mechanical ideas on atomic structure. As director of the Institute of Theoretical Physics in Munich, Sommerfeld had spent the past decade nurturing many of Europe's best scientists on a steady diet of cutting-edge research. As explained by Thomas Hager, "[Sommerfeld] knew everyone in theoretical physics, had collaborated with many of them and corresponded regularly with the rest."

Sommerfeld's lectures, famous in their time, were heralded for their innovative ideas, and it was Pauling's great fortune that his visit coincided with Sommerfeld's inaugural course on wave mechanics as applied to quantum theory. Historically significant as the first of their kind, Sommerfeld would later write that "My first lectures on this theory were heard by Linus Pauling, who learned as much from them as I did myself."

Though their connection would prove auspicious, Pauling's initial contacts with Sommerfeld were something of a disappointment. Rather than nourishing the work he had begun at Caltech, Sommerfeld instead chose to assign Pauling a mathematical problem relating to electron spin, an area that was only mildly appealing to him. After a period of half-hearted engagement with the task, Pauling convinced Sommerfeld to allow him to switch to a study of the motion of polar molecules. Core to Pauling's pitch was his belief that he could clarify portions of the Bohr-Sommerfeld model by introducing the effects of a magnetic field to the existing equations. This naturally piqued Sommerfeld's interest and Pauling was subsequently instructed to begin his investigation with the proviso that he pass along his results in advance of a presentation that Sommerfeld was scheduled to give in Switzerland. Pauling did so and, a few days after Sommerfeld had departed for the conference, Pauling also received an order to appear in Zurich to discuss his progress.

Once arrived, Pauling found himself surrounded by the leading physicists of Europe. Wolfgang Pauli, a young German investigator famous for his development of the revolutionary Pauli Exclusion Principle, was among those in attendance. On a whim, Pauling approached his distinguished colleague and began explaining his recent work on the Bohr-Sommerfeld model, but Pauli was unimpressed. As detailed in a 2001 talk by biographer Robert Paradowski,

> "Pauling told Pauli about this work he had been doing and it was a mixture of the old quantum theory and the new quantum mechanics, and Pauli listened to him and then just said 'not interesting.' It wasn't very interesting because it hadn't really sunk into Pauling what the revolutionary nature of quantum mechanics actually was, it would take him a while to absorb this."

In essence, the paradox-riddled Bohr-Sommerfeld model, and Pauling's work supporting it, was on its way out with the new ideas of quantum mechanics soon to take its place. Pauling's research was too late to be of any value and Pauli was not shy about telling him so.

***

After a summer vacation in Switzerland, Linus and Ava Helen returned to Munich for the fall semester. This turned out to be an important period for Pauling, who began to prove himself in the eyes of his colleagues and developed a reputation for extensive knowledge and concentrated enthusiasm.

Pauling's relationship with Sommerfeld was also strengthened during this time, in part through Pauling's discovery of a mathematical error in the work of Gregor Wentzel, a protégé of Sommerfeld's. The discovery and correction of this mistake garnered Pauling a great deal of respect in

Sommerfeld's eyes. But more importantly, it also inspired Pauling to apply Wentzel's work to the calculation of energy levels, which in turn created the foundation for a new investigation on the energy values of complex atoms. This approach to deriving atomic properties was entirely original and Pauling took full advantage of the breakthrough, publishing his findings in a paper titled "The Theoretical Prediction of the Physical Properties of many-Electron Atoms and Ions," which appeared in the *Proceedings of the Royal Society* in 1927. As noted by biographer Paradowski, "that was the beginning of his great success in quantum mechanics; it's still one of his most cited papers."

Throughout the fall, Pauling broadened his application of the new quantum mechanics to other problems including the calculation of light refraction and diamagnetic susceptibility. By then it was becoming clear that the 25-year-old American was emerging as a legitimate player in the European field of quantum mechanics. Looking back many years later, Pauling would recall that

"My year in Munich was very productive. I not only got a very good grasp of quantum mechanics — by attending Sommerfeld's lectures on the subject, as well as other lectures by him and other people in the University, and also by my own study of published papers — but in addition I was able to begin attacking many problems dealing with the nature of the chemical bond by applying quantum mechanics to these problems."

***

This attack on the nature of the chemical bond was beginning to feel more approachable for Pauling, but he would need more time overseas to work on it. As 1926 was coming to a close, he applied for an extension of his fellowship and, with the help of a particularly flattering cover letter from Arnold Sommerfeld, he was granted six more months of support.

Boosted by this news, Pauling quickly began planning visits to Copenhagen and Zurich, both home to some of Europe's finest research facilities. His first stop was the Danish capital, where he hoped to visit Niels Bohr's institute and discuss ongoing research with the renowned scientist. Unfortunately, he had arrived uninvited and found it almost impossible to secure a meeting. Bohr, with the help of Werner Heisenberg and Erwin Schrödinger, was at the time deeply engaged in research on the fundamentals of quantum mechanics. His specific focus during this period was a study of the physical realities of the electron, work which eventually resulted in a theory termed the "Copenhagen Interpretation."

Pauling did, however, make one valuable discovery in Denmark: a young physicist named Samuel Goudsmit. Born in the Netherlands in 1902, Goudsmit had already received acclaim for his 1925 collaboration with Eugene Uhlenbeck, in which the duo introduced the concept of electron spin to the scientific community. By the time that Pauling met him in 1926, Goudsmit was investigating complex spectra and the Zeeman effect. The two men quickly became friends and began discussing the potential translation of Goudsmit's doctoral thesis from German into English.

Their interactions eventually gained the notice of Niels Bohr, who finally granted Pauling and Goudsmit an audience. But sadly for the pair, Bohr was neither engaging nor encouraging. Nevertheless, the two rising figures continued to work together, their cooperation eventually culminating in 1930's *The Structure of Line Spectra*, the first book-form publication for either scientist. A year after the volume was published, Pauling described their month together in Copenhagen as "the happiest period of scientific cooperation in my life, and the most profitable for me."

***

Once finished in Denmark, Pauling made a quick trip to Max Born's institute in Göttingen before moving on to Zurich, where other advances in quantum mechanics promised an exciting stay. And yet again the man that Pauling was most interested in, Erwin Schrödinger, proved unavailable. The quantum mechanics revolution was consuming the time and thoughts of Europe's leading physicists and Pauling, though growing in reputation, still wasn't important enough to attract the attention of lions like Bohr and Schrödinger.

As a result, Pauling chose to converse and work with men of similar status in the scientific hierarchy. Two of them, German researchers Walter Heitler and Fritz London, had spent the past several months applying wave mechanics to the study of electron-pair bonding, and the relationship that Pauling built with them proved important.

Heitler and London's work was an outgrowth of their interest in Heisenberg's theory of resonance, which proposed that electrons are exchanged between atoms as a result of electronic attraction. Heitler and London determined that, under certain conditions, this process could cancel out electrostatic repulsion and result in the creation of electron bonds. Their investigations of hydrogen bonds likewise agreed with other leading theories, including Wolfgang Pauli's exclusion principle and G.N. Lewis' shared electron bond. In short, the Heitler-London model was well on its way to contributing to a new truth about the physics of the atom.

Pauling used his time in Zurich to absorb and experiment with Heitler and London's output. While he didn't produce a paper during his stay, the emerging model made a pronounced impression on him and he returned to the U.S. with a renewed sense of purpose. As he would write in 1929,

> "...the replacement of the old quantum theory by the quantum mechanics is not the overthrow of a dynasty through revolution, but rather the abdication of an old and feeble king in favor of his young and powerful son."

Pauling would embrace the spirit of this moment and use it to tackle the problem of atomic structure in a manner that would rapidly solidify his stature as a great scientific mind, one that could not be ignored.

***

Pauling came back to Pasadena in the fall of 1927, bursting with new ideas. While he was away, A.A. Noyes had sent word that a unique position had been created for his promising young faculty member, one that lined up nicely with Pauling's nascent agenda. Upon his re-arrival, Pauling was to begin working under the title of Assistant Professor of Theoretical Chemistry and Mathematical Physics. And though Noyes eventually dropped the physics course from the appointment, Pauling liked the idea of hybridizing his interests into one name and began referring to himself as a quantum chemist.

A top priority for Pauling was to use the knowledge gleaned from his Guggenheim experience to develop his own lecture series on quantum mechanics. (Among those who attended was none other than Albert Einstein who sat in on one of Pauling's talks in 1930.) The content of the course became the foundation for Pauling's first textbook, *Introduction to Quantum Mechanics with Applications to Chemistry*, which he developed with a former Ph.D. student named E. Bright Wilson, Jr.

In the fall of 1930, Pauling began work on a determination of the structure of the carbon tetrahedron, implementing a simplified version of the Schrödinger wave equation that had been developed by John C. Slater, an American physicist. Pauling's goal was to define the tetrahedron's atomic architecture, as it was understood by chemists, in terms of quantum mechanics. Success in this venture carried with it the potential to unify chemists and physicists in their understanding of molecular bonding.

Pauling worked through the fall without any major breakthroughs until finally, in December 1930, deciding to simplify the mathematics of the project by removing the radial function from his equation. After doing so, Pauling found that the resulting wave functions (which might be thought of as mathematical models of atomic structures derived from X-ray studies of substances), when mathematically combined, resulted in four hybrid orbitals oriented at the angles of a tetrahedron. Crucially, the calculations demonstrated that these bonds strengthened as the overlap between orbitals increased.

Pauling was elated to find that all variables could be accounted for using his new method, and it was immediately clear that this was a major discovery. In his own words,

"I was so excited and happy, I think I stayed up all night, making, writing out, solving the equations, which were so simple that I could solve them in a few minutes. Solve one equation, get the answer, then solve another equation about the structure of octahedral complexes such as the ferrocyanide ion in potassium ferrocyanide, or square planar complexes such as in tetrachloroplatinate ion, and various other problems. I just kept getting more and more euphorious as time went by."

***

In February 1931, Pauling mailed his results to the *Journal of the American Chemical Society (JACS)*. Confidently titled "The Nature of the Chemical Bond: Application of results obtained from the quantum mechanics and from a theory of paramagnetic susceptibility to the structure of molecules," the paper was published by *JACS* only six weeks after the manuscript arrived in the journal's offices. The incredible speed of this turnaround was abetted by the fact that the paper had not been refereed, as editor Arthur Lamb could think of no individual properly qualified to review the revolutionary content of Pauling's work.

Soon recognized as a classic of 20th-century scientific writing (more than 40 years later, Pauling would call it "the best work I've ever done"), the paper defined six rules for the shared electron bond and presented Pauling's findings in uncomplicated terms, thus enabling peers to examine the work without becoming lost in the mathematics. Between 1931–1933, Pauling would publish six more papers on the topic, the combined impact of which would push the field toward a new understanding of atomic structures and properties. As Pauling would reflect in 1977,

"It seems to me that I have introduced into my work on the chemical bond a way of thinking that might not have been introduced by anyone else, at least not for quite a while. I suppose that the complex of ideas that I originated in the period of around 1928 to 1933 — and 1931 was probably my most important paper — has had the greatest impact on chemistry."

In 1939, Pauling collected his research into an extremely influential textbook titled *The Nature of the Chemical Bond and the Structure of Molecules and Crystals: An Introduction to Modern Structural Chemistry*. The book was a huge success on many levels: not only did it describe research of fundamental importance to the study of chemistry, but it did so in a lucid style that was understandable to a wide range of users. In the estimation of Nobel laureate Max Perutz, the volume was proof that "chemistry could be understood rather than being memorized." A contemporary of Pauling's, Charles P. Smyth, would echo this perspective in a 1939 letter, writing,

"I have been very much interested by your new book and have assigned several of the chapters for reading in connection with a graduate course. As evidence of my interest in it, I can cite the fact that it is the first scientific book which I can remember reading during the course of a fishing trip, although I have carried many with me in the past."

***

Word that Pauling would receive the Nobel Chemistry Prize began to leak in late October 1954 and, on November 3, the news was officially confirmed. While 1954 proved his breakthrough year, subsequent research has revealed that Pauling had been nominated for the Chemistry Prize nearly every year beginning in 1940. Several times during this period he was put forth as a potential co-recipient of the prize, often with fellow scientists who had worked on similar projects rather than individuals with whom he had directly collaborated.

In 1954 Pauling's name was submitted for the Chemistry Prize by 13 different nominators. In 11 instances his was a solo nomination, and twice he was partnered with a second nominee: German organic chemist Hans Lebrecht Meerwein on one ballot and American organic chemist Robert Burns Woodward on another.

Alfred Nobel's will stipulated that one prize was to go to "the person who shall have made the most important chemical discovery or improvement." In 1954 Pauling was honored "for his research into the nature of the chemical bond and its application to the elucidation of the structure of complex substances." This decision by the Nobel Committee was important as Pauling's prize marked the first time that the group chose to recognize a collection of work rather than "the most important chemical discovery or improvement" made in a given moment.

The tenth American to win the Chemistry Prize, Pauling was honored by the Nobel Committee for his study of the structure of matter and of the seemingly invisible forces that hold its building blocks together. When asked for his thoughts on this work, Pauling first explained that it was the supportive environment that fellow scientists and collaborators had created at Caltech that helped him to win the prize. He likewise noted that he had been able to develop his theories because of many years' work — by him and others — on X-ray crystallography and the behavior of electronically irradiated chemicals.

But it is equally likely that this signature achievement would not have come to pass without the Guggenheim trip to Europe in 1926 and 1927. As he recalled in a 1975 exchange with author John H. Davis, who was researching a biography of the Guggenheim family, the opportunity that these travels provided to meet both leading and emerging physicists in Europe gave him more confidence in his own ideas on the application of quantum theory to chemical problems. Much of the theoretical work that followed was based on those months living overseas. That period of scientific exchange, propelled by the freedom afforded by the Guggenheim Fellowship, was precisely what he needed at the time.

Chapter 26

The Paulings Go to England

Peter, Linda, Ava Helen and Linus Pauling at Magdalen Great Tower, Oxford, England, 1948.

In 1946 the Second World War had come to a close and Linus Pauling was in transition from his wartime work back to the regular goings-on at the California Institute of Technology when he received an enticing invitation. That January, Frank Aydelotte, American Secretary for the Rhodes Scholarship Trust and director of the Institute for Advanced Study at Princeton, wrote to Pauling proposing that he be appointed as George Eastman Visiting Professor at the University of Oxford for the coming academic year.

The appointment would include a professorial fellowship at Balliol College, one of the oldest (founded in 1263) and most prestigious of Oxford's 38 colleges. It was an attractive offer; with only two or three lectures a week required of him, Pauling would have ample time to visit other European universities and steep in the vibrant culture of international chemistry.

Pauling felt deeply honored by the invitation and was anxious to return to Europe, which he had not visited since 1930, but the trip would have to wait a year while he remained in Pasadena to attend to administrative duties and finish his freshman text, *General Chemistry*. After much back-and-forth between Aydelotte and Pauling it was decided in early 1947 that he would serve as Eastman Professor for the winter and spring terms of 1948.

***

Though the professorship was postponed, Linus and Ava Helen managed to squeeze in an early summer visit to Europe in 1947, a mix of work and a few quiet days of rest and sightseeing. As was typical, the couple was kept busy by a wide range of activities including social affairs, preparations for the coming stay in Oxford, and Pauling's receipt of an honorary doctorate from the University of Cambridge.

Pauling's role at the forefront of American chemistry (he would learn in December that he had been chosen as President-Elect of the American Chemical Society) also garnered him a key role in scientific discussions abroad, and his July was filled to the brim with meetings and conferences. After three days at the International Congress of Experimental Cytology in Stockholm, he returned to England for the International Congress of Pure and Applied Chemistry. This event coincided with the International Union of Chemistry, where Pauling presided as Congress Lecturer. He also participated in the Centenary Celebration of The Chemical Society at the University of London.

At that final event, Pauling received another honorary degree and delivered an after-dinner speech on behalf of his fellow awardees. In it, he called on the scientific community to take the lead in advocating for an end to war and expressed his hope that there might soon be a "supra-national world government, and that we shall all be fellow citizens, citizens of the world."

***

Their summer escape primed the Paulings' excitement for their extended stay the coming year. However, the planning for this trip proved to be almost as difficult as the initial decision on when to go. With England still in the early stages of recovery from the war, travel inside the UK was less than ideal. Securing a house for Linus, Ava Helen and their three children still residing in Pasadena proved such an onerous task that the entire enterprise was on the verge of being canceled just a month before departure. Ultimately the family decided to make the sacrifice of staying in an Oxford hotel, Linton Lodge, for several weeks until a small flat was finally procured for them.

Aspects of the food rationing required during wartime also lingered during the post-war years and presented challenges for Ava Helen in preparing the very strict diet prescribed to keep her husband's kidney problems at bay. The doctor who was treating Pauling's nephritis, Thomas Addis, even wrote to the Ministry of Food to ensure that the visiting scientist would be able to receive the 40 grams of protein (from eggs, milk, cheese, cereals, vegetables and fruits — not meat, chicken or fish) required by his unique 2,500-3,000 calorie diet.

Indeed, Pauling left no stone unturned in his planning, even writing to a doctor friend in Maryland for advice on preventing seasickness during their transit across the Atlantic. Despite initial skepticism that adequate schools would be found for the children, Ava Helen managed to enroll nine-year-old Crellin in the Dragon School — where he was the "best man" in his form — and young teenager Linda in the Oxford High School for girls, which was mostly a positive experience (nice, well-fitting uniforms but difficult Latin classes). Peter Pauling had just started his first term at Caltech but was able to keep up while overseas by studying independently with tutoring from an American Rhodes Scholar. Oldest child Linus Pauling, Jr. had just married Anita Oser — the great-granddaughter of John D. Rockefeller and Cyrus Hall McCormick — and the young couple remained in the States while the rest of the family embarked on their adventure.

***

On December 26, excitement was at a high level for the Pauling children even before they boarded the *Queen Mary* and set sail for England. During the holiday period, New York City was experiencing its worst blizzard in years, and this was the first time the three southern California natives had seen snow. Despite the marvels of the winter wonderland, the family was in actual fact stuck, and it was only by a stroke of luck (and some extra cash up front) that the Paulings were able to convince a taxi driver to push his way through to the docks and get them to their ship on time. They celebrated New Year's Eve on the boat and, in a later interview, Linda recalled that members of the Canadian Ski Team, fellow passengers on the same Atlantic voyage, were keen on dancing with her all night — that is, until they found out that she was only 15!

Linus Pauling must have spent some time during the journey across the sea in introspective thought, for it was during this trip that he wrote his famous pledge, on the back of a piece of cardboard announcing one of his lectures, that:

> "I hereby make avowal that from this day henceforth I shall include mention of world peace in every lecture and address that I give."

***

There is another story from Pauling's crossing of the Atlantic that merits close examination; a story of an opportunity that was lost.

A fellow passenger on that *Queen Mary* trip was Erwin Chargaff (1905–2002), an Austria-born biochemist who became interested in DNA in the 1930s, far earlier than most other scientists. In 1944, after molecular biologist Oswald Avery published an important paper detailing the transforming principle of the *Pneumococcus* bacteria, Chargaff decided to devote his laboratory almost entirely to the study of nucleic acid chemistry. Experimenting with these delicate substances was not an easy task, but eventually a chromatographic technique was developed that would allow for the separation and analysis of the base rings in DNA. This work would later lead to the postulation of what became known as "Chargaff's rules."

DNA has two main structural components: a backbone made up of sugar and phosphate groups, and a series of nitrogenous bases in the middle of the molecule. There are four different bases found in DNA, Adenine (A),

Cytosine (C), Guanine (G), and Thymine (T), which are themselves classified into two categories, pyrimidines and purines. The pyrimidine bases, Cytosine and Thymine, contain only one nitrogen ring, while the purine bases, Guanine and Adenine, contain two rings. In the DNA structure, the bases pair complementarily, meaning that a purine base will bind with a pyrimidine base. More specifically, Adenine binds with Thymine, and Cytosine binds with Guanine.

Although this information is now considered fundamental biology, its importance wasn't fully understood until after James Watson and Francis Crick elucidated the structure of DNA in 1953. However, Chargaff's research in the 1940s was already indicating that the four bases paired in the manner described above. When Chargaff first decided to devote his laboratory to nucleic acids, he allowed a postdoctoral student named Ernst Vischer to choose his research topic from a list of suggestions compiled by Chargaff. Vischer decided to analyze the purines and pyrimidines in nucleic acids and went to work developing the chromatographic technique so crucial to isolating the bases. Although his approach was rather crude it ultimately proved effective, and Vischer achieved great success. His analysis showed that the amounts of Adenine and Thymine were about equal, and that the amounts of Guanine and Cytosine were also about equal. This led Chargaff to establish his primary rule: in a single molecule of DNA, Guanine/Cytosine = Adenine/Thymine = 1.

Chargaff's rules were officially announced in a lecture delivered in June 1949 and then published in May 1950. However, Pauling had heard about them much earlier — straight from Chargaff while on the *Queen Mary*. Oddly though, Pauling chose not to pay them any mind and thus missed out on a grand opportunity. On the contrary, Watson and Crick met with Chargaff in 1952 and combined his rules with Rosalind Franklin's beautiful X-ray crystallographic photos to correctly form a molecular model of the genetic material.

One can never know for sure why Pauling chose to ignore Chargaff, but it may have come down to the fact that he simply didn't like him. Chargaff was known to be a prickly figure. By way of example, in his 1978 memoir, *Heraclitean Fire: Sketches from a Life before Nature*, Chargaff called Watson and Crick "a variety act" and further described them as

"One [Crick] 35 years old, with the looks of a fading racing tout…an incessant falsetto, with occasional nuggets gleaming in the turbid stream of prattle. The other [Watson], quite undeveloped…a grin, more sly than sheepish…a gawky young figure. […] I never met two men who knew so little and aspired to so much. They told me they wanted to construct a helix, a polynucleotide to rival Pauling's [alpha] helix. They talked so much about 'pitch' that I remember I wrote it down afterwards, 'Two pitchmen in search of a helix.'"

In a 1995 talk, Crellin Pauling reflected on his father's chance interaction with this irritable character and wondered what might have been.

"Well, I'm sure many of you know that Chargaff had a reputation as a, well how do you put it politely, as a difficult personality. And what Daddy said to me was that he found Chargaff so unpleasant to be trapped on the *Queen Mary* with that he dismissed his work. And what he told me twenty-five years ago was that it may well be that if he had just read the papers, instead of having been trapped on the *Queen Mary* for five days with this guy, he would have thought more about what these numbers meant. It may well be […] that if he'd read Chargaff's papers in 1948 instead of crossing on the *Queen Mary* — if we'd crossed on the *Queen Elizabeth* — the story might have been different."

***

Arriving in Oxford at the start of 1948, the Paulings had ample time to settle in before Linus's first lecture on January 20. Shortly after unpacking, the family purchased bicycles for the whole group and took the opportunity "to peddle around the countryside." The Paulings also explored their new surroundings in England via a four-day car trip to Cambridge, Peterborough, Nottingham, Manchester, Chester, Shrewsbury, and then back to Oxford.

By February the family had assimilated well into their new temporary home, and Pauling's Oxford lectures were proving to be very popular — as he wrote to his Caltech colleague Carl Niemann, "My lectures have been going across well — there are 250 or 300 auditors still attending them."

His regular schedule — two lectures on the nature of the chemical bond at 5:15 Tuesdays and Fridays, plus a weekly afternoon seminar on inorganic chemistry each Wednesday at 2:30 — sounds light, but Pauling certainly wasn't taking it easy. Rather quickly, the visiting professor found that he was in high demand as a lecturer; not only were his Oxford presentations packed with standing room only, but he was also invited to speak all around England and Europe. Everyone from the Istituto Chimico to the Gesellschaft Deutscher Chemiker wanted to hear what this brilliant scientist and dynamic communicator had to divulge about their field.

The clamor for Pauling was such that he was forced to turn down many offers, but he still managed to give at least 39 invited lectures over six months, in addition to the three per week required by his Eastman Professorship. As he wrote to his close colleague Robert Corey, who was running the Pauling lab back in Pasadena,

"I am continuing to get along very well — perhaps being kept a little too busy, with so many extra lectures to deliver. However, I feel that when there is so much interest in what I have to say it is proper that I make the effort to say it."

Among those who had the opportunity to hear him speak were attendees of his three Scott Lectures in physics at Cambridge, as well as students and faculty at the British Undergraduate School of Medicine and many other universities across England and Scotland. Pauling notably gave three talks at University College London; the Sir Jesse Boot Lecture at the University of Nottingham (in which he correctly predicted both the basic manner in which genes act as templates for proteins as well as the means by which gene replication occurs); a Bedson Lecture at King's College London; the Liversidge Lecture for the Chemical Society; and the first Lyell Lecture at Oxford.

Famed molecular biologist Max Perutz remembered attending one of these talks, noting Pauling's ability to

"...reel off the top of his head atomic radii, interatomic distances and bond energies with the gusto of an organist playing a Bach fugue; afterwards he would look around for applause, as I had seen Bertrand Russell do after quoting one of his eloquent metaphors."

In short, he was the toast of the town. Priscilla Roth, Pauling's secretary during his time overseas, wrote in a letter that Pauling was "getting a royal welcome everywhere he goes." And despite the Paulings' initial disappointment in their lodging, which kept the couple from entertaining their English friends to the extent they had originally hoped, the pair nonetheless found themselves swept up in a social whirlwind, attending an event almost every night — be it a dinner, musical performance, or the ubiquitous English sherry party.

A prime example of this red-carpet treatment was the reception afforded Pauling during his Friday Evening Lecture at the Royal Institution in London on February 27. After a grand dinner, Pauling gave an hour-long talk on "The Nature of Forces between Large Molecules of Biological Interest" to an audience of men in tuxedos and women draped in furs and jewels. Biographer Thomas Hager described the evening as "an artifact from the days when the sciences were patronized like the arts...the scientific equivalent of playing Carnegie Hall."

***

Pageantry aside, the content of Pauling's major lectures was cutting edge. And as befitted his turn from structural chemistry to topics in molecular biology, his presentations typically brought to life new insights into the tiny world of molecules.

During this period, Pauling often began his lectures with a question: "What is it that defines living organisms as alive?" In developing his answer, Pauling would propose that it is molecular architecture that makes creatures unique and imparts upon them the properties that we identify as life. Expounding on this thesis, Pauling would speak of the wonders of the giant molecules that comprise living organisms and their special biological roles. In so doing, he would sometimes dig more deeply into specific topics, such as the oxygen-carrying proteins hemoglobin and myoglobin or the mechanics of catalysis by enzymes. Importantly, Pauling also remarked on the still mysterious question of the genetic material, echoing the common belief that "molecules of nucleoprotein" — instead of DNA itself — would "determine the characters of individual living organisms and [be] involved in the transmission of these characters to their progeny."

In a reflection on viruses, Pauling likewise questioned the boundaries of life. "Although these molecules may not ordinarily carry out the processes of respiration of air and metabolism of foodstuffs that we usually associate with life," he wrote, "they have one important property that causes us to regard them as living, the property of producing progeny." His focus returned to the triumph of the human body in a discussion of antibodies — the biological "police force" charged with identifying invaders, including viruses, and forming defenses against them.

At the end of this colorful journey through the body's biological and chemical networks, Pauling would typically conclude with a hopeful vision that an improved understanding of molecular architecture could be used to attack degenerative diseases, such as heart disease and cancer, in the same way that penicillin and the sulfa drugs had so effectively countered infectious diseases in the first half of the century.

***

Though most of Pauling's talks focused on scientific topics, he didn't forget about the peace pledge that he had made on the *Queen Mary*. Indeed, as he traveled around the UK, Pauling was frequently able to make forays into the world of politics and current events in speeches such as the "The Third Party Movement in the United States," presented in March to the English Speaking Union at Oxford University.

Somewhat radical in a time when Americans were fostering a mounting fear of communism and the Soviet Union, Pauling promoted a third political party outside of the Republicans and Democrats. Called the Progressive Citizens of America, this party identified itself as a "mild socialist organization" and supported Henry Wallace — Vice President and then Commerce Secretary during the Franklin Roosevelt administration — for President. Speaking to his British audience, Pauling's remarks painted a portrait of hysterical fears back home that were amplifying the emerging Cold War.

Socialism was lumped with communism in the minds of the American public, and Pauling noted that pressures on the third party were growing, with members and advocates of the movement being questioned and even losing their jobs.

Pauling was also critical of the Marshall Plan, which included $13 billion in U.S. aid for the European Recovery Program, and expressed worry about the fate of industry across the continent now that a decision had been "made that the Western European Union is to be capitalistic, patterned after past and present U.S., rather than socialistic, patterned after England." Clearly aware of rising tensions stateside, one might assume that Pauling found a degree of relief in being able to speak freely and share ideas with more sympathetic groups in the UK.

***

In his lab, a five-minute walk from his office at Balliol College (where he would sometimes boil eggs for lunch on an electric space heater), Pauling's research took a turn from the contents of his lectures — intermolecular forces and biological specificity — toward an increasing interest in the theory of metals. One of Pauling's to-do's for his Eastman Professorship was to revise the index for his new textbook, *General Chemistry*, but, as he wrote to Caltech colleague J. Holmes Sturdivant, "... it has turned out that I have devoted all of my time, and presumably shall continue to do so, to work on the theory of metals and intermetallic compounds."

In this he was aided by three other researchers: David Shoemaker and Hans Kuhn, both visiting from Caltech, as well as a young Ph.D. from the Netherlands, F.C. Romeyn. This group was proving to be highly productive, and in a March letter to Robert Corey, Pauling wrote of the impact that the change of setting was having in stimulating his thoughts.

> "I have been having wonderful success in my development of a theory of metals. I think that it has really been very much worthwhile for me to get away for this period of time, under circumstances favorable to my thinking over questions and trying to find their solution. The problem of metals has been on my mind for a number of years, and I haven't been able to leave it alone, so it is a good thing that I have now managed to get it solved."

This new theory of metals was an extension of Pauling's valence-bond approach to determining the structure of molecules, which he initially developed in the late 1920s and had expanded upon since then, to wide acclaim.

But during this time another chemist, Robert Mulliken (recipient of the 1966 Nobel Prize for Chemistry) had been steadily fostering and gaining traction with a rival approach, the molecular orbital theory. At the beginning of April, while his family enjoyed springtime in Paris, Pauling met head-to-head with Mulliken at a conference on Isotopic Exchange and Molecular Structure. An entire day of the meeting was devoted to a comparison of the two theories before a rapt audience of quantum chemists. Pauling had written earlier that Mulliken's molecular orbitals were confusing to students, but found at this meeting that, with more mathematics under their belts, advanced chemistry students were increasingly hungry for the more quantitative approach that Mulliken's theory offered. And indeed, as the years rolled by, consensus grew that a full understanding of the chemical bond would require engagement with both Pauling and Mulliken's ideas.

***

Sometimes flashes of insight come upon the great thinker at surprising times, and Pauling experienced just such a eureka moment in February during one of his regular Oxford lectures. As he wrote to J. Holmes Sturdivant,

> "I have just had a great stroke of luck. While giving my lecture on Tuesday I suddenly realized that a calculation about resonance energy of metals that I had just made and was reporting contained the key to the strange valence numbers and numbers of atomic orbitals and unused orbitals that have turned up in my theory of valency of metals."

Pauling subsequently worked out his ideas over several weeks and in pages and pages of careful handwritten calculations. In those notes, Pauling structures his thoughts by presenting a hypothesis about some aspect of metal theory, and then proceeds to calculate, revise, and recalculate until the theory either lines up with the X-ray diffraction data that he had on hand or is discarded. On one typical March day, Pauling was exploring several different approaches to intermetallic compounds. He began by writing,

"I shall now treat intermetallic compounds with my new ideas — resonance of bonds when an extra orbital is available, importance of n=1/2, 1/4 etc., concentration of bonding electrons into strong bonds (Zn-Zn, etc. as compared with Na-Na), transfer of electrons with increase in valence."

Hybrid orbitals, bond lengths, and the overall stability of structures were other items on Pauling's research agenda during that period. Some of this work would come together in an important paper that Pauling published in 1949, "A Resonating-Valence-Bond Theory of Metals and Intermetallic Compounds," which successfully extended Pauling's theory of resonance to the structure of, in particular, the iron-group transition metals.

Of course, not every idea is a winner, and a few theories led Pauling down the wrong path. In one set of notes, Pauling set out to, as he wrote, "consider sp hybridization — how can we set up a secular equation to give the results given by my bond-strength postulate?" After working through his process, Pauling found that "the ratio does not come out as desired. It is evident that my assumption that the energies can be taken proportional to 'bond strengths' is not right." But missteps such as these didn't deter Pauling from pressing on with his research. As he often said, "The best way to have good ideas is to have lots of ideas and throw away the bad ones."

***

Chemistry relies heavily on its own unique language and, as research advances, sometimes a brand new word is needed to describe an innovative concept. While tackling the nuances of metal theory at Oxford, Pauling wrote to Sturdivant about this problem, and in the process revealed the degree to which language truly mattered to him.

"By the way, I think that we should do something toward improving the nomenclature. For example, coordination number is an awkward and unwieldy expression — we need one short, precise word for this concept. Perhaps ligancy could be used. It would fit in well with ligand and the verb to ligate. We also need some general words to express the bonds between one atom and the surrounding atoms — we now use the word bond to refer both to the electron pair bond that is resonating around among alternative positions and to the fraction of an electron pair bond that is a portion to a particular position. I have also felt troubled about using the word position in this way — to mean the region between two atoms. If we do introduce any change in nomenclature, it must be very well thought out and must not involve too great a strain on the memory, or too great a departure from the past."

New advancements also call for innovations in instrument development and Pauling was in regular communication with his colleagues back home about tools that might be built to aid their work, including the specialized cameras used for X-ray diffraction of metallic crystals. He particularly admired the Cavendish's vast X-ray crystallography laboratory and was gaining new insights from reading British journals devoted to scientific instrumentation.

Pauling was similarly intrigued by the English system of graduate education, wherein students focused entirely on classwork during their first year and then spent practically 100% of their time on research during the two years that followed. As both a scientist and an administrator, Pauling was always looking to improve upon Caltech's existing programs. But as appealing as the English system was, Pauling acknowledged that, in implementing it, one would run the risk of not knowing whether a student was an apt researcher for their entire first year.

***

Though metals were consuming a good portion of his time during his fellowship at Oxford, Pauling's biological projects never strayed far from his thoughts. High on this list were the mysteries of proteins, whose structures and functions were slowly starting to be unraveled.

Pauling's interest in proteins was spurred in the 1930s when the Rockefeller Foundation announced that its grant money would mostly be devoted to investigations of the science of life. Early on Pauling set out to tackle hemoglobin, and though his affection for the molecule lasted for the remainder of his life, Pauling certainly didn't limit himself to the study of just one protein. Moreover, at a time when most were looking at proteins from the top down by trying to sort out the complicated data produced by their X-ray diffraction photographs, Pauling was working from the bottom up by seeking to determine the structures of individual amino acids, the building blocks of proteins.

One specific protein that kept coming back into view was keratin. In the 1930s the English scientist William Astbury had studied the structure of wool which, as with hair, horn and fingernail, is made up primarily of this enigmatic protein. Astbury proposed that the keratin structure was akin to a flat, kinked ribbon, but Pauling disagreed. "I knew that what Astbury had said wasn't right," Pauling recalled, "because our studies of simple molecules had given us enough knowledge about bond lengths and bond angles and hydrogen-bond formation to show that what he said wasn't right." Crucially though, Pauling didn't have a good alternative to offer. After reading Astbury's paper, he attempted to construct a model that matched the measurements dictated by the blurry X-ray diffraction images that Astbury had generated, but nothing seemed to work. Writing the project off as a failure, Pauling chose to set it aside and pursue other interests.

***

In 1945 Pauling found himself seated next to Harvard medical professor William Castle on a railroad journey from Denver to Chicago. Castle was a physician working on sickle cell anemia and the conversation that he shared with Pauling planted a seed in Pauling's mind about the cause of this debilitating disease. Castle mentioned that, in the bodies of those suffering from sickle cell anemia, red blood cells assume a sickled shape when they are in the deoxygenated venous system but retain their normal flattened disk shape in the oxygen-rich arterial system. Clearly this suggested that the oxygen content in sickle cell blood played a major role in its molecular architecture. By his own recollection, "within two seconds" Pauling deduced that the source of the problem must be a defect in the oxygen-carrying protein itself, hemoglobin. As he recalled in 1960,

> "...immediately I thought, 'could it be possible that this disease, which seems to be a disease of the red cell because the red cells in the patients are twisted out of shape, could really be a disease of the hemoglobin molecule?' Nobody had ever suggested that there could be molecular diseases before, but this idea popped into my head. I thought, 'could it be that these patients can manufacture a special kind of hemoglobin such that the molecules are sticky and clamp on to one another to form long rods, which then line up side by side to form a long needle-like crystal, which as it grows inside of the red cell becomes longer than the diameter of the cell and thus twists the red cell out of shape?'"

Amidst his residency in Europe, Pauling continued to develop this idea as a maestro from afar, directing scientists back at Caltech to search for differences in the hemoglobin of normal and sickled cells. In the meantime, he sought out and communicated new information gleaned from meetings such as the Barcroft Memorial Conference on Hemoglobin, held at Cambridge in June 1948.

Working in Pasadena, members of Pauling's research team, in particular Harvey Itano and S. Jonathan Singer, used electrophoresis and other methods to find, in the words of a 1950 Caltech press release,

> "...a difference — slight but still unmistakable — between normal hemoglobin and that of a sickle-cell anemia patient. Sickle-cell hemoglobin proved to have a greater positive electrical charge, under the proper chemical conditions, than did the hemoglobin from a normal person. Such a difference in electrical properties can only mean a difference in molecular architecture, in the way in which the hemoglobin molecules are constructed.

In other words, Pauling was right: sickle cell anemia was a molecular disease, and malformed hemoglobin was the cause.

In 1956 an English chemist named Vernon Ingram, using a new technique called fingerprinting, proved conclusively that sickle cell anemia was an inherited disease as well. Moreover, the illness was found to be caused by an astonishingly small change at the molecular level. In a 1976 film, physicist John Hopfield described it this way:

> "On the surface of the ten-thousand atom molecule, there is a slight change. A small group of a few atoms on the edge of the molecule is replaced by another small group of atoms. That's all that happens — an exchange of a few atoms. Yet it's enough to make people very ill. The effect of the change is to create a sticky point between an abnormal molecule and its neighbor, causing molecules to pile up on each other."

Although they weren't the first to think about diseases in terms of molecular aberrations, no one prior to the Pauling group had concretely demonstrated their existence. After their initial success, Singer and Itano continued to expand on the original research, eventually discovering a less severe form of the illness called sicklemia. The duo also described the manner in which sickle cell anemia is inherited. As such, not only did Pauling and his colleagues identify the exact source of the disease, but they also provided a link to genetics and confirmed Pauling's view that assessment on a molecular level can provide valuable information. Later, Itano would discover more abnormal hemoglobin molecules, and the molecular analysis of diseases would continue.

Since the Pauling group's breakthrough, many other illnesses have been categorized as molecular diseases: hemophilia, thalassemia, and muscular dystrophy to name a few. Thalassemia is also a disease of the hemoglobin molecule, but while sickle cell anemia is caused by the standard production of abnormal hemoglobin molecules, thalassemia conversely involves the abnormal production of standard molecules. More specifically, for patients suffering from thalassemia, the rate of production of a specific globin chain is decreased, which then results in the formation of abnormal hemoglobin molecules.

Pauling's conceptualization of sickle cell anemia as a disease of the hemoglobin molecule jump-started years of research on abnormal hemoglobins and opened many new doors in the study of inherited diseases. Although he wasn't directly involved in the discovery of the abnormal hemoglobin molecules specifically, Pauling's development of the concept of molecular disease was achievement enough to significantly raise his stature within the medical community and further cement his status as a major scientific figure.

***

While in England, Pauling had occasion to interact closely with a number of scientific greats. Among these were his close friend Dorothy Crowfoot Hodgkin, recipient of the Nobel Chemistry Prize in 1964 and known today as a pioneer in the development of protein crystallography. Likewise, Pauling conversed with Max Perutz, a protégé of Sir William Lawrence Bragg at the Cavendish Laboratory at Cambridge. Perutz would go on to discover the structure of hemoglobin and share the Nobel Prize for Chemistry in 1962. While fruitful in many respects, these interactions also served to heighten Pauling's feelings of urgency concerning the race to determine the structures of various proteins.

Bragg split the 1915 Nobel Prize in Physics with his father for their early development of X-ray crystallography and was a long-standing scientific rival of Pauling's. But it wasn't until Pauling saw, with his own eyes, the work being done at the Cavendish that he admitted to "beginning to feel a bit uncomfortable about the English competition." As he wrote to his colleague Edward Hughes back at Caltech,

"It has been a good experience for me to look over the X-ray laboratory at Cambridge. They have about five times as great an outfit as ours, that is, with facilities for taking nearly 30 X-ray pictures at the same time. I think that we should expand our X-ray lab without delay."

Seeing the Bragg group in action also prompted Pauling to start his researchers on a study of insulin, an arduous and complicated undertaking that required sample purification and crystallization prior to X-ray investigation. In relaying research findings from English scientists working on the topic to his partners back in Pasadena, Pauling intimated that

"It is clear that there is already considerable progress made on the job of a complete structure determination of insulin. However, there is still a very great deal of work that remains to be done, and I do not think that it is assured that the British school will finish the job. I believe that this is the problem that we should begin to work on, with as much vigor as possible..."

Little did Pauling know that, while lying in bed and using little more than a piece of paper, a pen and a slide rule, he would soon make a major breakthrough on his own.

***

It has been said that sometimes blessings come in disguise, and so it may be that we have the damp English weather to thank for the discovery of the alpha-helix — a fundamental and ubiquitous secondary structure pattern found in many proteins.

Pauling was plagued by sinusitis for much of his time in England, and for three days in March 1948 it became severe enough to put him in bed (later in life, Pauling would fondly recall that this was before he started taking vitamin C). Having grown tired of reading mystery novels, Linus asked Ava Helen if she would bring him some paper and his slide rule, at which point he started trying to figure out how polypeptide chains might fold up into a satisfactory protein structure.

His canvas an ordinary 8½ by 11-inch sheet of paper, Pauling's first step was to draw a series of bond angles and distances, as determined from previous X-ray crystallographic studies of polypeptides. Next, he folded the sheet along parallel lines into a sort of squared-off tube. Doing so allowed him to add in representations of hydrogen bonds, which the impromptu model suggested would form between amino acid residues and, as a result, hold the turns of the polypeptide together.

The model made sense and pretty quickly it was clear that Pauling had uncovered something important. As he later wrote, his folded creation "turned out to be the structure of hair and horn and fingernail, and [was] also present in myoglobin and hemoglobin and other globular proteins, a structure called the alpha-helix."

Pauling kept the breakthrough to himself until his return to the United States because something didn't match up quite right with the current laboratory data. Specifically, the turns of Pauling's helix didn't mirror the 5.1-angstrom repeat found in all of William Astbury's X-ray patterns. Pauling's structure came close, but it made a turn every 5.4 angstroms, or every 3.7 amino acid residues.

After his arrival home, and with the assistance of colleagues Robert Corey and Herman Branson, Pauling continued refining his alpha helix structure while also developing others, including the beta sheet. In the midst of this, the group's chief British rivals at the Cavendish Laboratory published a paper titled "Polypeptide Chain Configurations in Crystalline Proteins." The paper promised more than it delivered though, listing many possible structures, all of which Pauling found to be unlikely. The competition was still on.

Pauling was finally convinced to publish when he received word that a British chemical firm called Courtaulds had created a synthetic polypeptide chain that showed no sign of Astbury's 5.1-angstrom reflection in X-ray diffraction images. This was enough evidence for Pauling to decide that the 5.1-angstrom repeat was, perhaps, not a vital component of all polypeptide chains. And so it was that, in April 1951, Pauling, Corey and Branson published a landmark paper, "The structure of proteins: Two hydrogen-bonded helical configurations of the polypeptide chain," in the *Proceedings of the National Academy of Sciences*.

***

After devouring the Pauling group's results shortly after their publication, Max Perutz headed to the Cavendish lab to check the data himself. As he recounted later in life,

"When I saw the alpha-helix and saw what a beautiful, elegant structure it was, I was thunderstruck and was furious with myself for not having built this, but on the other hand, I wondered, was it really right?

So I cycled home for lunch and was so preoccupied with the turmoil in my mind that I didn't respond to anything. Then I had an idea, so I cycled back to the lab. I realized that I had a horse hair in a drawer. I set it up on the X-ray camera and gave it a two hour exposure, then took the film to the dark room with my heart in my mouth, wondering what it showed, and when I developed it, there was the 1.5-angstrom reflection which I had predicted and which excluded all structures other than the alpha-helix."

Subsequent trials with porcupine quills, synthetic polypeptides, hemo-globin and, for good measure, some old protein films that had been tucked away all resulted in the same conclusion:

"The spacing at which this [1.5-angstrom] reflexion appears excludes all models except the 3.7 residue helix of Pauling, Corey and Branson, with which it is in complete accord."

In communicating this to Pauling, Perutz added: "The fulfillment of this prediction and, finally, the discovery of this reflection in hemoglobin has been the most thrilling discovery of my life."

Perutz' recollection of this momentous experience also included a leg-endary interaction with his boss, whose retort lends insight into the scientific competition between the Cavendish and Caltech.

"So on Monday morning I stormed into my professor's office, into [William Lawrence] Bragg's office and showed him this [confirmation of the 1.5-angstrom reflection in the alpha-helix], and Bragg said, 'Whatever made you think of that?' And I said, 'Because I was so furious with myself for having missed that beautiful structure.' To which Bragg replied coldly, 'I wish I had made you angry earlier.'"

***

It wasn't until a year later that the mystery of Astbury's 5.1-angstrom reflection was fully solved. In 1952, on a visit to the Cavendish, Pauling met Francis Crick, at the time a graduate student soon to play a huge part in the discovery of the structure of DNA. The two maintained similar interests and during a taxi ride around Cambridge they found themselves discussing the matter of the alpha helix. "Have you thought about the possibility," Crick asked Pauling, "that alpha helixes are coiled around one another?" Whether Pauling had or had not considered this idea would later emerge as a point of contention, but Pauling remembered replying that he *had* because he had been playing with multiple higher-level schemes for his helixes including some in which the structures wound around each other.

When Pauling returned to Caltech he promptly set to work on the problem. Again with help from Robert Corey and Herman Branson, Pauling discovered a means by which the alpha helixes could wrap around each other in a coiled-coil "like a piece of yarn around a finger" to produce the problematic 5.1 angstroms found in Astbury's pictures of natural keratin. This addition to the alpha helix hypothesis built upon an undated idea of Pauling's that was written down in a travel journal that he kept during his stay in Europe. Those notes describe a structure that Pauling named "AB6" — six alpha helices (B6) coiled around a seventh (A).

Crick, in the meantime, was conducting a very similar study and learned from Peter Pauling — by then a graduate student at the Cavendish working alongside Crick and Watson — that the elder Pauling was also working on "coiled-coils" of the alpha helix. This news undoubtedly felt to Crick like Pauling had built upon the concept that Crick brought up in their conversation.

In a rush to be published first, Crick hurriedly finished his research and dashed off a note to the journal *Nature*, only to discover that Pauling's own manuscript had arrived just a few days before. However, in a surprise twist, Crick's manuscript was published before Pauling's, likely due to two factors: 1. his paper was shorter, and 2. it was sent with a cover letter from Perutz requesting high-speed publication.

The following month Pauling wrote a letter to Jerry Donohue, a former Caltech doctoral student who had worked with Pauling since the 1940s and was now stationed at the Cavendish on a Guggenheim fellowship. The communication was in reply to a letter that Donohue had written to Pauling reporting on Crick's *Nature* submission. In his response, Pauling explained that he remembered the taxicab conversation from the previous summer. Cognizant of the controversy brewing over the provenance of the coiled-coil idea, Pauling specifically wrote that the exchange with Crick was brief.

A few months later, in March 1953, Pauling wrote a similar letter to Perutz, but now containing more details on the matter. This time, Pauling explained that he had thought of coiled-coils prior to Crick's suggestion but had not fully fleshed it out, a claim perhaps supported by Pauling's travel journal.

Upon receiving a copy of Pauling's letter to Perutz, Crick offered his own version of the events that had transpired. By his recollection, the taxicab conversation was longer than Pauling remembered and coiled-coils were discussed in greater depth, thus leading him to the assumption that Pauling had built upon his ideas. This would have been fine, Crick wrote, had Pauling simply informed Crick, after which the two scientists could publish concurrently, giving credit where credit was due and bolstering each other's work.

Crick did, however, admit that Pauling's paper was more detailed and thorough than was his own and also came to different conclusions on key points. These factors were enough for both Caltech and the Cavendish to declare that Pauling and Crick had generated their ideas on coiled-coils independently of one other, if simultaneously.

***

In July 1948, Pauling's two fruitful terms as Eastman Professor were up. The family split their remaining time in Europe traveling in Amsterdam, Switzerland and Paris, and Pauling rounded out the summer by receiving yet another honorary degree, this time from the University of Paris. With eight months of productive and enlightening experiences in hand, Pauling and his family stepped aboard the *Queen Mary* once more on August 25, 1948, this time sailing westward toward home. Thus concluded what Pauling would later refer to as "one of the happiest years of my life."

Chapter 27

Visting Albert Schweitzer

Albert Schweitzer and Linus Pauling in Lambaréné, French Equatorial Africa, 1959.

Throughout his life, Linus Pauling was quick to acknowledge the work of other scientists, humanists and philosophers whose ideas had impacted his own. And though this list is legion, perhaps the one person — aside from his wife, Ava Helen — whom Pauling most frequently cited as being profoundly influential was fellow Nobel Peace laureate, Albert Schweitzer.

Born in Alsace in 1875, Schweitzer was a physician, theologian, musician and activist who is best known for having spent the bulk of his life leading a medical clinic in Lambaréné, French Equatorial Africa (present-day Gabon). Beginning in 1913 and excluding a hiatus during World War I and brief trips to Europe throughout his life, Schweitzer lived and worked at the compound, offering medical attention to the indigenous peoples of, in Pauling's words,

"...one of the most inaccessible areas of the world. An area heavily infected with sleeping sickness, elephantiasis, malaria, schistosomiasis, Framboesia, leprosy and many other terrible diseases, a large area without a single doctor."

In doing so, Schweitzer was motivated by a universal principle that he called "reverence for life," a basic ethical outlook that he applied to humans, plants and animals, and one that proved important to many thinkers including Pauling. As he would write in his 1923 book, *The Decay and Restoration of Civilization*,

"Reverence for Life affords me my fundamental principle of morality, namely that good consists in maintaining, assisting and enhancing life, and that to destroy, to harm, or to hinder life is evil."

Deeply troubled by the outbreak of World War I, Schweitzer pushed this idea in part because it was easy to understand but also because it was not attached to any form of religious dogma.

In 1953 Schweitzer was awarded the Nobel Peace Prize "for his altruism, reverence for life, and tireless humanitarian work which has helped making the idea of brotherhood between men and nations a living one." In addition to his work in Africa, Schweitzer was also a vocal critic of nuclear weapons testing and proliferation. This was an obvious point of intersection with Pauling, who would frequently refer to the idea of reverence for life — and, by extension, the minimization of suffering for all of Earth's inhabitants — in his lectures and writings.

Beginning in 1957 the duo began exchanging letters which, read today, serve as clear evidence of the respect that they held for one another. In the salutation to several of the letters the recipient is not specified by name but is called out simply as "Friend." Likewise, an early message from Pauling concludes, "Let me take this opportunity to express to you my deep feeling of affection and gratitude." Six years later, he would close a different letter with a similar sentiment: "It has been a great pleasure and an inspiration for me to be in touch with you over the years."

***

In the summer of 1959, Linus and Ava Helen took a trip around the world, visiting Norway, England, Sweden, Germany, Japan and, for the first and only time, Africa. The African spur of the journey was prompted by Pauling's desire to meet, dialogue with and pay homage to Schweitzer in Lambaréné, a community located in a rainforest on the river Ogooué less than 50 miles south of the equator.

The Paulings' visit to Schweitzer's clinic lasted ten days and is well-documented through two sources: a travel diary kept by Ava Helen and a type-written statement that Linus delivered at a press conference that he held upon returning to the U.S. Of these sources, Ava Helen's is far more evocative and provides a visceral sense of what it was like to travel in Sub-Saharan Africa in the late 1950s. In it, she notes that the last legs of their flight itinerary began in Douala (in present-day Cameroon) and included stops in "Bitan'" and Libreville before arriving at their final destination. Of the plane ride from "Bitan" she mentions that one passenger was loaded on a stretcher and that at Libreville "all got out and a new set came in including a number of French soldiers with frolicsome children." Once arrived, Lambaréné was a mix of beauty and chaos, replete with "goats, chickens, ducks, pigeons everywhere." Their first night's sleep was restless, "with babies crying and baby kids too. It is hard to tell which sometimes." The next day she observes beautiful birds and butterflies in one paragraph, and "a wee baby dying with froth on the lips" in another.

The days that followed were filled with long walks, theological discussions, singing and conversations with an international group of visitors who had also found their way to Schweitzer's compound. Halfway through their stay, Ava Helen mentions that Linus "gave a talk after supper on 'Biological Specificity' in German. Schweitzer slept most of the time. [...] [N]urses and doctors are always very tired at dinner."

Throughout her journal, Ava Helen also continued to note aspects of daily life that she found compelling. Some selections of what she recorded are as follows:

---

* Sic? It seems likely that Ava Helen was actually meaning to refer to the city of Bata on the coast of present-day Equatorial Guinea, a country known in 1959 as Spanish Guinea.

- "Dr. Catchpool says that the natives say that there is a telephone in the head which tells the legs, hands, etc. what to do."
- "Gave some washing to be done. Use charcoal irons — that is, an iron filled with charcoal."
- "Coffee is roasted over a fire in a log cylinder turned by hand by a native. 5-inch diameter, 2 feet high, 2 feet long."
- "Saw cut up crocodile (small) by hospital — being cut up and divided by natives."
- "One nurse takes the baby gorilla and baby white-nosed monkey in her room at night and the gorilla makes a frightful mess. I haven't discovered where the baby chimpanzee sleeps but no doubt in some kindhearted nurse's room for they say they are too young to sleep outdoors."

Elsewhere she documented their host's affectionate relationship with a "wild pig who slept on his porch, but Schweitzer had to put him to bed — fluffed his cushion and then played Brahms' lullaby! Schweitzer said pigs very intelligent."

***

Her husband's more formal remarks offer a less colorful view of the trip but are insightful in different ways. Reporting out on his activities, Pauling begins as follows.

> "I discussed the new developments in the field of the hereditary hemolytic anemias with Dr. Schweitzer and the members of his medical staff, and also presented a conference on molecular biology. In addition, I spent considerable time during the ten-day visit in discussions with Dr. Schweitzer about fundamental philosophical, ethical, and moral questions, and in particular about the problems of nuclear war and the testing of nuclear weapons."

Of his host's daily routine, Pauling observes,

"Dr. Schweitzer does not do very much himself in the way of giving medical treatment to the patients in his hospital in Lambaréné. He devotes part of his time to supervising his staff of six doctors and about fifteen nurses and to supervising the general activities of the hospital, with its hundreds of patients and their relatives. He also devotes several hours a day to his correspondence with people in nearly all the countries in the world. He puts in two or three hours a day reading newspapers and journals and making notes about news items and scientific items relating to nuclear weapons tests, radioactive fallout and its effects, and the nature of nuclear war, as well as to international problems in general."

One can conclude that Pauling also took advantage of his time with Schweitzer to more thoroughly interrogate his thoughts on the moral philosophy that both men found so important.

"Dr. Schweitzer clarified for me some questions about his system of philosophy and ethics, including his principle of reverence for life. He said that this principle did not require that every action that might cause loss of life to an animal need inevitably to be avoided. For example, the principle did not prevent the use of animals in medical research. He said, however, that according to this principle the investigator had the obligation to consider every single experiment, and to ask himself whether the experiment was important enough, in its promise of providing information that would decrease the amount of suffering in the world, to justify the sacrifice of an animal. [...] He said that it was important that each person think about these questions himself, and not accept decisions from people in authority, either in organized religions or in government. In application to medical research, he said that this principle required that the investigator think about each experiment, and not simply apply a rule that had been laid down by the director of the laboratory or some other authority.

***

Schweitzer would pass away in September 1965, but his influence on Pauling would last for many years more. One immediate outcome of the Paulings' trip to central Africa was the connection that they made with Frank Catchpool, a physician who spent several years at the compound and eventually became Schweitzer's chief of staff. In 1960, the year after the Paulings' visit, Catchpool relocated to Pasadena at Pauling's invitation, and worked at Caltech with Pauling on the theory of anesthesia. In later years, Catchpool was also involved with medical outreach activities sponsored by the Linus Pauling Institute of Science and Medicine.

Catchpool's friendship with Pauling was such that he was among the small number of speakers selected to offer remarks at Pauling's funeral in late August 1994. At a conference held a few months later, he shared these memories:

> "Ava Helen and Linus used to walk around hand-in-hand [in Lambaréné]. We at the hospital all thought that this was bizarre, a middle-aged couple walking around holding hands all the time. And yet I realized that that was how they had been and would be throughout their lives. They had this tremendous, affectionate, and warm loving relationship. They wanted to be in physical contact with each other all the time. On one occasion [in California], Ava Helen asked me to go to the airport and meet Linus, and she said, 'By the way, he is expecting me, and he won't be very pleased to see you.' At the airport Linus was quite put out. He did not want me to drive him up to Caltech. He wanted Ava Helen."

In 1964 Pauling and Catchpool began collaborating on a manuscript, "Albert Schweitzer, Physician and Humanitarian," that was penned to commemorate Schweitzer's 90th birthday and published shortly after he passed away. In it, they reaffirmed the esteem that they held for their mutual acquaintance, placing his name among the ranks of a few others who had profoundly impacted Pauling's world view. "Of the thousands of millions of human beings who have lived during the first half of the Twentieth Century," they wrote, "we may expect that the memory of only a few will be preserved in history."

"[O]f Einstein, whose new ways of looking at the world brought about a revolution in scientific thinking; of Bertrand Russell, who by application of his incisive intellect brought clarification to mathematics, philosophy, and politics; and, with little doubt, of Albert Schweitzer, who will be remembered as an outstanding musician and musicologist, philosopher and moralist, physician and humanitarian, and leader of and active participant in the effort to save civilization from destruction in a nuclear war."

A decade later, in September 1975, Pauling gave the keynote address at the Schweitzer Centennial Birth Anniversary Commemoration Symposium, which was held in Tokyo. After discussing the evolution of Schweitzer's work and principles, Pauling offered his thoughts on a number of contemporary issues including the depletion of the ozone layer, declines in whale populations, economic inflation, and the need to invest in solar, wind, wave and tidal energy sources. He concluded with a call that once again harkened back to a man and to ideas that had motivated his activism for many years.

"Now is the time for us to change from our immoral course, from our dedication to the archaic institution of war, to a policy of peace and rationality and morality. We must follow the teachings of Albert Schweitzer, and bend all our efforts toward achieving the goal of a world of justice and morality, a world in which the resources of the earth and the fruits of man's labor are used for the benefit of human beings, permitting each one to lead a good and full life, a world of freedom and dignity, a world in which idealistic young people will be glad to take part. I believe that we can achieve that goal."

Chapter 28

Pauling and the Soviet Union

Pauling in lecture after receiving the Lomonosov Gold Medal, Moscow, USSR, September 25, 1978.

Linus and Ava Helen Pauling visited the Soviet Union six times between the years 1957 and 1984. Generally speaking, these trips were motivated by scientific interests, though sometimes Linus also spoke on peace issues and the need to cease nuclear tests. Unlike many of his peers, Pauling did not see the Soviet Union as a monolithic threat, but chose to view it, at least in part, as a necessary partner in working for peace. In addition, Pauling regarded most of the Soviet scientists with whom he interacted as genuine colleagues who shared his passion for scientific inquiry. Unfortunately, Pauling's cordial relations with Russian academics caused others in the United States to be suspicious of his own beliefs and associations during the decades of the Cold War.

For those inclined to criticize Pauling, one early affiliation that raised eyebrows was the National Council of American-Soviet Friendship, which Pauling joined out of a hope that it might live up to its name. Pauling made this clear in a letter to the Council, in which he implored the group to assist in establishing scientific links between the two countries, particularly with respect to chemistry and medicine. Pauling was also invited to attend meetings of the Russian-American Club of Los Angeles. At one such gathering, in November 1945, he delivered a speech encouraging mutual cooperation to attain peace between all nations. Pauling additionally participated in events sponsored by Progressive Citizens of America, a group considered by some to be communist.

Whether or not this meant that he agreed with communist ideology was a matter of public debate during multiple periods of Pauling's life. In a November 1950 appearance before the California State Investigating Committee on Education, Pauling formally stated for the record that,

"I am not a Communist. I have never been a Communist. I have never been involved with the Communist Party."

But as he grew as an activist, Pauling became more interested in economic theory and increasingly came to denounce the ills of wealth disparity and the forces that drive it. By the late 1960s, Pauling was including the idea of a logarithmic "wellbeing index" in some of his talks and discussing the benefits of a transfer to the world's poor "of part of the income of the unconscionably rich 0.1 percent." One might reasonably assume then that Pauling was comfortable with left-leaning economic ideas more classically associated with socialism.

That said, Pauling harbored no illusions about a communist worker's utopia and was well aware of human rights abuses perpetrated by the Soviet state. On the world stage, he believed the government of the USSR to be largely recalcitrant but thought if the United States were to take the first step towards initiating peace, positive momentum would follow. More than anything though, Pauling firmly believed that differing national ideologies should not affect scientific relationships between the United States and the Soviet Union, and this was a point to which he would return again and again.

***

The nuances of Pauling's position were lost on many though, and over the course of the 1950s he increasingly became the subject of false public claims as well as a sharpening focus from the FBI. Near the end of 1952, a former Communist Party USA functionary name Louis Budenz declared before a Congressional committee that Pauling was a "concealed communist" who had made monetary contributions to the party. This charge certainly did nothing to dampen the FBI's existing interest in Pauling's activities, but the Bureau had trouble finding sources other than Budenz who would identify Pauling as a past or present Communist Party member. In the meantime, Pauling forcefully denied Budenz' allegations, calling his accuser a "professional liar" and stating that he was neither a member nor a contributor to the party, but that he was an advocate for the inclusion of Soviet scientists in international conferences and symposia. In the climate of the era, however, even this level of support was seen by some as grounds for reprimand.

Another action that contributed to suspicion of Pauling was his January 1953 appeal to the White House urging the commutation of the death sentences handed down to Julius and Ethel Rosenberg, two Americans found guilty of spying for the Soviet Union. Pauling was deeply interested in the Rosenberg case and closely studied the details underlying their sentencing. His actions on their behalf were based on his analysis of these details, one that led him to conclude that the death penalty judgement was extreme and unjust. But no matter the reason, this continued willingness to contradict the mainstream made it difficult for him to convince others of his trustworthiness and patriotism. As a result, institutions became increasingly shy about affiliating with Pauling and overseas travel became more and more difficult.

Though the climate in the United States was slow to change, Pauling's receipt of the Nobel Chemistry Prize in 1954 elevated his status as a public figure and afforded him a bit more space to voice his opinions and travel freely. Finally, in 1957, he made his first trip to the USSR where he was at last able to meet with many scientists who had themselves been restricted from participating in international meetings during this period of pitched international tensions.

**

The Paulings' first visit to the Soviet Union was initiated by Moscow State University biochemist A.I. Oparin, who invited Pauling to deliver a paper at the International Symposium on the Origin of Life on the Earth. At first, the Paulings were hesitant to say yes, due to high costs and questions about their ability to obtain the required visas. But ultimately these issues were resolved and they accepted the offer, excited by the prospect of the symposium and the opportunity to explore a new part of the world. In August they arrived in Moscow, where Pauling presented a paper titled "The Nature of the Forces of Operation in the Process of the Duplication of Molecules in Living Organisms."

During this stay in Russia, Ava Helen kept a diary in which she recorded her experiences, including trips to the Bolshoi Theatre to see a ballet, an opera, and an operetta. Other noteworthy excursions included the treasure house of the Kremlin, Cathedral Isaac, the Pushkin Museum and a Russian kindergarten. Of this last visit, Ava Helen noted that the children were presented in such a rigorously organized fashion — specifically in their music and gymnastics classes — that she had a hard time enjoying the event. Her impression of the Youth Festival parade was more positive, and she made particular note of spectacular performances and a breathtaking fireworks display.

The Paulings also made time to dine with Oparin and other colleagues at the Savoy Hotel in Moscow, and toured the Institute of Chemical Physics — housed in a repurposed monastery — to have a look at N.N. Semenov's laboratory. This was just one of several laboratory stops on their itinerary, which also featured outings to the nuclear physics lab in Moscow, Oparin's lab, the Orekhovich Lab, and the Tatyveskis Geo-Chemical Institute Lab.

**

Upon returning to the U.S., Pauling extended an offer to Academician V.N. Orekhovitch to come to Caltech and deliver a guest lecture on procollagen, which was Orekhovitch's subject of expertise. To pave the way, Pauling offered an honorarium of $250, which was a large sum in absolute terms but especially so for Soviets, for whom "hard currency" like U.S. dollars had tremendous spending power.

Orekhovitch readily accepted this invitation and the two initiated the process of requesting travel papers through the U.S. State Department. However, in November 1957, Pauling received an urgent telegram from Orekhovitch in which he stated that he was unable to obtain the mandatory documents and that he desperately needed Pauling's help. When Pauling received the telegram, he immediately began trying to figure out why the visa had been denied. One clue was a recent article that he had read that hinted that Pasadena — among other cities — had been declared to be off-limits for Soviet travelers. Pauling was not certain of this information though, or if it was the reason for Orekhovitch's denial.

Seeking answers, Pauling contacted multiple colleagues across the country asking whether or not they had encountered similar difficulties. At Harvard University, Paul Doty replied that he had recently become aware that travel by Soviets was not permitted to the entire state of Massachusetts *except* for Cambridge, where Harvard was located. This detail seemed to confirm that Pauling's initial fears were true, that the State Department had established certain areas of the United States as out of bounds for Soviet visitors. A subsequent letter from the government confirmed that Pasadena, San Francisco, and Los Angeles were officially closed to anybody holding a Soviet passport. Outraged by this action, Pauling called State Department official Lawrence Mitchell, urging him to make an exception for Orekhovich. Mitchell denied the request, noting that a deviation of this sort would have "little effect in applying pressure on the Russian Government."

Pauling next appealed to the Secretary of State, claiming that the office's policy "gives the Russian scientists who come to the United States a false impression — the impression that we are a police state, where scientists are not free to talk with other scientists, but are ruled by the Department of State." Bothered by a nagging feeling that Pasadena's inclusion on the list of banned cities was directed specifically at him, Pauling also formally objected to "the discriminatory action" allowing Soviet visitors to visit Harvard or UC-Berkeley, but not Caltech. A Harvard colleague echoed Pauling's feelings, describing the situation as "embarrassing and frustrating" in his own letter to the State Department.

About a month later, Pauling received a reply from Frederick Merrill, Director of the East-West Contacts Staff. In it, Merrill reiterated Lawrence Mitchell's original argument and added that it was U.S. policy to increase contacts with peoples of Eastern Europe, but that this aim had been rejected by the Soviet Union. As such, and until talks on the matter could be revived, the United States had to restrict Soviet travel as a way of pressuring the USSR into negotiations. Merrill also explained that Pasadena was deemed to be a strategic city because of its geographic location; that Caltech was off-limits simply because it was a part of Pasadena; and that all Soviet citizens, not just scientists, were excluded from traveling there.

What the letter did not do is clarify why Pasadena's location made it a strategic city, nor did it provide any details on why Berkeley or Cambridge were not geographically strategic. Rather, the suggestion was merely that Orekhovitch had been denied a visa as a matter of routine and that the ban on travel to Pasadena was logical due to the declared importance of the city. Orekhovitch did eventually made it to the U.S. but was unable to visit Pauling or Caltech.

***

On June 20, 1958, almost exactly 25 years after being inducted into the National Academy of Sciences, Pauling was unanimously approved for inclusion in the Akademia Nauk (Academy of Sciences) of the USSR, becoming the second American to receive this honor. Founded in 1724 during the reign of Peter the Great and charged with conducting national research and overseeing scientific publications, the Academy had attained a position of major importance in Soviet society and its domestic members were among the country's highest paid workers.

While the responsibilities of his membership were purely honorary and the Academy insisted that he was being recognized for his scientific accomplishments, many media outlets, including the *New York Times*, suspected that the decision had been politically motivated. Responding to these insinuations, Pauling noted that the Soviets "have been strongly critical of my work in the past," and pointed out that, in 1951, the Academy had deemed his theory of resonance to be "reactionary" and "bourgeois." In the years since, Pauling supposed that the Soviets had "learned that you can't mix politics up with science."

Pauling was well-aware that his acceptance of the Academy's nomination would garner criticism, but for him it was worth it to take a stand in favor of academic freedom. In a statement to the Associated Press, he affirmed his strong belief "in the importance of improving international relations in every way" and expressed enthusiasm at the idea of "becoming better acquainted with the scientists in the USSR." The letters of congratulation that he received from his colleagues indicate that this point of view was shared by many.

Pauling did not travel to the Soviet Union to accept his award, but he did address the topic of his membership in several lectures delivered during the summer of 1958. One talk, given at Antioch College on the day of his nomination, used the honor as a rhetorical starting point for a deeper discussion of a path toward reducing the risk of nuclear war. Afterward, the president of the college sent Pauling a follow-up letter indicating that local journalists had mostly accepted Pauling's ideas on merit, though the *Dayton Daily* had refused to report on the event at all due to Pauling's membership in the Soviet Academy.

In addition to Pauling, one other American, scientist and former Johns Hopkins University president Detlev Bronk, was added to the Soviet Academy in 1958. So too was Bruno Pontecorvo, a highly regarded Italian-born physicist, who was living in the U.S. and working on atomic research when he disappeared in 1950. Considered missing for several years, Pontecorvo eventually appeared on Russian television, at which point it became clear that he had defected and that he had, in fact, risen to a position of authority within the Soviet nuclear development program. Confirmation of Pontecorvo's defection came as a shock, and some feared that Pauling would follow suit. Needless to say, this did not come to pass. Pontecorvo, on the other hand, remained in the USSR and worked under its auspices until his death in 1993.

***

The Paulings made their second visit to Moscow in November 1961, where Linus gave a speech titled "World Cooperation of Scientists" at a conference hosted by the Soviet Academy in celebration of the 250th birth anniversary of famed Russian scientist and writer, M.V. Lomonosov. In his remarks, Pauling discussed the approaches taken by Lomonosov and other Russian scientists in their investigations on the structure of matter. He also commented on contributions that Soviet scientists had made toward world peace and urged the Soviet state to reconsider its official stance on resonance theory.

Pauling expounded on the resonance controversy at a later talk given at the Academy's Institute for Organic Chemistry. Pauling's theory used quantum mechanics and wave functions to model a hypothetical structure of a molecular system as expressed as a sum of wave functions. His presentation of this idea during the 1961 trip was particularly important because, ten years earlier, the Institute of Organic Chemistry of the Academy of Sciences, USSR had formally rejected the concept as "pseudoscientific" and "hostile to the Marxist view."

In response, Pauling had written to the Akademia Nauk arguing in support of his work and asking the organization to reconsider its decision. In 1954 the group eventually consented to a written debate between Pauling and Professor N.D. Sokolov, but this exchange never took place. By 1961, when Pauling gave his lecture, the political controversy over the theory had begun to die down and Pauling was able to constructively engage with his Soviet colleagues on some of its more technical details.

During this trip, Pauling also gave a talk in which he urged the Soviet Union to end its nuclear testing program and reduce its stockpiles of nuclear weapons. He also attended a panel discussion at which he once again called on the Soviet government to halt all nuclear tests.

Ava Helen attended these events with her husband, while finding time for adventures of her own. As before she kept a diary, portions of which are dedicated to the "wild rides" that she experienced with her guide, Angella Gratcheva. Apparently Gratcheva drove very erratically, and while navigating the Russian roads commonly recited poetry, sang songs and engaged in animated conversations with Ava Helen. Her driving was so unpredictable that, at one point, the police stopped them, a "misunderstanding" that the guide cleared up with more animated speech. From scientific controversy to peace activism to white knuckle transportation, Russia was proving to be an interesting place indeed.

***

In 1967 Pauling was invited back to the USSR by the Akademia Nauk to join their general special meeting session in commemoration of the 50th anniversary of the Great October Socialist Revolution. He was not able to attend but did agree to participate in the publication of a text being sponsored by the Academy, *Functional Biochemistry of Cell Structures*, for which he submitted a piece titled "Orthomolecular Methods in Medicine." The paper discussed Pauling's growing interest in the molecular basis of health and disease and provided both a rationale and examples that supported an orthomolecular approach to medicine. The piece was published in 1970, and the volume was edited by Pauling's old friend, A.I. Oparin.

That same year, Pauling was awarded the International Lenin Peace Prize, the Soviet Union's most prestigious award for humanitarian efforts. Pauling was the fifth American to receive the prize since its inception in 1949, following the likes of civil rights activist W.E.B. DuBois and artist Rockwell Kent. The award was presented to Pauling by physicist Dmitry V. Skobeltsyn at a public ceremony held inside the Soviet Embassy in Washington, D.C. In his acceptance address, Pauling emphasized the need to achieve global peace through international law and expressed growing confidence in the world's ability to facilitate international relations without reliance on nuclear weapons. Despite the prestige of the occasion, only a small number of people were invited to attend the ceremony itself. But that did not mean that the prize went unnoticed, and Pauling received a great many letters of congratulation once word of his accomplishment circulated through the media.

In addition to an engraved medal bearing the image of Vladimir Lenin in profile, the prize came with a 25,000-ruble honorarium. But because rubles were valueless outside of the Soviet Union at the time, an interesting conversation quickly ensued about how the monetary award could be converted into usable currency. The situation was eventually sorted out when Linda Kamb, Pauling's daughter, visited the Soviet embassy shortly before the award ceremony was to take place. Upon arriving, Linda spoke with Henry Kissinger, who was serving as the U.S. National Security Advisor at the time, and who also happened to also be at the embassy that day. Linda met as well with the Soviet Ambassador to the United States.

In these conversations, Pauling's daughter asked the two men about her father's unusual problem of not being able to spend the rubles being given to him. The officials subsequently conferred and decided that the prize money could be converted into U.S. dollars at a rate of one ruble to $1.10, with the exchange happening within the embassy. This quote was apparently satisfactory, and a delve into Pauling's financial documents for the year 1970 indicates that he did in fact utilize the currency conversion option that his daughter had helped arrange.

***

In 1975 Linus and Ava Helen made a return visit to the Soviet Union to help celebrate the 250th anniversary of the Akademia Nauk. Linus was one of 27 Americans invited to participate in the event which, according to the Associated Press, had been delayed for more than a year "to head off embarrassing discussions on intellectual freedom and Jewish emigration."

While in Moscow, Pauling was asked to appear on a television program, *The 9th Studio*, alongside Bulgarian scientist Angel Balevski, Soviet physicist Nikolay Basov, and Soviet philosopher Dzermen Gvishiani. Broadcast to a potential audience of 80 million people throughout the Eastern bloc, the roundtable was asked to discuss modern science, the prohibition of nuclear weapons, and the struggle for peace.

By now the bulk of Pauling's interactions with Soviet colleagues focused on the pursuit of world peace and disarmament, but many were also intrigued by his work with vitamin C. His prominence in this area sparked an invitation to return to Moscow in 1978 to give talks on ascorbic acid and chemistry. While there, he spoke to the Shemyakin Institute of Bioorganic Chemistry on his growing interest in using vitamin C in the treatment of cancer. He presented again on vitamin C to the USSR Academy of Sciences and attended the International Symposium of Frontiers in Bioorganic Chemistry and Molecular Biology.

During this visit, Pauling was also awarded the Lomonosov Gold Medal, the highest award conferred by the Academy of Sciences of the USSR. Officially, the prize was given for his outstanding achievements in chemistry and biochemistry though, as stated in a letter from Soviet poet Mikhail Vershinin, the award also sought to recognize Pauling's work as a "knight of peace and progress." Pauling accepted the medal by giving an address that detailed the specifics of his most current work. Titled "Ortho-molecular and Toximolecular Medicine Compared," Pauling's lecture was delivered to an audience of more than 300 people, including 70 scientists visiting from other countries.

Later, Pauling gave another talk on a completely different area of interest: "The Nature of the Bond Formed by the Transition Metals in Bioorganic Compounds and other Compounds." And as usual, the Paulings did their best to take in as much culture as possible, including a sightseeing trip to Uzbekistan, where they visited the cities of Tashkent, Samarkand, Bukhara and Khiva.

***

Just days before accepting the medal in Moscow, Pauling was handed an untranslated letter written by Andrei Sakharov, the famed Soviet dissident who had received the Nobel Peace Prize in 1975 "for his struggle for human rights in the Soviet Union, for disarmament and cooperation between all nations." In the letter, Sakharov urged Pauling to use the Lomonosov cere-mony to speak out against the wrongful imprisonments of Soviet physicist Yuri Orlov, mathematician Alexander Bolonkin, and biologist Sergei Kovalev. "I am convinced that today you share the concern of many Western colleagues over violations of human rights in the whole world," Sakharov wrote, "and particularly in the Soviet Union."

Kovalev's case was representative of the persecution suffered by many scientists who spoke out in favor of reform. A member of the organization Action Group for the Defense of Human Rights in the USSR, Kovalev's activ-ities had garnered a sentence of seven years in a hard labor camp followed by another three years in a standard prison.

Pauling was caught off-guard by Sakharov's communication which, unbeknownst to him, had also been released to the media. Pauling did not address the contents of Sakharov's request during his trip and when he returned to the U.S., he found that his reputation had suffered for this inaction.

In a letter to the editor of *Physics Today* authored a month later, Pauling defended himself, noting that,

"I had signed statements and had written letters about scientists and other people whose rights have been reported to have been violated by the USSR government and other governments, although I could not remember with confidence whether or not I had taken action about these three men. I added that all governments are immoral, and cited the example of the United States government, which in 1952 refused me a passport and thus prevented me from participating in the two-day symposium in London that had been organized by the Royal Society..."

A response to Pauling's letter by I.I. Glass of the University of Toronto called him to task for comparing "what happened to him during the McCarthy twilight era with the darkness in which many of our colleagues in the USSR are living today." Pauling offered this reply:

"All governments are immoral. But I agree with Glass that the immorality of the government of the US is different from that of the government of the Soviet Union. Also, I am concerned about Sakharov and other scientists in the Soviet Union. My letter to *Physics Today* expressed my concern, although only briefly, and expressed also another concern, about how the Sakharov problem is being handled. I wish that I knew more about the whole matter."

Time moved forward but Sakharov refused to let the issue fade. Two years later he sent several letters — including a handwritten message handed to Pauling via his son-in-law, Yefrem Yankelevich — repeating the same urgent call to support the three Soviet scientists. Some of these letters even included personal statements from the scientists themselves, and Yankelevich appears to have added updates on their lives. For Sergei Kovalev, the situation seemed to be deteriorating rapidly as he was reportedly suffering from tuberculosis as well as partial paralysis.

In addition to the personal handwritten notes, Sakharov once again released a separate public letter to Pauling, which was published in translated form in *Freedom Appeals* magazine. This time, Sakharov sought to enlist

Pauling's support for the release of Kovalev as well as his daughter-in-law, Tatiana Osipova.

While Sakharov's initial correspondence had been fairly dry, this latest public letter was more emotional. Addressing Pauling, Sakharov wrote,

"I know neither your political views nor the extent to which you may be sympathetic to the Soviet regime. But what I am asking of you is not politics. To save honest and courageous people who are about to perish is the duty of humaneness and a question of honor. Please make good use of your prestige; appeal to Soviet leaders and to the leaders of Western countries. Please do what you can."

This new approach seems to have made an impact, if in an oblique way. Even though Pauling once again did not publicly act to free the imprisoned Soviet scientists — Sergei Kovalev was eventually released by Mikhail Gorbachev in 1986 — he did eventually come to the aid of a different Soviet intellectual: Andrei Sakharov himself.

***

In 1980, just five years removed from his receipt of the Nobel Peace Prize, Sakharov was sent into exile in the city of Gorky and was routinely subjected to isolation and harassment in the years that followed. In April 1981, Pauling and Gerhard Herzberg, a fellow Nobel Chemistry laureate, sent a letter to Soviet Premier Leonid Brezhnev and the Canadian Ambassador to the Soviet Union demanding the "end of [Sakharov's] confinement." In this message, Pauling and Herzberg explained that their letter was not a publicity stunt, and that there would be "no communication about it to the 'media.'"

Instead, the authors put forth that, "every society needs its critics if it is to diagnose successfully and overcome its problems [...] Surely your nation is mighty enough to tolerate a patriotic critic of the stature of Andrei Sakharov." The authors concluded by harkening back to the dark years of gulags and secret police, exhorting to Brezhnev that "Surely you do not want a return to Stalinism."

In 1981, after a year in exile, Sakharov began a hunger strike to demand that his daughter-in-law, Liza, be permitted to move to the U.S. to be with her husband, Sakharov's son Alexei. In addition to an open letter, which was broad and impersonally written, Sakharov sent a direct message to Pauling, imploring him specifically to support his daughter-in-law's emigration.

Ultimately the campaign worked and before the year had concluded Liza was granted an exit visa to live in the United States. But the victory did not come without cost: as a penalty for having gone on the hunger strike, Sakharov was stripped of all his accolades by the Soviet government. In reaction to this, an international campaign initiated by the Norwegian Helsinki Committee — a non-governmental organization dedicated to ensuring that human rights are respected and practiced worldwide — solicited prominent scientists to urge Premier Brezhnev to release Sakharov from exile and allow him to return to his home in Moscow. Pauling promptly complied, arguing for Sakharov's release on the grounds of human rights violations. Delivered in August 1981, the letter apparently fell on deaf ears.

By 1983 Sakharov had been in exile for three years and his health was starting to decline. Shifting tactics in an attempt to secure his release, Pauling sent a telegram to the Soviet Academy of Sciences and to new Soviet Premier Yuri Andropov, offering Sakharov a job as a research associate in theoretical physics at the Linus Pauling Institute of Science and Medicine in Palo Alto. In discussing this offer, Pauling told news reporters that "I feel sympathy for Sakharov as a person who gets into trouble for criticizing his own country." Sakharov subsequently announced that he was willing to accept the position and emigrate, but his application for an exit visa was declined, with officials citing the need to protect "state secrets" connected to his scientific work on the hydrogen bomb during World War II.

In 1986 Sakharov was finally given his freedom, a beneficiary of the Gorbachev regime's policies of *glasnost* and *perestroika*. The famed scientist and activist promptly returned to Moscow, and in 1989 he died in his home. While there is no documentary evidence that their relationship advanced in the years following his release, it is worth mentioning that Pauling received an advance copy of Sakharov's memoir prior to its posthumous publication in 1990. It is not clear if Pauling requested the copy, but his receipt of the volume is a suggestion that, even in death, Sakharov would not soon be forgotten by his American comrade.

***

Pauling visited Moscow again in 1982 for ten days to attend the 60th anniversary celebration of the founding of the USSR. This time, he was the only American invited. The trip came near the end of a long run of international travel that had been scheduled, in part, to distract him from the death of Ava Helen, who had passed away the previous December. Pauling's pocket diary from these travels is suggestive of a wistful mindset; of his arrival in Moscow, he notes only the landing time and a "Russian girl with a Barbie doll."

In between this visit and his next trip in 1984, Pauling continued to think about the political and cultural norms developing in Moscow, writing a support notice for the book *Give Peace a Chance: Soviet Peace Proposals and U.S. Responses* and attending a conference, "What About the Russians?" that took place at his alma mater, Oregon State University. As a past recipient, he also nominated two of his colleagues, Dorothy Hodgkin and Joseph Rotblat, for the 1983 Lenin Peace Prize.

Pauling returned to the Soviet Union for the final time in June 1984, touring the national biological research center and attending the opening session of another "Frontiers in Bioorganic Chemistry and Molecular Biology" conference. In this symposium he and others discussed agreements jointly proposed by the Union of International Research and the Akademia Nauk.

On this trip Pauling also attempted to arrange a meeting with Andrei Sakharov, then confined to his apartment in the city of Gorky. In his diary, Pauling recorded that this request was denied and that a Soviet official had told him that "he was sure I could understand that a person with secret information might have to have his travel restricted." So ended Pauling's in-person encounters with the USSR, a nation whose enchantments and flaws had revealed themselves like few others over many years of acquaintance.

Chapter 29

Travels in Latin America

The Paulings dancing the samba in Brazil, September 1980.

Linus Pauling's travels included frequent visits to countries in Latin America, during which he typically gave speeches on familiar topics including hemoglobin, structural chemistry, nuclear weapons and, of course, vitamin C. He also consistently advocated for human rights in the region and was the recipient of all sorts of awards. A review of these trips surfaces many notable details that are personal to Pauling, while also providing insight into his rhetoric, his research interests, and his geopolitical preoccupations as they evolved over four busy decades.

***

In September 1949, on one of his first professional trips south of the U.S. border, Pauling visited Mexico City to attend a somewhat ill-fated gathering, the Western Continental Congress for Peace. Pauling had been designated as the United States delegate for the conference, and his participation began with the delivery of his delegate's address.

In it, Pauling pointed out that the purpose of the conference was to work towards "permanent, worldwide peace" and to foster more effective cooperation between the citizens of the Americas. Acting in this spirit, Pauling shared his feeling that order was evident everywhere in the natural world except for the seeming impulse toward self-destruction so consistently expressed by the human race. Pauling believed that scientists must play a special role in combating this impulse, suggesting that "the world looks to science for the ultimate solution of the threatening natural problems that menace it." He further stressed the need to strengthen the United Nations such that it could not be dominated by one or two great powers. To do so, Pauling proposed that participating nations transfer part of their sovereignty to the United Nations to form a democratic world government.

At the end of his delegate's address, Pauling again emphasized that movement toward world peace must be a collective, democratic undertaking, proclaiming that

> "It is we, the people, who now have the duty of working for peace, for the welfare and happiness of human beings everywhere. If another devastating world war comes, it will be because we, the people of the world, have failed. We must not fail."

Later this same day, Pauling gave a second speech titled, "Man — An Irrational Animal." In this talk, he reiterated his "deep interest in the structure of the material world," and his appreciation for the harmony and workings of nature, and again proposed that the world of man was an anomaly to nature's preference for balance. Pauling lamented that "we see groups of men, who make up the nations of the world, devoting the material wealth of the world and the intellectual powers of man, the 'rational' animal, not for the welfare of mankind, but for destruction."

Pauling attributed many of the era's problems to the growing struggle between the Soviet Union and the United States, pointing out that nearly ten percent of the world's income was being used for the preparation or execution of war. He also underscored that, in the U.S., suspicions of communist or radical ties were inhibiting many scientists from finding work in universities and the private sector alike. One corrective that Pauling proposed was the channeling of more resources toward UNESCO's (United Nations Educational, Scientific and Cultural Organization) peace efforts, and spending less on military stockpiles.

Interestingly, Pauling's participation in the Mexico City assembly managed to cause resentment in both the United States and within the conference itself. As it turns out (and unbeknownst to Pauling at the time), the Western Continental Congress for Peace was a Communist-organized gathering and was accurately labeled as such in the American media. In biographer Thomas Hager's words,

"...that, of course, did not bother the Paulings. They loved Mexico City — Ava Helen was becoming an admirer of folk art from around the world and spent time combing the *mercados* for pieces to add to her collection — but were less enthusiastic about the meeting, which seemed to consist of speech after long-winded speech defending the Soviet Union and attacking the United States. [Pauling's] keynote address ranged from standard socialist anti-imperialism...to a purposeful and carefully evenhanded denouncement of both the United States' and USSR's policies of curtailing freedom and preparing for war. The audience, expecting another one-sided attack on the Yankees, responded with lukewarm applause."

Pauling's next visit to Latin America came about in May 1955, when Linus and Ava Helen were invited to a conference at the University of Puerto

Rico by the Chancellor of the university, Jaime Benitez. At the meeting, Pauling gave three speeches: "The Hemoglobin Molecule in Health and Disease," "The Structure of Proteins," and "Technology and Democracy."

"Technology and Democracy" was the only talk of the three that Pauling did not give on a regular basis to many other groups. In it, Pauling commented that it was impossible for people to consider themselves "cultured" if they did not have a reasonable grasp of the sciences. To this end, Pauling urged that science curricula be expanded from the elementary level on up, in part because "knowledge of the nature of the world in which we live contributes to our happiness."

***

These trips to Mexico and Puerto Rico were just the beginning of an extensive political and scientific relationship that Pauling built with Latin American countries. His next trip came in January 1962, when he visited Chile to address the Seventh International Summer School at the University of Concepción and to accept a certificate of honorary membership in the Chilean Society of Chemistry. While there, both Linus and Ava Helen delivered multiple talks during visits to four other universities and several additional scientific organizations.

The theme of the Concepción Summer School was "The Man of Today, His Problems and His Future." Pauling gave the opening address, in which he expressed his belief that humanity had accumulated enough knowledge to control the world instead of being controlled by it, but that with this knowledge came the power to destroy civilization. His thesis was a familiar one to those who had followed Pauling's activism:

> "I believe in the philosophy of humanism — that the chief end of human life is to work for the happiness of man upon this earth, to work for the welfare of all humanity, to apply new ideas, scientific progress, for the benefit of all men — those now living and those still to be born."

Pauling also commented on improvements in the understanding of diseases caused by gene mutation, such as sickle-cell anemia and phenylketonuria. Some gene mutations, he hastened to add, had been caused by the circulation of radioactive materials released by nuclear bomb tests. More pressingly, the U.S. was now in possession of 100,000 megatons of bombs, while only 20,000 megatons would be needed to decimate Russia. Similarly, Pauling estimated that the Soviets had stockpiled 50,000 megatons of bombs, with just 10,000 necessary for destroying the U.S.

Pauling reinforced to his Chilean audience that the impact of a nuclear war would not be confined to the U.S. and Russia, but would affect the Southern Hemisphere as well, in the form of catastrophic nuclear fallout. The surest way to safeguard the future was through complete disarmament. "The survival of the whole human race now depends upon whether or not we can work together for the common good," he concluded.

***

When Hurricane Flora hit Cuba in 1963, pounding the country for four days and killing more than 1,700 people, Pauling attempted to visit to help provide emergency disaster relief. After his request to travel to the Communist country was rejected, he and Ava Helen did what they could to support the Cuban people from afar. As part of this, the couple lent their support to Fair Play for Cuba, an organization that protested the trade embargo that the U.S. had placed on its Caribbean neighbor.

That same year, Pauling was invited by Professor N. Matkovsky of the Vienna-based International Institute for Peace, to visit the leaders of various Latin American countries and support their declaration that the entire region remain a nuclear-free zone. This statement had been signed by five countries — Brazil, Bolivia, Ecuador, Chile and Mexico — on May 1, 1963, and would lead to the ratification of the Treaty of Tlateloloco in 1967, which included signatures from a total of 33 parties. In August, accompanying Matkovsky as guest observers, Linus and Ava Helen met with leaders in Brazil and Chile, and also with Arturo Illia, the President-elect of Argentina.

A month after the Paulings returned home, they received word that more than 50 women peace workers had been arrested in Rosario, Argentina. Linus and Ava Helen had met some of these individuals during their visit to Buenos Aires and, upon learning of their detainment, Pauling wrote directly to Illia, imploring him to take action.

"I have been hoping that, after a period during which the authorities of the Republic of Argentina suppressed the rights of individual human beings and carried out many oppressive actions, your nation would take its place among the civilized nations of the world, would recognize the rights of individual human beings, and would abandon the dictatorial and oppressive policies that are characteristic of governments in backward nations."

He echoed his appeal in letters to the current President at the time, Arturo Mor Roig, and to Raul Andrada, a judge in Argentina's federal court, but his entreaties went ignored.

***

Pauling's next journey south came just ten months later, in May 1964, to help celebrate the centenary of the National Academy of Medicine in Mexico City. Invited as a guest of honor, Pauling gave a speech titled, "Abnormal Hemoglobin Molecules and Molecular Disease." He began this talk by expressing his view that the molecules that make up our DNA are the most important molecules in the world, since "[t]he pool of human germ plasm is a precious heritage of the human race."

Pauling also used this platform to discuss advances in the scientific understanding of various molecular diseases including phenylketonuria, which was responsible at the time for one percent of individuals institutionalized in the U.S. as "mentally defective." Pauling explained that the disease occurs when both parents carry a gene for phenylketonuria and then pass it along to their child. Thus afflicted, the newborn would suffer from a reduced ability to manufacture the enzyme that catalyzes the oxidation of phenylalanine to tyrosine. As a result, if the baby ate food containing protein, phenylalanine would build up in the bloodstream and interfere with the growth and function of the brain.

The only way to treat this disease, Pauling continued, was to eat a diet of protein hydrosylate from which most of the phenylalanine had been removed. This protocol would have to be initiated within the first year of life or intellectual disability would occur. The diet would also have to be followed for the rest of the patient's life. It was not an easy existence to be sure, but brighter days were on the horizon. Concluding his presentation, Pauling offered that

> "...the continued study of the molecular structure of the human body and the nature of molecular disease will provide information that will contribute to the control of disease and will significantly diminish the amount of human suffering. Molecular biology and molecular medicine are new fields of science that can be greatly developed for the benefit of mankind."

Pauling rounded out his trip to Mexico with one more talk, "Molecules and Evolution," sponsored by the National School of Anthropology. Pauling also spent a great deal of his time in Mexico discussing the devastating effects of nuclear war, repeating his conviction that the United Nations should have custody and control of radioactive substances produced by the United States and Russia. This work done, the Paulings left Latin America behind for a while, not returning until the beginning of the next decade.

***

On multiple occasions, Pauling found himself visiting volatile places at dangerous times. One such example was a 1970 trip to Chile, taken when he and Ava Helen were invited to the Universidad Técnica del Estado for the university's Summer School.

The Paulings were asked to attend by Enrique Kirberg, the rector of the university, who had visited Pauling in the States and was very enthusiastic to host him as a guest speaker. During this time, Chile was still under the leadership of President Eduardo Frei Montalva, who had been elected in 1964 but had increasingly become the focus of opposition from both conservatives and leftists. By 1970 tensions were such that, as Pauling noted in his diary, he and Ava Helen were escorted everywhere by three armed detectives.

The Paulings arrived in Pudahuel, Chile, on January 8 and the inauguration of the Summer School took place the next day. Pauling spoke at this opening, delivering his 40-minute lecture, "Science and the Future of Humanity," entirely in Spanish. In this speech, which he gave often (albeit in English), Pauling reiterated his stance on the necessarily close connection between science, morality and ethics, and expressed frustration that scientists were not pressing governments and the body politic more forcefully in the direction of informed decision-making.

Another issue that Pauling surfaced in this talk was concern over the size of the world's population, which by 1970 was about four billion people. Pauling believed that the Earth was already too crowded, and that growing malnutrition was one result of the stress being placed on the planet. His solution was to reduce population growth little by little, until it eventually reached what he believed to be the ideal number of one billion people sometime around the year 2200. At this population level, Pauling reasoned, all humans could expect to lead a good life.

Later in the trip, Pauling had the opportunity to meet Salvador Allende, who would be elected President of Chile in September 1970. Allende's government was overthrown by a military coup in 1973 after which General Augusto Pinochet assumed power. Amidst this upheaval, Enrique Kirberg, the university rector who had prompted Pauling's 1970 visit, was arrested. He was then taken to Dawson Island, a land mass in the Tierra del Fuego archipelago that is subject to Antarctic weather and that was used, at the time, to house political prisoners suspected of being communists. For more than a year, Kirberg lived on the island in a camp before being returned to Santiago, where he was found guilty of tax fraud and given a long prison sentence.

When Pauling caught wind of his friend's plight in 1974, he wrote a letter to General Pinochet inquiring about Kirberg's whereabouts and asking that he be permitted to leave the country if he wished. Kirberg was eventually freed, and, in 1975, Pauling received a letter of gratitude thanking him for being a part of the effort to secure his release.

Although Pauling would not return to the country again, he did serve as a sponsor for the National Coordinating Center in Solidarity with Chile, which contributed to the struggle for democracy during the military dictatorship. He also supported the Office for Political Prisoners and Human Rights in Chile during the late 1970s and co-sponsored the Madrid World Conference in Solidarity with Chile in 1978.

***

On a rare solo trip, Ava Helen visited Bogotá, Colombia in July 1970 to participate in the Third Congress of Women of America. The gathering was organized by the Women's International League for Peace and Freedom (WILPF) and lasted for five days.

WILPF was founded in 1915 by a group of women from 12 countries and has worked for peace and gender equality ever since. Key objectives for the Colombian League in 1970 included rights related to marriage and divorce, and support for the education of women. The Third Congress also featured conversations on the relationship between the U.S. and Soviet Union, the Colombian economy, population control, the equitable use of resources, and balancing the distribution of wealth. The group likewise addressed social issues including sexual taboos and family planning.

The Paulings next went to Tijuana, Mexico, in March 1972 for a conference sponsored by the Chemistry Association of Tijuana, where Linus received a certificate of appreciation and attended various meetings. While there he gave his speech "Science and the Future of Humanity," a version of which he had delivered two years prior in Chile.

Ava Helen also gave a lecture during this trip, titled "The Liberation of Women." In her remarks, she noted that the last 15 years had seen an increase in the struggle for the liberation of oppressed people all over the world, including women, and that "[t]he Women's Movement has developed so rapidly that it is difficult to keep up with their various activities." She then touched on a small grievance, the difference in titles available to women versus men: "Miss," if a woman is unmarried and "Mrs." if they are, with men simply being called "Mr." Though admittedly a minor problem, Ava Helen made clear that she would always prefer to be called "Ms."

From there Ava Helen listed four demands that had attained primacy within the women's liberation movement. The first was that women should receive equal pay for equal work. (According to Ava Helen, in a 1965 survey, women were found to be receiving only 60 percent of the salary that men earned for the same work.) The second demand was equal opportunity in employment, without discrimination. Third, working women needed to have access to 24-hour childcare centers "[i]n order to do their jobs well." The fourth and final demand was free and ready access to abortion. "Women are demonstrating in all countries for the repeal of abortion laws," she said, specifically citing the 1971 Women's National Abortion Action Coalition demonstration in Washington D.C., in which 3,000 women participated.

Along with these four demands, Ava Helen also presented a collection of major concerns being discussed within women's liberation circles. These included "nutrition in general, nutrition for the pregnant woman, free lunches for school children, nursery schools, adequate housing, and a guaranteed income for everyone." She finished her speech by stressing that, "[women] are becoming politically sophisticated and ever more aware that they, in working for their own freedom from discrimination and oppression, are working for the freedom of all humankind."

***

The setting for the Paulings' next trip to Latin America was a novel one: the Preventive Medicine Cruise to Mexico. Departing from Los Angeles on April 13, 1977, the ship's ports of call included stops in Puerto Vallarta, Acapulco and Mazatlán. The cruise included 60 passengers and lasted for ten days, from April 13th to the 23rd, although the Paulings only took part until April 18, owing to prior engagements.

Linus agreed to give two lectures while onboard the S.S. *Fairsea*, and used them to discuss biochemical specificity in nature, the use of massive doses of vitamin C to alleviate cancer distress, and topics in immunology. Other lecturers on the cruise included an allergist, Theron Randolph, and Virginia Livingston Wheeler, a physician who specialized in cancer research. Participants could also avail themselves of a 30-hour educational program in the sub-specialties of preventive and orthomolecular medicine, clinical ecology, and cancer immunology.

A year later, in 1978, the Paulings finally made their long-desired visit to Cuba to take part in the Fifth Cuban Congress on Oncology, which was held in Havana from March 19–27. Pauling gave a talk at the gathering titled, "Nutrition and Cancer," in which he discussed the benefits of ingesting vitamin C and other nutrients to increase cancer survival times. His hopeful remarks included a report that

"[a]s much as 75 grams of vitamin C per day has been administered, both intravenously and orally, to patients with advanced cancer, and there is some evidence that the larger intakes are considerably more effective than the usual intake of 10 grams per day."

After giving his lecture, Linus and Ava Helen immersed themselves in Cuban culture, attending a performance of the National Ballet of Cuba, enjoying the music of a Cuban Folklore Ensemble, and even going to the Tropicana nightclub. Tourism at that level was rare for the Paulings, but this trip was the culmination of a long-held ambition to explore a country that had previously been off limits.

Later that same year, the Paulings made a return trip to Brazil, their final visit to Latin America in the 1970s. The impetus was an invitation to Linus that he speak as guest of honor at the Second International Vitamin C Symposium in Rio de Janeiro. Pauling gave the opening address on August 24 and voiced his belief that the world was entering into the Megavitamin or Orthomolecular Age. While there, Pauling also acted as chairman of a workshop on Vitamin C and cancer research and delivered the symposium's closing address as well.

***

In 1980, Pauling once again traveled to Brazil for the third iteration of the International Vitamin C Symposium, which took place in Sao Paulo and Rio de Janeiro. This time the event ran from September 2–13, with Pauling again seated as the guest of honor and delivering the gathering's opening address.

Session titles at the conference were familiar ones to those who had followed Pauling's recent career: "Vitamin C in Immunology," "Vitamin C in Lipid Metabolism" and "Vitamin C in Cancer." Pauling coordinated and participated in the latter program, presenting a paper titled "The Incidence of Squamous Cell Carcinoma in Hairless Mice Irradiated with Ultraviolet Light in Relation to the Intake of Ascorbic Acid."

The paper communicated the results of a study that Pauling and three other investigators had conducted, in which they observed the development of large malignant skin tumors in 700 hairless mice. The mice, divided into groups, were "intermittently exposed in a standard way to long-wavelength ultraviolet light over a period of 110 days," with each group being given a consistent diet but differing amounts of Vitamin C. At the end of the study, it was determined that a strong correlation existed between the number of tumors that the mice developed and the amount of Vitamin C that they consumed.

Pauling went to yet another symposium in January 1981, this time in Mexico City and focusing on the subject "Metabolic Treatment of Heart Conditions." The conference took place at Juarez Hospital and was coordinated by Dr. David Contreras, Chief of Cardiology. At this short, two-day meeting, Pauling only gave one lecture, "Treatment of Old Age."

***

After Ava Helen's death in December 1981, Pauling did not travel to Latin America again until November 1982, when he was invited to a symposium held in Caracas, Venezuela and marking the 100th anniversary of Charles Darwin's death. He gave a lecture on November 8 at the Central University titled "The Joy of Research" in which he talked about his capacity to find pleasure in scientific discoveries made by others, specifically citing his excitement in learning that new clues related to the extinction of dinosaurs had been found in layers of clay. But even more joy could be felt through one's own process of discovery. As he recounted on a different occasion,

> "When Ernest Lawrence got married...I was an usher at the wedding, in 1931. I drove back in the car with some people and I said that I was happy because I had in my pocket a crystal of sulvanite, Cu_3VS_4. And I had just determined the structure of this and it was a very striking structure, anomalous, it didn't fit in with my ideas about sulfide minerals. But I knew what the structure was, nobody else knows, nobody in the world knows what the structure is and they won't know until I tell them. This is an example of the feeling of pleasure I had on discovering something new."

Pauling's next trip to the region was for the International Symposium on Vitamins in Nutrition and Therapy, held in Cartagena and Bogotá, Colombia in November 1983. During this visit, Pauling met with President Belisario Betancur, who led the country from 1982 to 1986. Pauling also gave a speech, "The Necessity of World Peace." In it he discussed the terrifying possibility of a third World War and how it might result in the extermination of the human race. Echoing his refrain from decades before, Pauling underlined the need for cooperation between the world's superpowers, since retaliation would be suicide.

In making this point, Pauling referred to Korean Airlines Flight 007, which had been shot down by Soviet interceptors over the Sea of Japan that previous September after entering Soviet airspace around the time of a planned missile test. All of the flight's passengers and crew were killed in the tragedy, including Lawrence McDonald, a member of the United States House of Representatives. The Soviet Union eventually claimed that the aircraft was on a spy mission but, according to Pauling, "President Reagan saved the world by not taking retaliatory military action, as was urged on him by the right-wingers in the U.S."

***

In 1984 Nicaragua was struggling to rebuild itself under a new government, the Sandinista National Liberation Front (FSLN), which had come to power after 43 years of dictatorial rule by the Somoza family. Five years removed from the overthrow of Anastasio Somoza Debayle, the nation was enveloped in a brutal civil war in which the left-leaning Sandinistas struggled to hold power over a U.S.-backed guerrilla corps collectively known as the Contras. By some estimates, more than 30,000 people died in this conflict, which lasted for all of the 1980s.

Amidst the chaos, the government of Norway, along with a small group of Nobel Laureates, decided to help by stocking a ship with 13,000 tons of humanitarian aid and delivering it to the suffering Nicaraguan people. The Peace Ship, as it was called, began its journey with a press conference held in Panama City on July 23, after which it set sail to Port Corinto, Nicaragua. Passengers on board the Norwegian cargo ship (formally named the W.V. *Falknes*) included Pauling; Adolfo Perez Esquivel and Betty Williams, winners of the Nobel Peace Prize; George Wald, winner of the Nobel Prize in Medicine; and the leaders of various religious groups.

Those on board sent a message "To People of Conscience, From the Peace Ship," which stated,

"[this ship] carries instruments for health and life, not implements of war; medicines, educational materials, fertilizers, small fishing boats and paper donated by the governments of Norway, Sweden and non-governmental organizations to facilitate Nicaragua's forthcoming [November 1984] national election."

Pauling and Wald also sent a telegram to President Reagan informing him of their mission and noting their intent to issue a statement in Managua backing "the right of self-determination, support for the efforts of the Contadora [conflict mediators] Group to bring peace to the region, the cessation of all foreign intervention, and the withdrawal of all foreign advisers from the region."

After the Peace Ship arrived in the country, Pauling and Wald rode to Managua in a Land Cruiser driven by Daniel Ortega, a member of the Junta of National Construction that operated the FSLN. At the time Ortega was running for President of Nicaragua, and he would eventually win that election, the first democratic, multi-party race held in the country's history. As they drove through the countryside, Pauling recalled that Ortega and his men kept "machine guns on the floor of the car because they expect action from the Contras who often carry out assassinations."

Accompanied by Ortega and Wald, Pauling visited a small hospital in Managua in which wounded soldiers who had been injured by the Contra forces were being treated. The Nobel laureates also received glimpses of a nascent socialized heath care network, wherein "most doctors do private practice and also work for the national health system," a situation that Pauling, in a later interview, judged to be an improvement over what had existed previously. From what Pauling had seen, "before the Sandinista fighters won...the poor people just didn't get much in the way of health care or physician services. Now they do."

Pauling was also intrigued by the Sandinistas' mixed approach to creating markets for agricultural goods. His notes mention a visit "to one of the privately owned big farms," where he "talked to the owner about his relationships with the government and how, when crops were poor a couple of years ago, the government had helped him. The government sets prices but the owners make a profit."

Most material conditions, however, were desperate. Following a visit to a medical school, Pauling reflected that "my high school chemistry class in 1913 was better equipped than this college." Of Nicaragua as a whole, Pauling retained an overwhelming sense of "a miserably poor country. I felt about as bad concerning conditions there as I had about India...[in 1973]." And ever-present throughout the trip, there existed the specter of the ongoing civil war. Again from his diary:

> "We met another large landowner, Gladys Volt, whose husband had been kidnapped and killed by the Contras just eight days before."

Thankfully the tour also included a certain amount of lighter fare — one particular highlight was a trip to the countryside to visit an active volcano, which Pauling found to be home to flocks of green parakeets. This excursion was hosted by Humberto Ortega Saavedra, Daniel Ortega's brother and the commander-in-chief of Nicaragua's armed forces.

Pauling's transit home was marked by a stark reminder that there is sometimes a price to be paid for acting in opposition to official government policy. Journeying north through Mexico City, Pauling's passport was confiscated for six hours. After a period of confusion, the passport was returned to Pauling, though not until he arrived at his final destination. Pauling believed, perhaps from hard experience, that he was being sent a message. As he recorded in a letter to his children,

> "I think that an order was issued to take my passport away from me. George [Wald] and I had sent President Reagan a telegram from the ship and said we were on this ship taking material from Norway to Nicaragua. So I started thinking, this is what's happening. They are taking my passport and I won't get it back and I wondered what to do. Should I call Meet the Press, or Face the Nation?
>
> "When we got to San Francisco and after I stepped off the plane, the stewardess came with an envelope and gave it to me. It contained my passport and I went through customs without any trouble. I think that some higher official in the U.S. government had decided that it was better not to take my passport (which had, of course, been denied me from 1952 until 1954, when I was given the Nobel prize in chemistry.)"

***

After his return to the U.S., Pauling continued to act on behalf of Nicaragua's struggle for peace and freedom. He supported the International Committee of the Support of War Victims of Nicaragua, and endorsed a resolution authored by two doctors, Robin W. Briehl and Kenneth Barnes, that opposed the "U.S.-directed violations of human rights and interference with scientific development in Nicaragua." This resolution was submitted for consideration to the American Association for the Advancement of Science and urged the U.S. government to stop funding the Contras and to aid in the safe release of a kidnapped medical brigade.

While Pauling continued to advocate on behalf of these and several other Latin American causes, his voyage on the peace ship marked his last major trip to the region. So concluded a long string of memorable activities and experiences that had begun some 35 years before.

Chapter 30

Japan and the Japanese

Pauling at an event in Hiroshima, Japan, August 6, 1959.

Linus Pauling's lengthy and fruitful association with Japan and its citizens dates to at least the 1940s and possibly begins with his connections to two Japanese Americans who worked under him at the California Institute of Technology.

Over the years, Pauling forged close relationships with many of his graduate and doctoral students, offering guidance that, in numerous cases, changed the course of a student's life. During World War II, he fought particularly hard for two of his research assistants, Miyoshi Ikawa and Carol Ikeda. In both cases, Pauling's intervention prevented these young colleagues from being forcibly incarcerated. Instead, Ikawa and Ikeda each moved on to graduate studies and fruitful careers in science.

Miyoshi "Mike" Ikawa was born in California in 1919 to first-generation immigrant parents. He pursued undergraduate studies at Caltech, where he was a member of the Chemistry Club and the Tau Beta Pi engineering honor society, and also competed on the Fleming House wrestling team. When he graduated in 1941, he was already working in Pauling's lab, preparing compounds and providing other support for research on the serological properties of simple substances. Pauling subsequently served as his graduate advisor.

Carol Ikeda came to Caltech in 1939, having begun his higher education at Paris Junior College in Texas. Ikeda transferred to Caltech with the intent to study chemistry and find employment as an organic chemist. At Caltech, he stood out among many other very bright students, with Pauling describing him as "one of the top men in the class." Not in the habit of giving compliments lightly, Pauling recognized Ikeda's potential from his performance in the classroom and his work in organic research labs. Before Ikeda had even decided to continue on to graduate studies, Pauling had recruited him for the serological project as an assistant in the Immunochemistry department. Both Ikeda and Ikawa show up as co-authors for two of Pauling's earliest serological papers, which is noteworthy in part because much of the work that the papers communicate was conducted while the pair were still undergraduates.

Up until World War II, it appeared that Ikawa and Ikeda were each moving well down the path toward successful careers in immunology, organic chemistry, or biochemistry. This all changed when President Franklin Roosevelt issued Executive Order 9066 and declared Pasadena to be a military zone. Even before the attack on Pearl Harbor, U.S. citizens of Japanese descent faced discrimination on the basis of race as well as suspicions that their first loyalties were to Japan, even if they were second- or third-generation Americans. Acutely aware of these mounting tensions, Pauling was determined to support students bearing this burden and to make sure that they could find positions at Caltech for which they were suitably qualified.

That said, Pauling was likewise aware that colleagues at other universities did not necessarily share his point of view. In the recommendations that he wrote, he provided full disclosure and acknowledged potential discomforts regarding race. In one illuminating instance, Pauling wrote,

> "...the two best men scholastically in our graduating class are American-born Japanese, Ikawa, and Ikeda. Although one of them has, I think, a satisfactory personality for teaching work, I doubt that you would be interested in appointing him because of his racial handicap."

Some colleagues responded positively to recommendations of this sort; one at the University of Iowa, for instance, confirmed that race wouldn't be a problem at all. Rather, Pauling's Iowa contact assured that the institution was aligned with Pauling's stance and would consider the qualifications of all applicants regardless of their ethnicity. The reply went on to state,

> "While we have not had any American-born Japanese on our teaching staff, I see no reason why they would not get along satisfactorily, if they have the necessary intelligence and ability."

***

Ikawa and Ikeda had been working on the serological project for more than a year when Pasadena was declared a military zone. Cognizant of the need to help his assistants relocate to a safer area, Pauling had a relatively easy time finding Ikawa a position as a graduate student at the University of Wisconsin, where he worked under Karl Paul Link. This move ultimately changed the trajectory of Ikawa's career. Before receiving his doctorate, Ikawa, along with Link and Mark Stahmann, synthesized a compound called warfarin and obtained a patent for it to be used as a rat poison. By 1950, warfarin (now commonly referred to by its brand name, Coumadin) was being used to treat blood-clotting disorders such as thrombosis, because it was a strong anticoagulant, and it still serves this purpose today.

With the war over, Ikawa was free to return to the West Coast, where he conducted postdoctoral research at Caltech and UC Berkeley before moving on to the University of Texas. In the early 1960s he became a professor at the University of New Hampshire, where he focused on marine biotoxins.

In 1972, he and his colleagues established the Paralytic Shellfish Monitoring Program for the state of Maine, a project that emerged after the first red tide was observed in the southern Gulf of Maine. Ikawa taught at New Hampshire for 20 years and spent much of his later career advising technical panels and participating in peer review committees for federal research grants. He passed away in 2006.

***

Pauling had a harder time finding a position for Ikeda. In February 1942, shortly after Roosevelt released his executive order, Pauling sent a letter to Robert Millikan — the chairman of the Caltech Executive Council — with an update on Ikeda's progress and position. In it, Pauling tried to make the case that, while his serological research wasn't directly related to defense work, its results could be valuable for their medical application. He also pointed out that finding someone as competent as Ikeda to continue these studies would be nearly impossible.

Pauling soon discovered that this approach had backfired due to pervasive nervousness over Japanese American involvement in war work of any kind. In his response, Millikan asked Pauling to vouch for Ikeda's loyalty, after which he might be allowed to continue "to undertake, under special arrangement, research work which may involve defense matters." Pauling affirmed that Ikeda's research skills were strong but hesitated to comment on his loyalty, writing that someone with a more personal working relationship with Ikeda could give a better answer. He also suggested that Ikeda could be transferred to a teaching position if the issue of loyalty could not be resolved to the Institute's satisfaction.

As Millikan deliberated, Pauling began to feel that Caltech might not be the best environment for Ikeda, even if he was reclassified into a teaching role. In short order, Pauling contacted Michael Heidelberger, a faculty member at Columbia University's College of Physicians and Surgeons. In doing so, Pauling offered Heidelberger a quid pro quo of sorts, suggesting that Heidelberger accept Ikeda into his program at Columbia in exchange for Pauling hosting one of Heidelberger's researchers in Pasadena. This plan broke down when the Columbia researcher whom Pauling had in mind wrote back to say that he could not accept an appointment at Caltech and that he wished to stay in New York instead.

(The situation was not improved much by Heidelberger's blasé attitude toward the internment camps. Recognizing that "wholly" patriotic people would be unjustly punished, Heidelberger remained unconvinced that there was much that he or Pauling could do to alleviate the issue, an opinion shared by many.)

Fortunately for Ikeda, circumstances worked out at the last minute. In April 1942, just two weeks before Ikeda was scheduled to report to a camp, Pauling secured him a graduate position at the University of Nebraska, where he ultimately completed his Ph.D. in 1945. In 1947 Ikeda accepted an offer of employment from the DuPont company in Delaware and later moved within the corporation to Philadelphia. In 1962, he received a patent for a resinous coating material that he developed while working for DuPont. He passed away in Phoenix, Arizona in 1996, having enjoyed a successful career.

***

Though Pauling was modestly active in world affairs before the onset of World War II, he tended to keep his views private. If sometimes too caught up in his work to spare much attention for politics, he fundamentally valued principles of neutrality and objectivity, qualities that stemmed from his training as a scientist.

Pauling's political inclinations remained relatively mild until March 1945, when his family hired a returned Japanese American evacuee, George Nimaki, to work as a gardener at their home for a brief period prior to his scheduled reporting for duty in the U.S. Army. A few days following Nimaki's hire, the Paulings woke up to graffiti on their garage door. Someone had painted, in bright red, "Americans die but we love Japs — Japs work here Pauling" alongside an image of the rising sun flag.

A series of subsequent hate letters and death threats followed; chillingly, they appeared to be the work of people who lived nearby:

> "We happen to be one of a groupe [sic] who fully intend to burn your home, tire [sic] and feather your body, unless you get rid of that jap…the more publicity you give this matter, the sooner we will take care of you just like Al Capone did some years ago… [signed] A neighbor."

Pauling, still deeply engaged in a vast program of scientific war work, was outraged that his loyalties might be questioned and his family threatened. Speaking to the *Pasadena Independent* newspaper, Pauling expressed his frustrations:

"I do not know who is responsible for this un-American act. The people in Pasadena and the surrounding region are, in general, intelligent and patriotic. I have, however, come in contact with a few people who do not know what the Bill of Rights is and what the Four Freedoms are and what the principles are for which the United Nations are fighting. I suspect that the trespass on our home was carried out by one or more of these misguided people who believe that American citizens should be persecuted in the same way that the Nazis have persecuted the Jewish citizens of Germany and the conquered territories."

With aid from the American Civil Liberties Union, he was able to prod the largely unsympathetic local sheriff's office into providing a guard to protect his wife and children against violence. And while none of the threats were carried out, the "Japanese Gardener Incident" proved to be an important moment in Pauling's life, as it marked a move towards more active and open involvement in public affairs. Another event a few months later would cement this shift.

***

On August 6, 1945, the United States dropped an atomic bomb on Hiroshima, Japan. Another nuclear weapon was detonated over Nagasaki on August 9. The day after the first bomb devastated Hiroshima, a day that he never forgot, Pauling read of the attack in a local newspaper. He was immediately interested in the physics of the bomb, but did not share in the euphoria that was sweeping the nation.

During the war Pauling had been offered a spot in the chemistry division of the Manhattan Project, where the atomic bomb was developed, but he was busy with his own scientific war work and had little personal interest in the opportunity. Following the bombings in Japan, groups of concerned scientists who had contributed to the project began discussing the world-altering implications of their work and later began distributing information about the role that atomic weapons might play in a rapidly changing world. Pauling received much of this material and began to attend informal and formal meetings where issues such as civilian control of atomic weapons and technology were the main topic of discussion.

After hearing what many other scientists had to say, and reflecting on his own beliefs, Pauling became openly supportive of sharing atomic secrets with the USSR, and of increased cooperation generally. While on a trip in September, Pauling wrote to Ava Helen about his growing concerns, noting that,

> "[Samuel] Allison [a physicist who worked on the Manhattan Project] has made a strong public statement against keeping the A-bomb secret from Russia... I think that Union Now with Russia is the only hope for the world."

From there, Pauling learned more about the science of the bomb and began giving talks around southern California, starting with a lecture hosted by the Rotary Club in Hollywood. As time went on, he increasingly came to incorporate thoughts on politics and international relations into his presentations, but many in his audience found his non-science discourse dry and unconvincing. After one of these early speeches, Ava Helen gave her husband an honest assessment of what she was seeing as a spectator and concluded that he should stop discussing war and peace. He later wrote that her comments changed his life.

> "I thought 'What shall I do? I am convinced that scientists should speak to their fellow human beings not only about science, but also about atomic bombs, the nature of war, the need to change international relations, the need to achieve peace in the world. But my wife says that I should not give talks of this sort because I am not able to speak authoritatively. Either I should stop, or I should learn to speak authoritatively.'"

From that point on, Pauling devoted up to half his time to the study of peace and the abolition of war.

***

Beginning in the 1950s, Japan emerged as a favored spot for research, cultural engagement, and public speaking for the Paulings. As their fame grew, the couple were held in consistently high esteem by the Japanese public and Pauling maintained close contact with many of the country's leading scientists. These connections resulted in the extension of numerous invitations to deliver lectures and attend conferences in Japan, which Linus visited at least nine times over 31 years.

The Paulings' first proposed trip to East Asia was scheduled for 1953, with travel to Tokyo scheduled from February to March, but the visit was cancelled due to Pauling's chronic passport difficulties. Instead, 1955 marked the first venture to Japan. Dedicated to delivering talks on the chemical bond, hemoglobin and proteins, Pauling spoke at Tokyo University and Osaka University. In between, Pauling also attended seminars on proteins. Of this first trip, Pauling remarked,

"As a scientist I am interested in Japan and primarily in the universities… [I am] greatly impressed by the natural and cultural richness of the country…[where] scientific work is of the highest quality…Science of the modern world has been accelerated here by the atom-bomb and radiation…Because of this, hopefully steps will be made towards the goal of permanent world peace."

Linus and Ava Helen returned to the country in August 1959 to attend the Hiroshima Fifth World Conference against Atomic and Hydrogen Bombs. The Paulings' arrival came in the midst of an incredibly heavy travel schedule. They had started their summer with a two-week scientific trip to England and a nine-lecture *No More War!* tour through West Germany. From there they visited with Albert Schweitzer in Lambaréné, French Equatorial Africa, and then returned to Europe for the Triennial World Congress of the Women's International League for Peace and Freedom. In early August they jetted to Japan and began their trip by participating in a march at the Hiroshima Peace Park, after which Pauling gave a brief lecture titled "Physical and Biological Aspects of Radon" at Hiroshima University.

The Fifth World Conference came about at an interesting time for the peace movement. Though millions had been spurred toward an anti-war stance by the atomic bombings of Hiroshima and Nagasaki, by the 14th anniversary of those attacks, attitudes were beginning to shift, even in Japan. In an exchange of letters with the Austrian philosopher Günther Anders, Pauling noted that the biggest change between the Fifth World Congress and those that had preceded it was the attitude of the Japanese government, which,

> "under [Prime Minister Nobusuke] Kishi, has now changed its policy from that of the new Japanese constitution, which is opposed to military armaments. The government is attempting to revise the security pact with the United States (and the United States government is also in favor of this revision) in such a way as to lead to the rapid remilitarization of Japan. In consequence of this change in the Japanese government, the government withheld its subsidy of the Hiroshima Conference this year."

As one might expect, the text of the conference's finalized Hiroshima Appeal reacts in direct opposition to the posture that the Kishi government was assuming. Specifically, the document calls for

> "An international agreement that would result in the permanent neutralization of West and East Germany and adjacent countries and also of Japan...No revisions of present treaties, [and] no new military alliances... that would permit arming nations with nuclear weapons or prevent this eventual demilitarization."

This emphasis on international agreements to control the development of nuclear weapons programs, as well as specific support for the three-nations nuclear testing treaty being hammered out in Geneva, form the heart of the Appeal.

Given that the conference took place in 1959, it should come as no surprise that the gathering was plagued by suspicions of communist activity. Indeed, four delegates to the event — Wayland Young and Arthur Goss of

Britain, and Rolf Schroers and Carola Stern of West Germany — walked out in protest of what they perceived to be heavy communist infiltration. In particular, the boycotters felt that the conference was failing to condemn any nuclear ambitions that may have been emerging in China, a charge that Pauling (who did favor the admission of China to the United Nations) flatly denied.

In his report to Anders, Pauling detailed his perspective on the conflict:

"The appeal was written after Schroers walked out of the conference. There seemed to be a general feeling that he and Wayland Young, from England, were determined to make trouble. I was not present at the preliminary meeting, with which Schroers was dissatisfied. I was present, however, at all of the following meetings, and I participated vigorously in the discussions. I think that the arguments that I presented were effective. It seems pretty clear, from the Hiroshima Appeal, that the conference did not suffer much from the walkout by Schroers and his associates. On the other hand, if there had been a really serious difference of opinion, with real domination by communists, then it might not have been possible to get a good appeal accepted, and the walkout by Schroers and his associates might have been a very serious matter."

After the conference concluded, Pauling gave a talk in the Grand Lecture Hall at Hiroshima University that was starkly titled, "Our Choice: Atomic Death or World Law." In it he again advocated for a world government — calling it a "path of reason" — and condemned the dysfunction of "insensate militarism." These ideals were extended in additional meetings with the Japanese Committee of the Pugwash Conference, a collection of other scientists and academics. In these conversations, Pauling reiterated his stance that it is the scientist's duty to understand the physical reality of nuclear war and to relay its horrors to the world. In his last few days in Japan, he met with colleagues in Hiroshima, Osaka, and Kyoto to further stress these points.

***

Sixteen years passed between Pauling's participation in the 1959 Hiroshima Conference and his next visit to Japan in fall 1975. And while the latter trip dealt largely with his findings on Vitamin C, some of his time was devoted to peace-related activities.

Notably, Pauling attended a symposium of the Keidanren Kaikan Memorial Lecture in Tokyo, and a symposium of the Memorial Lecture at Hiroshima-Ishikaikan in Hiroshima. He also attended an event sponsored by the Albert Schweitzer Fellowship of Japan where he presented a paper titled "Reverence for Life and the Way to World Peace." His short "peace tour" additionally included a guest appearance on a talk show with Dr. Soichi Iijima and a lecture delivered at a high school in Hiroshima, titled "The Development of Science and the Future of Mankind."

Then the vitamin C engagements began. Throughout the 1970s, Pauling's efforts to spread the word about ascorbic acid had become a decidedly international project and certainly informed his stops in Japan. Partly because of this, Pauling developed close working relationships with two Japanese colleagues, one of whom was Fukumi Morishige. A surgeon at Fukuoka Torikai Hospital for over 30 years, Morishige had introduced himself to Pauling via letter not long before the 1975 trip. In this initial communication, the Japanese physician informed Pauling of his own research on vitamin C and asked to meet with him during his visit. Pauling was happy to comply and, at Morishige's request, also gave a talk on the value of vitamin C in health and disease. Thus began a collaboration and friendship that would last for the remainder of Pauling's life.

Fukumi Morishige was born in Fukuoka, Japan in 1925. He attended Kurume University where, in 1961, he received his medical degree. His career advanced quickly and within six years of completing his studies, he had become the chief surgeon of the Fukuoka Torikai Hospital. After visiting the Tottori Sakyu Hospital and witnessing some inspiring work being led by the resident surgeons there, Morishige began to take into consideration the importance of vitamin C. As he later recalled,

"I knew that giving vitamin C to patients helps them to heal quicker for some reason, but I didn't know why. I decided to do more research on how vitamin C impacts human bodies and made up my mind to explore vitamin C's effect and stay in this field."

Morishige's initial vitamin C studies concentrated on the prevention of serum hepatitis in patients receiving blood transfusions, but over time his interests moved in the direction of Pauling's focus on cancer. Through a correspondence that spanned nearly 20 years, Morishige would frequently relay information about new ideas on cancer research and Pauling would unfailingly reply with enthusiasm and support, often voicing his desire to bring Morishige to the U.S. to discuss his progress.

Spurred by Pauling's encouragement, Morishige conducted several experiments involving vitamin C and other therapies for cancer. In 1983, Morishige, Pauling and three additional Japanese scientists published a paper in the journal *Cancer Research* titled "Enhancement of Antitumor Activity of Ascorbate against Ehrlich Ascites Tumor Cells by the Copper:Glycylglycylhistidine Complex." The publication details the group's efforts to increase the antitumor activity of ascorbate by use of an "innocuous form of cupric ion complexed with glycylglycylhistidine." While it did not significantly "oxidize ascorbate," the researchers found that the compound "killed Ehrlich ascites tumor cells" in high concentrations of ascorbate. They further reported that glycylglycylhistidine "prolonged the life span of mice inoculated with Ehrlich tumor cells."

In 1986 Morishige was introduced to a cancer patient who seemed to be controlling her illness by drinking reishi tea and, in short order, the scientist launched his own program of research on reishi mushrooms. Through his investigations, Morishige came to believe that the fungus acted as both a cancer preventive and a tumor suppressant. He then began to combine the reishi treatments with vitamin C therapy and found that the two appeared to complement one another.

Though Morishige successfully used this approach with several cancer patients, it is still looked upon as an alternative healing remedy rather than a medically accredited technique. Indeed, for both Pauling and Morishige, it became common for their work to be rejected by the medical community, yet they both doggedly continued to research the topic, determined to demonstrate the benefits of ascorbic acid in medicine. Amidst their shared struggles, Morishige remained extremely grateful for Pauling's support and continually expressed his appreciation to Pauling for the interest and advice that he imparted.

***

Even prior to learning about Morishige's work, Pauling was paying close attention to research being conducted by another Japanese colleague, Akira Murata, who was studying the inactivation of viruses by vitamin C. Over the years that followed, Morishige and Murata often worked in partnership, and as with Morishige, Murata became a close colleague of Pauling's, hosting him on numerous visits to Japan and, on at least a few occasions, traveling across the Pacific to work with him in California.

Akira Murata was born in Shimonoseki, Japan in 1935 and later attended Kyushu University, receiving his Ph.D. in microbiology in 1963. In 1966 he accepted the position of Associate Professor at Saga University, where he remained for the bulk of his career. From early on, Murata was interested in vitamin C and, in particular, the impact that it could make on viruses. In 1975 Murata summarized much of this work in a paper written for the Intersectional Congress of the International Association of Microbiological Societies and titled "Virucidal Activity of Vitamin C: Vitamin C for Prevention and Treatment of Viral Diseases." In it, he outlined a series of clinical trials that he had conducted with Morishige, which used phages and their host bacteria as model systems to demonstrate the impact of vitamin C on viruses. A year later, in 1976, Murata went to the United States to study vitamin C and the immune system at the Linus Pauling Institute of Science and Medicine.

A parallel track of research conducted by Murata and Morishige in the 1970s focused on the impact of vitamin C on hepatitis. The duo notably authored an important paper titled "Vitamin C for Prophylaxis of Viral Hepatitis B in Transfused Patients." The article reported on a series of tests where patients who had received blood transfusions were also given specific dosages of vitamin C, after which data was collected on hepatitis contraction among the transfusion patients. The researchers found that, between 1967 and 1976, no hepatitis B cases were recorded for those who had received large doses of vitamin C following a blood transfusion. The paper accordingly concluded that, when given in large amounts, vitamin C has a "significant prophylactic effect against post-transfusion hepatitis, especially type B." Prior to its publication, Pauling annotated and edited Murata and Morishige's text, adding his suggestions for how the manuscript could be improved.

In 1976, the year of his residency at the Pauling Institute, Murata also published observations made with Morishige on the effect of increased doses of ascorbic acid on various viral and bacterial diseases. In their study, the researchers found that ascorbic acid showed a therapeutic effect on infectious hepatitis, measles, mumps, viral orchitis, viral pneumonia and certain types of meningitis.

Murata continued this line of research through the 1980s, continually seeking out new ways to test the effects of vitamin C on human health. Like Pauling and Morishige, Murata was also highly interested in vitamin C's possible therapeutic use with cancer. Several papers arose from this area of investigation, including one titled "Prolongation of Survival Times of Terminal Cancer Patients by Administration of Large Doses of Ascorbate." Together, Pauling and Murata also served as chairmen and panel members for at least one workshop on vitamin C, immunology and cancer.

By the late 1980s, Murata had contributed upwards of 25 publications on vitamin C and its effects upon various diseases, and Pauling continued to keep in contact. Murata typically hosted at least a portion of Pauling's many visits to Japan, often helping to arrange scientific meetings as well as social gatherings. Murata also translated a few of Pauling's books into Japanese. Among them was *Vitamin C, the Common Cold, and the Flu*, the preface to which contains Pauling's note of thanks to Murata and the observation that "it is important that everyone know about the great value that vitamin C has in improving health and in protecting against disease." Murata also translated Pauling's best-seller, *How to Live Longer and Feel Better*. The pair remained friends and collaborators throughout the last two decades of Pauling's life, both benefiting greatly from their cross-cultural exchange.

***

Most of Pauling's trips to Japan came in the 1980s. At the beginning of the decade, Linus and Ava Helen traveled across the Pacific for a busy stay that was mostly focused on vitamin C. Pauling began the visit by giving a talk to the general public on the health benefits of ascorbic acid. He then attended the general meeting of the Society of Japan Agricultural Chemistry at Fukuoka University — the topic of which was vitamin C and cancer — and was named an honorary member at its conclusion. Next, at Kyoto University, he gave a talk titled "What Can We Expect for Chemistry in the Next 100 Years?" after which he attended another symposium on vitamin C at

Gakushi Kaikan. Prior to returning home, Pauling gave one final lecture, "Prevention from Disease: Vitamin C, the Common Cold and Cancer," and also received an unusual award: an honorary blackbelt (7th dan) from the Federation of All Japan Karate-Do Organizations.

In 1981 Pauling returned to Japan for two short trips. The reason for the first was to attend the International Conference on Human Nutrition. During the second he appeared on Japanese television to discuss orthomolecular medicine with Drs. Kitahara and Morishige. A few days later he gave a lecture on the same topic to the Japanese Pharmacist Association.

Upon his return home, Pauling reached out to Fukumi Morishige to ask if he had tested his protocol on patients suffering from gastrointestinal cancer, noting a very personal motivation for doing so: this was the type of cancer from which Ava Helen was currently suffering. Morishige wrote back with a treatment plan that he thought might aid in slowing down the disease. Pauling attempted to act on this recommendation, but a variety of barriers arose to its implementation and, less than three months later, Ava Helen passed away.

***

In the years following, Pauling visited Japan three more times. These trips were spurred, at least in part, by the continuing need to secure financial support for the Linus Pauling Institute of Science and Medicine. Most notably, in concert with his travels in 1981, Pauling wrote to an industrialist named Ryoichi Sasakawa, asking his permission to establish a Ryoichi Sasakawa Research Professorship in Cancer Research at the Institute. Pauling further requested that Sasakawa endow the position, knowing of his support for cancer research in general and of Pauling's efforts to explore vitamin C in particular. Though Sasakawa did not fulfill this specific request, he did subsequently gift large sums to the Institute.

Ryoichi Sasakawa was among the more controversial of Linus Pauling's many acquaintances. To this day, opinions on Sasakawa tend to polarize: a politician, successful businessman and generous philanthropist, he was also considered by some to be a war criminal. Many Japanese additionally referred to him as a *kuromaku* — a shadowy force behind the visible power of a nation — because he had a hand in selecting two prime ministers and, as a result of his immense wealth, was a strongly influential player within Japanese politics. Sasakawa was also an avowed anti-communist and one-time admirer of Benito Mussolini.

Sasakawa himself admitted that he was a *kuromaku*; he thought them useful in a society where laws were ambiguous and enforcement was weak. This impulse toward power was, however, couched in the rhetoric of equality — words that were backed up by vast amounts of charitable giving. Especially in his later years, Sasakawa publicly espoused the notion that "the world is one family; all mankind are brothers and sisters," an idea that guided him in his philanthropy. Bringing peace to the masses became a stated life goal, and as a rich and powerful figure, Sasakawa saw himself as well-equipped to redistribute resources to the poor and needy of the world.

***

Sasakawa was born in 1899 to a sake brewer and grew up in a Buddhist household. As a young man he was fascinated by airplanes to such an extent that he ran away from home to learn to fly and was eventually drafted into the Japanese Air Force. After injuring his shoulder while working on an airplane, Sasakawa was discharged from the service early. His military career ended, he returned to his hometown and founded the *Konnichi Shimbun* newspaper.

Coming from a respected family, having experience in the Air Force and professing a zeal for making things right in the world, Sasakawa became the councilor of his village at the age of 22. Quickly, he reformed the local government and eradicated a major drinking problem that had impacted the community's leadership. At the same time, Sasakawa began accumulating his fortune by investing in the rice exchange. As his wealth grew, a business rival became jealous and had Sasakawa arrested for "charges unknown." Anticipating a possibility of this sort, Sasakawa had effectively sheltered his money before his arrest and emerged from the incident unscathed.

Not long after Sasakawa was released from custody, World War II engulfed the Pacific. Already a successful regional leader, he decided to involve himself further in the realm of national politics. Rather quickly, Sasakawa and Isoroku Yamamoto — the Commander-in-Chief of the Combined Fleet during World War II and, over time, a friend of Sasakawa's —

came to be known as the rightist leaders of the era. Both favored the ideals of fascism.

While Sasakawa and Yamamoto both spoke out against the outbreak of war, the two men were also strong patriots and did all they could to contribute to the success of Japan and its war effort. Notably, in 1932, Sasakawa became the leader of the nationalistic Volunteer Air Corps and became connected to a supply trade that included military goods and opium. Sasakawa was also a strong advocate for Yamamoto, and because the United States viewed Yamamoto as a warmonger and a threat, the Americans put Sasakawa on its watch list. With the defeat of Japan in 1945, Sasakawa's support of Yamamoto, coupled with his extreme nationalism and his contributions to "developing Japan's totalitarianism and aggression," earned him the label of war criminal. He was ordered by Douglas MacArthur's General Headquarters to face the Tokyo War Crimes Tribunal and did so with pride and cheering supporters. During his stint in prison that followed, he became an icon for a segment of the Japanese public.

As punishment for his war activities, Sasakawa had actually been sentenced to death. In 1948, however, he was released for unknown reasons. By his own account, during his period of incarceration Sasakawa made a vow that, if he survived his ordeal, he would dedicate his life to preventing war and seeking world peace. Once freed, he vowed to stay out of politics for the remainder of his life and wrote a resolution that begins,

"The most horrible sin on earth is killing, with war being the paramount example. Despite the dedicated efforts of numerous people in the cause to end all wars, human history has shown us nothing but a repetition of wars. We cannot possibly account for all the victims of wars to date, but the number would be unimaginable. The only way to allow the souls of the war dead to rest in peace is to bring about everlasting world peace and rid the earth forever of the horror of war, building a heaven on earth where all people can live in harmony as brothers and sisters. There is no doubt in my mind that anyone dedicated to this worthy cause is abiding by the will of Heaven and will enjoy eternal life. May God protect and lead us in our efforts to achieve an early realization of our goal."

While in prison Sasakawa also developed an interest in powerboats, and after his release he introduced motorboat racing and gambling to Japan, eventually founding the Japan Ship Promotion Company. This venture proved successful, and he was able to accumulate trillions of yen annually as a result.

Before he died in 1995, Sasakawa stepped up his efforts to help others, particularly by promoting good health. Over time, Sasakawa's Nippon Foundation, also known as the Japanese Shipbuilding Foundation, donated substantial sums to a wide variety of health-related projects. Working with the United Nations, the World Health Organization and a host of other groups, the foundation allocated tens of millions of dollars towards efforts to cure smallpox and leprosy, to control parasites and hunger in impoverished nations, to study population control worldwide, and to provide relief to areas hit by natural disasters.

Sasakawa also worked with former U.S. President Jimmy Carter to promote amicable relations among the world's people through a project called The Friendship Force. He likewise created the B & G Foundation, which built exercise facilities in hopes of fostering sound minds and healthy bodies for young people. For his efforts he was, in 1975, awarded the presidency of the Japanese Science Society and, in 1980, given the Golden Heart Presidential Award by the President of the Philippines.

It was in this light that Sasakawa also chose to support Linus Pauling and his research on vitamin C. Having heard of Pauling's work on the common cold, the flu and cancer, Sasakawa traveled to the U.S. in June 1980 to meet Pauling in person. While there, Sasakawa introduced the possibility of starting a program to fight leprosy with vitamin C. Pauling suggested contacts elsewhere who might be able to pursue this line of work, confiding that research of the sort was not something that the Institute was equipped to take on.

But the meeting planted the seeds of a relationship and over the next decade, Pauling and Sasakawa corresponded frequently and visited one another on multiple occasions. As their relationship deepened, Sasakawa became a generous supporter of the Linus Pauling Institute of Science and Medicine. Beginning in 1981, the Japanese Shipbuilding Foundation pledged five million dollars over ten years to support the Institute's research, primarily on vitamin C and cancer. The Institute also parlayed the foundation's support to establish the Sasakawa Aging Research Center, which used fruit flies to test theories of antioxidant protection against stress and in support of extending life span.

It is apparent from his records that Pauling was aware of Sasakawa's past and reputation in Japan. It is also clear that Sasakawa's funds were crucial to the Institute's ability to remain financially viable during some very difficult years in the 1980s. In acknowledgement of his support, the Institute bestowed upon Sasakawa the 1983 Linus Pauling Medal for Humanitarianism, an award that was usually given to important financial backers. And on at least six occasions in the 1980s, Pauling nominated Sasakawa for the Nobel Peace Prize. While Pauling often nominated multiple individuals for the award in a given year, and while his nominations of Sasakawa tended to be relatively brief, his expression of formal support is an important detail for those seeking to understand the contours of the two men's relationship.

***

Over the course of the 1980s, Pauling also cultivated a warm connection with author and educator Daisaku Ikeda. A resident of Tokyo and the son of a seaweed farmer, Ikeda witnessed first-hand the devastation that two nuclear bombs wrought upon his homeland. This experience instilled in him an insatiable yearning to understand and eliminate the sources of war.

In pursuing this ambition, Ikeda studied political science at what is now Tokyo Fuji University and committed himself to the pacifist lifestyle of a Nichiren Buddhist. Ikeda's chosen faith, named after a 12th-century priest who emphasized the Lotus Sutra as the authoritative text for adherents of Buddhism, was becoming extremely popular among East Asians following World War II. Fundamental to the practice's message was a strong call to treat others with respect and compassion, recognizing that all will become Buddhas in the end.

Ikeda also joined a new religious organization called Soka Gakkai, which followed the teachings of Nichiren, and he ultimately became the group's president in 1960. Working in this capacity, Ikeda focused intently on opening Japan's relationship with China and establishing the Soka education network of humanistic schools from kindergarten through university. He also used historical fiction to document the early years of the organization in a series of newspaper articles that were eventually published as a book.

As his tenure as leader advanced, Soka Gakkai grew into an international network of communities dedicated to peace and to cultural and educational activities. In 1975 Ikeda founded an umbrella organization known as Soka Gakkai International (SGI) to fund, direct resources, and help facilitate communication between the dispersed Soka Gakkai membership. In the 1980s, he turned his attentions toward anti-nuclear activism and citizen diplomacy, and it was in this capacity that he came into close contact with Linus Pauling.

***

Pauling's first interaction with SGI came in the early 1980s, by which time the non-governmental organization was already actively cooperating with the United Nations' department of public information to mobilize citizens demanding peace. Seeking to increase SGI's influence in propelling the peace movement, Ikeda reached out to Pauling, who was by now splitting his time between the family ranch in Big Sur, California and the Linus Pauling Institute of Science and Medicine in Palo Alto. It was at the latter location where Ikeda's associate, Mr. Tomosaburo Hirano, first met and interviewed Pauling.

This meeting proved to be the first step in a lengthy "courtship" that involved extensive correspondence between Pauling's secretary, Dorothy Munro, and Ikeda's assistant, Hirano. In fact, more than six years would pass before Pauling communicated directly with Ikeda and, a bit later on, finally met Ikeda in person.

Over the course of those six years, Hirano interviewed Pauling two more times, focusing primarily on Pauling's views on peace and, to a lesser degree, his scientific work. Extracts from these sessions were often published in the *Seikyo Shimbun Press*, Soka Gakkai's daily newspaper in Japan, for which Hirano served as associate editor. The pieces typically highlighted Pauling's anti-nuclear activism and were often published in tandem with Ikeda's release of strategic articles bearing titles like "A New Proposal for Peace and Disarmament" and "Toward A Global Movement for a Lasting Peace."

Finally, during a 1987 trip to Los Angeles, Ikeda requested a personal meeting with Pauling, which the scientist obliged. Face to face at last, the esteemed acquaintances developed an easy rapport with one another, quickly exhausting the time allotted for their meeting and discussing (with a translator) a wide range of subjects including science, peace, childhood and adult life. The conversation even drifted into Pauling's hobby of collecting different editions of the *Encyclopedia Britannica*.

Ikeda was fascinated by Pauling's fond recollections of major figures like Albert Einstein, Albert Schweitzer, Bertrand Russell and, of course, Ava Helen Pauling, whose influence Pauling cited as having been directly responsible for his peace activism. The two also talked about Pauling's Nobel Peace Prize lecture, in which he discussed the future of war and the need for disputes to be solved by world law.

In that same lecture, Pauling emphasized that, were it up to him, he would prefer to be remembered as the person who discovered the hybridization of bond orbitals, rather than through his work to reduce nuclear testing and eliminate war. Nonetheless, Pauling considered the Nobel Peace Prize to be the highest honor that he had ever received, in particular because of the onus that it placed upon him to continue that work. By contrast, Pauling felt that his Nobel Chemistry Prize, awarded in 1954, had been earned for work already completed.

As they continued their conversation, Ikeda also learned more about Pauling's ideas on the minimization and prevention of suffering for all human beings. In this, Pauling's point of view as a humanist matched up well with Ikeda's Buddhist philosophy. Specifically, Ikeda's faith taught that one should regard others' sufferings as their own and should seek to eliminate it — a principle also expressed in the teachings of Christ, Kant's Categorical Imperative, and the Analects of Confucius, and more generally known in the U.S. as the Golden Rule.

Though Pauling was an avowed atheist, Ikeda pointed out that he did not feel his own religion to be an impediment to his rationality — the same rationality that Pauling believed guided his own desire for peace. Rather, Ikeda argued that

"Religions must make every effort to avoid both bias and dogma. If they fail in this, they lose the ability to establish a sound humanism and can even distort human nature. The twenty-first century has no need of religions of this kind."

***

Pauling and Ikeda's 1987 conversation clearly made an impression on both men. Not long after, Ikeda's assistant wrote to Pauling again, thanking him for the meeting and asking if he might consider authoring a manuscript for publication in Japan.

In 1988, about a year after their first visit, Ikeda wrote to Pauling directly to express interest in co-authoring a dialogue that might "provide some suggestion for the young generation who are to shoulder the responsibility in the 21st century, as well as serve the cause of peace and prosperity of humankind." The dialogue would be published in an interview format and would be based on the transcript of their meeting in Los Angeles and supplemented by additional material. The first step toward completion was for Pauling to answer a series of 73 questions probing his outlook on life. Pauling was interested and promptly responded.

Many of the project's supplemental questions concerned the evolution of Pauling's views on war and peace over the course of his life. Pauling began by explaining that he was only 13 years old when World War I started and had few thoughts about international relations at the time. He did recall the conclusion of the war in 1918 and participating in a victory parade held in Corvallis, Oregon, where he marched alongside other cadets serving in the Oregon Agricultural College Army Reserve Officers Training Corps.

By the dawn of the Second World War, Pauling was well-established at the California Institute of Technology. During the war years he directed much of his energy toward projects sponsored by the explosives division of the National Defense Research Committee, where his expertise was used to support the killing and maiming of enemy soldiers, including the Japanese. Though he would spend much of his life working to limit the amounts of human suffering on Earth, Pauling commented that he felt satisfaction at the conclusion of the war, heartened that Hitler "and his associates" had been denied their goal of gaining control of the planet. Afterward though, the emerging development of nuclear weapons and the ongoing threat of nuclear war prompted Pauling to ramp up his peace activism. Over time, this point of view evolved into a desire to eliminate all war from Earth.

In 1990 the Japanese version of the Pauling-Ikeda dialogue, titled *In Quest of the Century of Life*, was published. Not long after, Pauling delivered a commemorative lecture at the second Soka University Pacific Basin Symposium, held at the Los Angeles campus of Soka Gakkai University.

The next year, Pauling and Ikeda, along with Johan Galtung, the Norwegian founder of the discipline of Peace and Conflict Studies, signed the Oslo Appeal. This document urged the United Nations to require that nuclear member states agree to a joint Nuclear Test Ban Treaty and Nuclear Non-Proliferation Treaty; outlaw the production and stockpiling of chemical and biological weapons; prohibit the international weapons trade; and sponsor an international conference to discuss the redirection of resources released through disarmament to support development in the Third World.

Pauling subsequently received the Daisaku Ikeda Medal for Peace, awarded by SGI in 1992. Later that year, the English translation of his and Ikeda's dialogue was published in the West under the title, *A Lifelong Quest for Peace.*

Following Pauling's death in 1994, Ikeda expressed a desire to honor his friend with a travelling exhibit that would be funded by Soka Gakkai. Initially meant to educate the public on ideas in chemistry and introduce children to Pauling as a role model, the project's scope later shifted toward honoring all facets of Pauling's career as a humanitarian, activist, scientist, and medical researcher. Once finalized, the exhibit toured the world for six years. Millions of people saw it in Europe and Japan, as well as multiple locations in the United States, including Washington D.C., San Francisco, Boston, and Pauling's birthplace, Portland, Oregon.

Chapter 31

Two Trips to China[1]

Ava Helen and Linus Pauling at the Great Wall of China, fall 1973.

[1] Names of Chinese individuals in this chapter are written with family name placed first, as per Chinese nomenclature. For example, in Wu Yu-Hsun, Wu is the family name (or last name, as Westerners prefer to call it).

In March 1973, little more than one year after U.S. President Richard Nixon's historic visit, Linus Pauling received a letter inviting him to travel to China for three weeks in the coming summer. He was invited by Wu Yu-Hsun, Vice President of the Scientific and Technical Association of the People's Republic of China, who informed Pauling that his accommodations and transportation would be provided by the Association. "It is my belief that your visit will contribute to the promotion of the traditional friendship and scientific exchanges between the scholars of China and America," Wu wrote.

Following Wu's instructions for obtaining a visa, Pauling wrote to the Embassy of the Chinese People's Republic in Ottawa, Canada, requesting papers for him and Ava Helen. (At that time, there was no Chinese embassy in the United States as diplomatic relations between the two countries had not yet been formalized.) Two months later, he received a reply from the embassy, accompanied by applications for the travel documents. On August 8 Pauling wrote to Vice President Wu to let him know that the trip details had been finalized and informing him of plans to arrive in Hong Kong on September 16 and to leave on October 8.

In his letter, Pauling mentioned his book *Vitamin C and the Common Cold*, stating his belief that vitamin C not only decreases the severity and instances of the common cold, but does the same for other diseases. As such, he expressed a desire to engage with relevant Chinese medical authorities and members of the Ministry of Public Health about this matter. Pauling also communicated his interest in talking with physicians and scientists about oxypolygelatin, a blood plasma substitute that he had developed during World War II. Finally, it was Pauling's wish to see his former student, Chia-si Lu, a chemist and crystallographer, and also Professor Tsien, an authority on rockets whom he knew from Caltech. On August 9, Pauling returned the applications along with a letter stating that "I do not travel without my wife, and I have assumed that the invitation includes her also."

After receiving their visas, the Paulings departed San Francisco for Hong Kong on Friday, September 14, 1973. They spent that night in Honolulu, and arrived in Hong Kong on Sunday, September 16.

***

Much of what we know about the Paulings' first visit to China comes from Linus's travel log. The document is somewhat dry and very detail-oriented — so much so that one wonders to what extent the specifics were owed to Pauling's insatiable scientific appetite versus his knowledge that the U.S. government was historically suspicious of him and likely maintained a particular interest in his activities while traveling through communist China.

The Paulings arrived in Hong Kong on Sunday night, stayed an extra day, and went by train to Canton on Tuesday. They spent the night there at a guest house, where Pauling noted that it was "very hot during the day, very humid, and humid and hot during the night, too." After visiting Sun Yat-sen University and having lunch, the couple flew to Shanghai, where Pauling judged the humidity to be less oppressive. They toured the Shanghai Industrial Exhibition on the morning of September 20 and the Institute of Biochemistry in the afternoon. In his travel log, Pauling recorded the details of research being conducted by an Institute staff member who was working on nucleotides and nucleosides. One investigation in particular focused on the effectiveness of nucleotides in increasing the yields of different plants such as rice. Pauling also spoke with a Mr. Kung who, in 1965, was among the first scientists to synthesize insulin, and a man named Lee, who was conducting work on liver cancer.

Pauling was particularly interested in a screening of 150,000 Shanghai residents that was described to him by Mr. Lee. In the screening, 158 people were found to have an embryonic globulin in their blood that is manufactured in large amounts by people who have liver cancer. All of these 158 subjects either already had cancer or developed it later. Pauling suggested that the people who tested positive for this embryonic globulin be given 10 grams of vitamin C per day to stave off further progression of the cancer.

While in Shanghai, the Paulings frequently saw members of the Philadelphia Philharmonic at mealtimes, as they were staying at the same hotel. Pauling noted that he and Ava Helen had tickets to hear the orchestra on September 21. Earlier that same day he visited the Peking Institute of Organic Chemistry of the Academia Sinica, where he saw the laboratories and learned about the Institute's work on steroids. Meanwhile, Ava Helen went shopping and visited the zoo where, Pauling's record shows, she saw "three giant pandas and several small ones."

In the afternoon, the Paulings toured a commune, likely one of many established by Mao Zedong in the late 1950s with the aim of turning China into an industrialized nation. At the commune, Pauling recorded his impressions of the work and lifestyle of its 24,000 inhabitants, who mostly made tools or farmed the land.

On Saturday, September 22, the Paulings visited the Shanghai Institute of Pharmacology and later the Shanghai Psychiatric Hospital, where they observed a selection of treatments being given to patients, including acupuncture. Pauling presented the director of the hospital with a copy of his book, *Orthomolecular Psychiatry*, and discussed megavitamin therapy with the hospital's staff. Afterwards the Paulings watched an acrobatics performance. The next day they continued to enjoy China's culture by visiting the Children's Palace and the Palace of History.

When the couple went on a sightseeing tour with Chia-si Lu, Pauling's former student, Chia-si told them of the hardship that had existed before China's "liberation." By liberation, Pauling's guide was referring to the 1949 Chinese Revolution in which Mao and his supporters took over China's government and installed communist rule. According to Chia-si, who had been in the U.S. for five years during the 1940s, only about ten percent of the money that he periodically sent home to his wife and son actually reached them; the rest was taken by the Bank of China. For a few years, Chia-si's wife and son were close to starvation, as were many other people in China. However, after the revolution, the new Maoist government controlled the price and distribution of food and, in Chia-si's estimation, the quality of life improved. (Note: this perspective is contrary to other scholarly analyses of Chinese food security under Mao.)

Although a few days are excluded from his formal travel log, Pauling wrote notes in his daily planner about activities related to hemoglobin and orthomolecular medicine on September 24, and a trip by train to Hangzhou that evening, where he and Ava Helen took in the local attractions. Their tourism included the Ling Yin Temple, the Tiger Spring, the Jade Fountain, a tea ceremony, a boat ride on West Lake and a visit to a brocade factory. They also attended another in a long string of banquets and saw the Dragon Well Spring before leaving by train.

Pauling next wrote in his journal on Friday, September 28 to record his and Ava Helen's tour of a petroleum refinery. The day before, on Thursday, they had visited the big bridge across the Yangtze, after which Pauling gave a lecture on vitamin C and good health at Nanjing University. On Thursday night, the Paulings went to a song and dance performance staged by children from the district of Nanjing. Then that weekend, the couple attended the National Day banquet in Beijing, held in the dining room of the Hall of Ten Thousand, to which 1,000 guests were invited by the Minister of Foreign Affairs. Before the banquet, the Paulings visited the Forbidden City once in the morning and again in the afternoon.

Pauling's log does not contain any entries after September 30, which leaves the next week in the country unaccounted for. However, other records indicate that the Paulings did get their desired opportunity to speak with scientists Ma Hai-teh and Rewi Alley on the subject of oxypolygelatin, a substance that Linus hadn't touched in 30 years but had clearly been thinking about. Pauling wrote to Ma a few months later, in February 1974, to tell him about further work being conducted on the serum. In his letter, Pauling intimated that he had the idea that the properties of gelatin as a plasma extender would be improved if the long thin gelatin molecules could be tied together into rosettes using hydrogen peroxide, such that the molecules would not escape into the dilate urine through pores in the glomerular filter. This would improve the substance's time of retention in the body.

***

In the summer of 1981, Pauling participated in the First International Conference on Human Nutrition, which took place in Japan and China. The meeting lasted from May 31 to June 8 and was sponsored by the China Medical Association and the Foundation for Nutritional Advancement (FNA), the latter of which Pauling was president. The conference was held sequentially in two locations, Tokyo, Japan followed by Tianjin, China, travels to which would comprise the first part of a trip that would also take the Paulings to Germany and London.

Pauling made the opening remarks at the beginning of the conference on June 1. After the first set of sessions was completed three days later, the Paulings flew to Peking, traveled in an official vehicle to Tianjin (a "red flag limousine," as recorded by Pauling in his journal) and stayed in the State Guest House in the same suite used by Richard Nixon during his iconic 1972 visit. From June 4–8, Pauling participated in the China leg of the conference, which was jointly planned by the FNA and Professor Chou Pei-yuan, the President of Peking University.

A day after arriving in China, the Paulings toured Tianjin Medical College, Tianjin Hospital and Tianjin Children's Hospital before attending a formal reception given by Li Xiannian, who eventually became the Chinese Head of State in 1983. The conference formally opened on June 6, again with Pauling delivering the plenary remarks.

In his talk, Pauling explained that he had decided to learn more about organic chemistry in an effort to better understand how molecules are built and how they interact with each other. During this time, Pauling studied hemoglobin, antibodies, immunology, sickle cell anemia, and other hereditary anemias. In 1954 he decided to look at other groups of diseases to see if they could be classed as molecular diseases and chose to study mental illness over cancer because he felt that many people were working on cancer already.

After researching mental illness for ten years, he became interested in vitamins in part because of work that Canadian scientists Abram Hoffer and Humphry Osmond were doing to treat schizophrenia patients with megadoses of niacin. Around this time, a different researcher, Gerald Milner, had been giving large amounts of ascorbic acid to mentally ill patients, with positive results. Pauling later observed that vitamin C had value in the control of cancer and turned his focus in that direction. Near the end of his address, Pauling remarked, "As I look back on my life, I see that I have enjoyed myself very much and a good bit of this enjoyment has come from the continued recognition of something new about the universe."

The conference closed on June 8 and the next day the Paulings took part in a tour of the Great Wall and the Ming tombs. Later that week, Pauling gave a talk on chemical bonds in transition metals at Peking University and continued to meet with various scientists throughout the rest of his time in China.

***

The trip took a dramatic turn for the worse when, on the afternoon of June 19, Ava Helen had a heart attack and was taken to the hospital. Though she left the facility the next day, she remained medicated and too sick to travel for a few days following, causing the Paulings to change their plans. She felt weak for the rest of their time in China but recovered enough to complete their planned itinerary through Germany and England.

When the couple returned to California and Ava Helen underwent exploratory surgery, it was determined that she was facing a recurrence of the stomach cancer that had plagued her for the past five years. Following close colleague Ewan Cameron's advice, she took 10 grams of vitamin C daily. She also opted against receiving chemotherapy.

Throughout her treatment, Linus clung to the belief that megadoses of vitamin C would work for Ava Helen, just as it had for Cameron's success stories in Scotland. "Daddy was convinced that he was going to save her," remembered daughter Linda. "And that was, I think, the only reason he was able to survive… He said to me after she died that until five days before, he thought he was going to be able to save her."

After more surgeries and numerous medical complications, Ava Helen died in her home on December 7, 1981, three weeks shy of her 78th birthday. The devastation that her husband felt was palpable and long-lasting. In a handwritten letter to his children written on March 23, 1982, Pauling provided "a report about how I am getting along after your mother's death," adding that, "I find it hard to talk about it, but possible to write about it."

"I am getting along moderately well, and also moderately poorly. I still enjoy doing my work, especially the theoretical work on molecular structure and nuclear structure. I am able to handle the burden of work at the Institute.

"My evaluation of my situation is that I get along most of the time by managing to forget that Ava Helen has died. I think that when I am traveling, I think that she is at home in Portola Valley waiting for me. When I am in the Institute, I think that she is at home and that I shall soon see her. When I am in our house in Portola Valley, working at a table, I think that she is in another room. When I am at the ranch too, I think that she is in another room.

"But then, several times a day, something happens to make me realize that these thoughts are not true, and that my dear companion is indeed dead. I usually then experience a paroxysm of grief, such as I have never experienced before. I usually begin to cry (as now, on the plane from Oklahoma City to San Francisco). I usually also begin to moan. So far as I can remember, I have never before moaned. Indeed, I have been moaning here on the plane, but not very loudly. I think that no one has noticed my tears or my moaning — the seat next to me is vacant."

Nearly two years later, on February 27, 1984, the day before his 83rd birthday, Pauling penned another tear-stained manuscript.

"Just before she died, my wife said to me, 'Our molecules are together.' She was, I am sure, happy with this thought — the thought that she and I, as unique individuals, had not lost completely the uniqueness of our association by her death — by the fact that her body, like that of Julius Caeser, would soon be turned to clay, or, rather, for her to molecules of carbon dioxide, water, dinitrogen, and inorganic ash. Instead, some of the molecules of DNA that had determined her nature and some of those that had determined my nature were present together in the bodies of our four children and had cooperated in determining their nature; and they had cooperated also, somewhat diluted, in determining the nature of our fifteen grandchildren and our great-grandchildren, then three in number. She was very very very very, very very very very, very very very fond of me, just as I was very very very very, very very very very, very very very very fond of her, through the sixty years of our association, and I am sure that despite her feeling of sadness that our life together was ending, she found some consolation in the knowledge that our cooperation was not being ended by her death."

Sourcing

Much of the work presented in this book relied upon two secondary sources to provide basic context: *Force of Nature: The Life of Linus Pauling* (1995) by Thomas Hager, and *The Pauling Chronology* (2001) by Robert Paradowski. From there, original research was conducted primarily in the Ava Helen and Linus Pauling Papers (MSS Pauling) at the Oregon State University (OSU) Libraries Special Collections and Archives Research Center (SCARC), with primary attention paid to the components of the collections identified below.

Preface

- Images: Pauling at the Grand Canyon (MSS Pauling — 194?i.65); Clifford Mead, Linus Pauling and Tom Hager at OSU, 1991. (courtesy of Clifford Mead)
- For more on the history of the Pauling Papers see: Petersen, Chris. "The Ava Helen and Linus Pauling Papers at Oregon State University: A Short History," *Archivoz*, September 2, 2019. Available online at: https://www.archivozmagazine.org/en/the-ava-helen-and-linus-pauling-papers-at-oregon-state-university-a-short-history/
- Referenced works by or related to Robert Paradowski:
  - Davenport, Derek. "The Many Lives of Linus Pauling: A Review of Reviews," *Journal of Chemical Education*, 73, 9, September 1996.
  - Paradowski, Robert J. *The Structural Chemistry of Linus Pauling.* Doctoral dissertation. University of Wisconsin-Madison, 1972.
  - Paradowski, Robert J. "The Pauling Chronology." Originally published in *Linus Pauling — A Man of Intellect and Action*, Cosmos Japan International, Tokyo, 1991. Republished in English, Spanish and German on the OSU Libraries Special Collections/SCARC website. See, for example, https://scarc.library.oregonstate.edu/coll/pauling/chronology/page1.html.

- o Paradowski, Robert J. "An American in Munich: Truth and Controversy in the Life and Work of Linus Pauling during the Golden Years of Physics." Presentation delivered at *A Liking for the Truth: Truth and Controversy in the Work of Linus Pauling (1901–1994)*, [Pauling centenary conference], Oregon State University, Corvallis, Oregon, February 28, 2001. Available online at: https://scarc.library.oregonstate.edu/events/2001paulingconference/video-s2-2-paradowski.html.
- o "Panel Discussion and Closing Remarks," *A Liking for the Truth: Truth and Controversy in the Work of Linus Pauling (1901–1994)*, [Pauling centenary conference], Oregon State University, Corvallis, Oregon, February 28, 2001. Available online at: https://scarc.library.oregonstate.edu/events/2001paulingconference/video-s2-4-panel.html.
- Referenced works by Thomas Hager:
  - o Hager, Thomas. *Force of Nature: The Life of Linus Pauling*, New York: Simon & Schuster, 1995.
  - o Hager, Thomas. *The Demon Under the Microscope: From Battlefield Hospitals to Nazi Labs, One Doctor's Heroic Search for the World's First Miracle Drug*, New York: Harmony Books, 2006.
  - o Hager, Thomas. *The Alchemy of Air: A Jewish Genius, a Doomed Tycoon, and the Scientific Discovery that Fed the World but Fueled the Rise of Hitler*, New York: Harmony Books, 2008.
  - o Hager, Thomas. *Ten Drugs: How Plants, Powders, and Pills have Shaped the History of Medicine*, New York: Abrams Press, 2019.
- Other Referenced Works
  - o Linus Pauling Jr.'s reflections on his father in the 1960s can be found in OSU SCARC, History of Science Oral History Collection (OH 017). Oral History Interview of Linus Pauling, Jr. by Chris Petersen, June 7, 2012.
  - o Petersen, Chris, ed. *Visions of Linus Pauling*, Singapore; Hackensack, New Jersey: World Scientific, 2023.

Part I: Pauling in Oregon

1. Herman Pauling: Striving for a Better Life

- Blog Posts:
  - o **The Paternal Ancestry of Linus Pauling** (September 23, 2008), by Chris Petersen

- o **Herman Pauling's Letter to The Oregonian** (January 20, 2009), by Chris Petersen
- o **Condon, Oregon: Pauling's Wild West** (February 3, 2009), by Trevor Sandgathe
- o **Snapshots of Pauling's Childhood in Condon** (March 3, 2009), by Trevor Sandgathe
- o **Herman Pauling's Condon Pharmacy** (March 20, 2009), by Ben Jager
- o **The Life of Herman Pauling — Linus Pauling's Father** (October 15, 2014), by Andy Hahn
- o **Herman Pauling: Striving for a Better Life** (October 22, 2014), by Andy Hahn
- Major Sourcing:
  - o Image: Herman Pauling with young children Linus and Pauline, 1903. (MSS Pauling, 1903i.5)
  - o Much of what we know about Herman Pauling, Belle Pauling and Linus Pauling's early years comes to us via Thomas Hager's research and writing. In addition to the primary and secondary sources that he itemized in his biography, a major resource for those interested in this topic are the many interviews that he conducted in support of the project. SCARC is home to Hager's papers and these interviews have been cataloged into his collection as follows: MSS Hager, Series 2: *Force of Nature*: Interviews — Audio Cassettes and Transcriptions. Nearly all of Hager's interviews with Pauling have been digitized, transcribed and made available online here: https://scarc.library. oregonstate.edu/omeka/exhibits/show/oacvoices/pauling/.
  - o MSS Pauling, Series 1: Correspondence. Box 305. Paradowski, Robert J.
  - o MSS Pauling, Series 7: Newspaper Clippings. 1910n.1 "Reading for 9-Year-Old Boy"; "Advice to a Bright Boy," *Portland Oregonian*, May 13, 1910.
  - o MSS Pauling, Series 13: Biographical, Sub-Series 5: Personal Materials and Family Correspondence. Box 5.001. Pauling Family Genealogical Materials; Box 5.003. Materials re: the early life of Linus Pauling; Box 5.004. Publications: *The History of Gilliam County, Oregon*, 1981; *A Pictorial History of Gilliam County*, 1983.

- Herman and Belle Pauling's correspondence remains with the Pauling family and is not a component of MSS Pauling.

2. **Belle Pauling: Hard Times**
- Blog Posts
  - **The Darlings: Maternal Ancestors of Linus Pauling** (October 9, 2008), by Chris Petersen
  - **Condon, Oregon: Pauling's Wild West** (February 3, 2009), by Trevor Sandgathe
  - **The Life of Belle Pauling — Linus Pauling's Mother** (October 29, 2014), by Andy Hahn
  - **Belle Pauling: Hard Times** (November 5, 2014), by Andy Hahn
- Major Sourcing
  - Image: Portrait of Belle Pauling, early 1900s. (MSS Pauling, 190?i.28)
  - MSS Hager, Series 2: *Force of Nature*: Interviews — Audio Cassettes and Transcriptions.
  - MSS Pauling, Series 1: Correspondence. Box 305. Paradowski, Robert J.
  - MSS Pauling, Series 13: Biographical, Sub-Series 5: Personal Materials and Family Correspondence. Box 5.001. Pauling Family Genealogical Materials; Box 5.003. Materials re: the early life of Linus Pauling; Box 5.004. Publications: *The History of Gilliam County, Oregon*, 1981; *A Pictorial History of Gilliam County*, 1983; Box 5.050. Family Correspondence: Lucy Isabelle Pauling, Herman Pauling and Herman W. Darling.

3. **Pauline and Lucile Pauling**
- Blog Posts
  - **Pauline Pauling (1902–2003)** (July 16, 2009), by Jamee Asher
  - **Lucile Pauling (1904–1992)** July 21, 2009), by Jamee Asher
- Major Sourcing
  - Image: Lucile, Linus, Belle and Pauline Pauling, 1922. (MSS Pauling, 1922i.27)
  - MSS Hager, Series 2. Interview with Lucile and Pauline Pauling, March 26, 1990.

- o MSS Pauling, Series 13: Biographical, Sub-Series 5: Personal Materials and Family Correspondence. Box 5.051. Family Correspondence: Pauline Pauling; Box 5.052. Pauline Pauling: Scrapbooks of Assorted Photos and Newspaper Clippings; Box 5.053. Family Correspondence: Frances Lucile Pauling.

4. **Pauling's Adolescence**
- Blog Posts
  - o **Portland, OR: Pauling's Teenage Years** (April 23, 2009), by Trevor Sandgathe
  - o **Working in Oregon: The Blue Collar Adolescence of Linus Pauling** (May 7, 2009), by Trevor Sandgathe
  - o **A Prominent High School Dropout** (June 23, 2009), by Ben Jager
  - o **Pauling's Freshman Diary, Part 1** (September 15, 2009), by Ben Jager
  - o **Pauling's Freshman Diary, Part 2** (September 17, 2009), by Ben Jager
  - o **Think Independently: Pauling's Years at OAC** (June 19, 2014), by Jindan Chen
- Major Sourcing
  - o Image: Linus Pauling as a teenager, ca. 1910s. (MSS Pauling, 191?.158)
  - o MSS Hager, Series 2: *Force of Nature*: Interviews — Audio Cassettes and Transcriptions.
  - o MSS Pauling, Series 1: Correspondence. Box 305. Paradowski, Robert J.; Folder 441.1 Washington High School.
  - o MSS Pauling, Series 13: Biographical, Sub-Series 1: Academia. Box 1.001, Folder 1.1 "Diary (so-called)", by Linus Pauling, 1917–1918; Sub-Series 5: Personal Materials and Family Correspondence. Box 5.003. Materials re: the early life of Linus Pauling; Sub-Series 6: Scrapbook Pages Mounted by Linus Pauling. Item 8.287 "Dr. Pauling, Non-Grad of WHS", Washington High School (Portland, Oregon) *Washingtonian*, May 11, 1962; Item 8.293 "Worldly Physicist Gets Only Honorary Washington Degree", Washington High School (Portland Oregon) *Washingtonian*, June 8, 1962, and "New Honor for Pauling: A High School Diploma", *New York Times*, June 19, 1962.
  - o MSS Pauling, Series 16: Personal Safe, Sub-Series 2. Item 25.31 Student record for LP, Washington High School, 1916.

5. **Pauling's Freshman Year at Oregon Agricultural College, 1917–1918**
- Blog Posts
    - o **Pauling in the ROTC** (November 13, 2008), by Chris Petersen
    - o **Working in Oregon: The Blue Collar Adolescence of Linus Pauling** (May 7, 2009), by Trevor Sandgathe
    - o **Becoming Dr. Pauling** (March 5, 2015), by Luis Marquez Loza
    - o **Pauling's OAC** (September 20, 2017), by McKenzie Ross
    - o **Study and Social Life at the Oregon Agricultural College** (September 27, 2017), by McKenzie Ross
    - o **Rook Life** (October 4, 2017), by McKenzie Ross
    - o **Pauling's OAC: Completing the Freshman Year** (October 11, 2017), by McKenzie Ross
- Major Sourcing
    - o Image: Pauling wearing his OAC freshman beanie, ca. 1917. (MSS Pauling, 191?i.160)
    - o MSS Hager, Series 2: *Force of Nature*: Interviews — Audio Cassettes and Transcriptions.
    - o MSS Pauling, Series 13: Biographical, Sub-Series 1: Academia. Box 1.001. Oregon Agricultural College: Diary, Academic Work, Biographical Information, etc.; Box 1.002. Oregon Agricultural College: Academic Work, Biographical Information, etc.; Sub-Series 5: Personal Materials and Family Correspondence. Box 5.044, Folder 44.1, Letter from Linus Pauling to Peter Pauling, June 10, 1960.
    - o OAC *Beaver* Yearbook, 1919 (1917/1918 academic year). Available online: https://oregondigital.org/concern/documents/9w0323217.
    - o OAC General Catalog, 1917–1918. Available online: https://oregon-digital.org/concern/documents/fx719v040.
    - o *The Daily Barometer* digital collection. Available online: https://oregondigital.org/collections/daily-barometer.

6. **Pauling's Sophomore Year at Oregon Agricultural College, 1918–1919**
- Blog Posts
    - o **Working in Oregon: The Blue Collar Adolescence of Linus Pauling** (May 7, 2009), by Trevor Sandgathe
    - o **Fraternity Life** (September 29, 2009), by Ben Jager

- o **Pauling's OAC: Sophomore Year** (September 19, 2018), by Sarah Litwin
- o **Pauling's OAC: Life During Wartime** (September 26, 2018), by Sarah Litwin
- o **Pauling's OAC: Sophomore Social Life** (October 3, 2018), by Sarah Litwin
- o **The "Spanish Flu" at OAC, 1918–1919** (October 7, 2020), by Anna Dvorak
- • Major Sourcing
  - o Image: Pauling in his ROTC military dress, 1918. (MSS Pauling, 1918i.33)
  - o MSS Hager, Series 2: *Force of Nature*: Interviews — Audio Cassettes and Transcriptions.
  - o MSS Pauling, Series 1: Correspondence. Folder 97.2 Delta Upsilon.
  - o MSS Pauling, Series 13: Biographical, Sub-Series 1: Academia. Box 1.001. Oregon Agricultural College: Diary, Academic Work, Biographical Information, etc.; Box 1.002. Oregon Agricultural College: Academic Work, Biographical Information, etc.
  - o MSS Pauling, Series 16: Personal Safe, Sub-Series 2. Item 39.102 "Petition of the Gamma Tau Beta Fraternity of the Oregon State Agricultural College to the Delta Upsilon Fraternity, October 12, 13, 14, 1916."
  - o OAC *Beaver* Yearbook, 1920 (1918/1919 academic year). Available online: https://oregondigital.org/concern/documents/xd07gt046.
  - o OAC General Catalog, 1918–1919. Available online: https://oregondigital.org/concern/documents/fx719v058.
  - o *The Daily Barometer* digital collection. Available online: https://oregondigital.org/collections/daily-barometer.

7. The Boy Professor, 1919–1920

- • Blog Posts
  - o **The Paving Inspector Job** (November 25, 2009), by Will Clark
  - o **Think Independently: Pauling's Years at OAC** (June 19, 2014), by Jindan Chen

- o **Pauling's OAC, 1919-1920: The Boy Professor** (September 18, 2019), by Sarah Litwin
- o **Pauling's OAC, 1919-1920: The Campus Scene** (September 25, 2019), by Sarah Litwin
- Major Sourcing
  - o Image: Portrait of Linus Pauling, 1920. (MSS Pauling, 1920i.1)
  - o MSS Hager, Series 2: *Force of Nature*: Interviews — Audio Cassettes and Transcriptions.
  - o MSS Pauling, Series 1: Correspondence. Folder 121.5 Fulton, John.
  - o MSS Pauling, Series 2: Publications. 1920p.1. The manufacture of cement in Oregon. *The Student Engineer* (The Associated Engineers of Oregon Agricultural College, Corvallis, Oregon) 12 (June 1920): 3-5.
  - o MSS Pauling, Series 13: Biographical, Sub-Series 1: Academia. Box 1.001. Oregon Agricultural College: Diary, Academic Work, Biographical Information, etc.; Box 1.002. Oregon Agricultural College: Academic Work, Biographical Information, etc.
  - o OAC *Beaver* Yearbook, 1921 (1919/1920 academic year). Available online: https://oregondigital.org/concern/documents/7d278t39s.
  - o OAC General Catalog, 1919–1920. Available online: https://oregondigital.org/concern/documents/fx719v22x.
  - o *The Daily Barometer* digital collection. Available online: https://oregondigital.org/collections/daily-barometer.

8. **Pauling's Junior Year at Oregon Agricultural College, 1920–1921**

- Blog Posts
  - o **Think Independently: Pauling's Years at OAC** (June 19, 2014), by Jindan Chen
  - o **Children of the Dawn** (June 5, 2015), by Luis Marquez Loza
  - o **Pauling's OAC, 1920-21: A True Junior Year** (October 14, 2020), by Miriam Lipton
  - o **Pauling's OAC: A Maturing Relationship with Chemistry** (October 21, 2020), by Miriam Lipton
  - o **Pauling's OAC: A New Decade** (October 28, 2020), by Miriam Lipton

- Major Sourcing
  - Image: Pauling at the OAC Interfraternity Smoker, cropped from a larger group photograph. (MSS Pauling, 1920i.27)
  - MSS Hager, Series 2: *Force of Nature*: Interviews — Audio Cassettes and Transcriptions.
  - MSS Pauling, Series 4: Speeches. 1921s.1 "Children of the Dawn," Oregon Agricultural College, Corvallis, Oregon, Spring 1921.
  - MSS Pauling, Series 13: Biographical, Sub-Series 1: Academia. Box 1.001. Oregon Agricultural College: Diary, Academic Work, Biographical Information, etc.; Box 1.002. Oregon Agricultural College: Academic Work, Biographical Information, etc.; Box 1.004. Oregon Agricultural College: Lab Book from Metallography 426.
  - OAC *Beaver* Yearbook, 1922 (1920/1921 academic year). Available online: https://oregondigital.org/concern/documents/3x816m87j.
  - OAC General Catalog, 1920-1921. Available online: https://oregondigital.org/concern/documents/fx719v236.
  - *The Daily Barometer* digital collection. Available online: https://oregondigital.org/collections/daily-barometer.

9. Ava Helen in Oregon

- Blog Posts
  - **The Ancestry of Ava Helen Pauling** (October 16, 2008), by Chris Petersen
  - **Ava Helen in Oregon** (December 8, 2009), by Will Clark
  - **The Miller Farmhouse** (September 7, 2011), by Olga Rodriguez-Walmisley
  - **A Trip to North Salem High School** (October 23, 2013), by Chris Petersen
  - **Ava Helen's OAC** (September 29, 2021), by Miriam Lipton
  - **Ava Helen Miller in Corvallis, 1921–1923** (October 6, 2021), by Miriam Lipton
- Major Sourcing
  - Images: Ava Helen Miller with her sisters and mother (MSS Pauling, 1914i.2); Portrait of Ava Helen Miller at age 18. Photo by Ball Studios, Corvallis. (MSS Pauling, 1922.1)

- Carson, Mina. *Ava Helen Pauling: Partner, Activist, Visionary*, Corvallis: Oregon State University Press, 2013.
- MSS Hager, Series 2: *Force of Nature*: Interviews — Audio Cassettes and Transcriptions.
- MSS Pauling, Series 15: Ava Helen Pauling. Sub-Series 3: Biographical. Box 3.001. Ava Helen Pauling: Biographical Materials; Box. 3.023. Ava Helen Pauling's Parents: George Richard Miller and Elnora Ellen Gard Miller — Genealogy, Biographical Information, and Correspondence; Box 3.028. *Genealogy of the Phillip E. Linn Families*, 1965.
- OAC *Beaver* Yearbook, 1923 (1921/1922 academic year). Available online: https://oregondigital.org/concern/documents/5t34sj58c.
- OAC General Catalog, 1921–1922. Available online: https://oregondigital.org/concern/documents/fx719v06j.
- *The Daily Barometer* digital collection. Available online: https://oregondigital.org/collections/daily-barometer.

10. Pauling's Senior Year at Oregon Agricultural College, 1921–1922

- Blog Posts
  - **The Paulings' Wedding Anniversary** (June 17, 2008), by Trevor Sandgathe
  - **Working in Oregon: The Blue Collar Adolescence of Linus Pauling** (May 7, 2009), by Trevor Sandgathe
  - **Fred Allen's Notebook** (October 27, 2009), by Ben Jager
  - **The Paving Inspector Job** (November 25, 2009), by Will Clark
  - **Pauling's Failed Rhodes Scholarship Application** (December 1, 2009), by Will Clark
  - **Ava Helen in Oregon** (December 8, 2009), by Will Clark
  - **Young Love** (June 19, 2013), by Chris Petersen
  - **Becoming Dr. Pauling** (March 5, 2015), by Luis Marquez Loza
  - **Pauling's Senior Class Oration** (June 10, 2015), by Luis Marquez Loza
  - **Pauling's OAC: Super Senior Year** (September 22, 2021), by Miriam Lipton
  - **Ava Helen Miller in Corvallis, 1921–1923** (October 6, 2021), by Miriam Lipton

- Major Sourcing
  - Images: Ava Helen Miller, Linus Pauling and two OAC classmates, 1922 (MSS Pauling, 1922i.12); Linus Pauling on OAC graduation day, June 1922. (MSS Pauling, 1922i.21)
  - MSS Hager, Series 2: *Force of Nature*: Interviews — Audio Cassettes and Transcriptions.
  - MSS Pauling, Series 1: Correspondence. Box 278. Noyes, A.A.
  - MSS Pauling, Series 4: Speeches. 1922s.1. "Senior Class Oration," Oregon Agricultural College, Corvallis, Oregon, May 31, 1922.
  - MSS Pauling, Series 13: Biographical, Sub-Series 1: Academia. Box 1.001. Oregon Agricultural College: Diary, Academic Work, Biographical Information, etc.; Box 1.002. Oregon Agricultural College: Academic Work, Biographical Information, etc.; Box 1.005. Oregon Agricultural College: Fred Allen Notebook.
  - MSS Pauling, Series 15: Ava Helen Pauling. Sub-Series 3: Biographical. Box 3.001. Ava Helen Pauling: Biographical Materials; Box 3.002. Ava Helen Pauling: Biographical Materials.
  - MSS Pauling, Series 16: Personal Safe. Pauling's correspondence with Ava Helen Miller during the summer of 1922 and beyond is held in Sub-Series 1: Linus Pauling's Safe — Drawer 1.
  - OAC *Beaver* Yearbook, 1923 (1921/1922 academic year). Available online: https://oregondigital.org/concern/documents/5t34sj58c.
  - OAC *Beaver* Yearbook, 1924 (1922/1923 academic year). Available online: https://oregondigital.org/concern/documents/tq57nr47g.
  - OAC General Catalog, 1921–1922. Available online: https://oregondigital.org/concern/documents/fx719v06j.
  - OAC General Catalog, 1922–1923. Available online: https://oregondigital.org/concern/documents/fx719v07t.
  - *The Daily Barometer* digital collection. Available online: https://oregondigital.org/collections/daily-barometer.

11. An Honorary Doctorate from Oregon State Agricultural College

- Blog Posts
  - **Pauling and Einstein** (September 1, 2010), by Will Clark
  - **Pauling's OSAC Honorary Doctorate** (June 13, 2018), by Dani Tellvik

- Major Sourcing
  - Image: Honorees at the 1933 OSAC Commencement. (OSU SCARC, Harriet's Photograph Collection, HC 1578)
  - MSS Pauling, Series 1: Correspondence. Box 298. Oregon State College/Oregon State University.
  - MSS Pauling, Series 7: Newspaper Clippings. 1931n.2. "Young Chemist Wins The Langmuir Award", *New York Times*, August 20, 1931; 1933n.1. "At Only 32", *Oregon Daily Journal*, 1933.
  - MSS Pauling, Series 8: Awards. 1931h.1 American Chemical Society, Prize for meritorious work in pure chemistry, [A. C. Langmuir Prize] September 2, 1931; 1933h.1 National Academy of Sciences of the United States of America, Certificate of Membership, April 26, 1933; 1933h.2 Oregon State College, Honorary Doctor of Science, June 5, 1933.

12. The Story of Ralph Spitzer

- Blog Posts
  - **The Story of Ralph Spitzer** (March 13, 2012), by Madeline Hoag
  - **Ralph Spitzer: The Firing** (March 21, 2012), by Madeline Hoag
  - **Spitzer: The Aftermath** (March 28, 2012), by Madeline Hoag
- Major Sourcing
  - Images: Ralph and Therese Spitzer at Oregon State College, 1949 (OSU SCARC, Harriet's Photograph Collection, HC 1308); Ralph and Therese Spitzer in the living room of their Vancouver home. (MSS Pauling, 197?i.98)
  - This book chapter makes heavy use of a manuscript written after the original posts on the Pauling Blog were published. See "Life History of Ralph W. Spitzer" by Chris Petersen, October 2013. Available online: https://ir.library.oregonstate.edu/concern/articles/fb494f55r. Source interviews with Ralph Spitzer, Eloise Spitzer and Hisako Kurotaki gathered for this manuscript have been cataloged by SCARC into the History of Science Oral History Collection (OH 017).
  - See also Robbins, William G. "The Academy and Cold War Politics: Oregon State College and the Ralph Spitzer Story." *Pacific Northwest*

Quarterly 104 (Fall 2013): 159-175; and Clark, Suzanne. *Cold Warriors: Manliness on Trial in the Rhetoric of the West*, Southern Illinois University Press, 2000.

o MSS Pauling, Series 1: Correspondence. Box 298. Oregon State College/Oregon State University.

o MSS Pauling, Series 2: Publications. 1949p.18. "Dr. Linus Pauling's Letter — President A. L. Strand's Reply." Letter to the Editor. *The Daily Barometer*, (Oregon State University) March 5, 1949, 2; 1949p.19. "Spitzer Case." Letter to the Editor. *Chem. Eng. News* 27, no. 13 (March 1949): 935.

o MSS Pauling, Series 13: Biographical, Sub-Series 2: Political Issues. Box 2.034. Ralph Spitzer: Academic Freedom and Passport Difficulties.

o MSS Pauling, Series 15: Ava Helen Pauling, Sub-Series 1: Correspondence. Box 1.006, Folder 6.15. Spitzer, Ralph and Terry.

o MSS Pauling, Series 16: Personal Safe, correspondence between Pauling and Spitzer written in October and November 1950 can be found in folder 1.032.

o MSS Pauling, Series 17: Pauling Personal Library. Spitzer, Therese. *Psychobattery: A Chronicle of Psychotherapeutic Abuse*; with Medical Discussion by Ralph Spitzer; foreword by Joseph Needham. Clifton, New Jersey: Humana Press, 1980.

o *The Daily Barometer* digital collection. Available online: https://oregondigital.org/collections/daily-barometer.

13. The Black Student Union Walkout

- Blog Posts
  o **The 1969 Black Student Union Walkout** (March 23, 2016), by Anna Dvorak
  o **A Lecture Interrupted and a Campus Torn Apart** (March 30, 2016), by Anna Dvorak
- Major Sourcing
  o Image: Rich Harr speaking at the OSU Centennial lecture walk-in, February 25, 1969. (Gwil Evans Papers, Folder 1.1, OSU SCARC)

- This book chapter makes use of a detailed journalistic account of the Black Student Union Walkout authored by the staff of *The Oregon Stater* alumni magazine, which devoted its entire April 1969 issue to the subject. The issue is available online here: https://oregondigital.org/concern/documents/fx71bs44q.
- *The Daily Barometer* also covered the walkout in real time. See in particular the issues published from February 25 to March 12, 1969, as available in this digital collection: https://oregondigital.org/collections/daily-barometer.
- *The Scab Sheet* underground newspaper is also available online at https://oregondigital.org/collections/osu-student-protest-underground-pubs.
- MSS Pauling, Series 4: Speeches, 1969s.1. "The Advancement of Knowledge — Orthomolecular Psychiatry," Oregon State University Centennial Lecture, Corvallis, Oregon, February 25, 1969.
- OH 017, Oral History Interview of Ken Hedberg by Chris Petersen, October 20, 2011.

14. A Sentimental Trip

- <u>Blog Post</u>
  - **A Sentimental Trip** (December 29, 2009), by Will Clark
- <u>Major Sourcing</u>
  - Image: Lucile Jenkins, Linus Pauling, Paul Emmett and Pauline Emmett at the 60th anniversary reunion of the OAC Class of 1922, June 12, 1982. Cropped from a larger group photograph. (MSS Pauling, 1982i.48)
  - MSS Pauling, Series 1: Correspondence. Folder 189.1 Jeffress, Lloyd A.; Box 305. Paradowski, Robert J. [Pauling discussed this trip at length in a letter to Paradowski dated August 6, 1982.]

Part II: Caltech Administrator

Most of the content that has been compiled into Part II was written by Andy Hahn, who worked as a student archivist in SCARC while a doctoral candidate in the History of Science at OSU. The blog posts itemized below were the result of original research that Hahn conducted in a set of 17 boxes of

Pauling's Caltech administrative records that have been arranged into MSS Pauling, Series 13: Biographical, Sub-Series 1: Academia. Also utilized were three more boxes held in MSS Pauling, Series 11: Science, Sub-Series 14: Scientific, Research and Grant-Funding Organizations. The specific boxes that were consulted are itemized in the following. Other materials used for given chapters are identified as well.

Major Sourcing for Part II

Series 11

o   Box 14.037. Rockefeller Foundation, 1928–1939
o   Box 14.038. Rockefeller Foundation, 1936–1946
o   Box 14.039. Rockefeller Foundation, 1943–1983

Series 13

o   Box 1.017. California Institute of Technology: Materials re: Teaching and Advising of Graduate Students by Linus Pauling, 1929–1964.
o   Box 1.018. California Institute of Technology: Materials re: Teaching and Advising of Graduate Students by Linus Pauling, 1932–1964.
o   Box 1.019. California Institute of Technology: Materials re: Division of Chemistry and Chemical Engineering, 1915–1945.
o   Box 1.020. California Institute of Technology: Materials re: Division of Chemistry and Chemical Engineering, 1946–1956.
o   Box 1.021. California Institute of Technology: Materials re: Division of Chemistry and Chemical Engineering, 1946–1957.
o   Box 1.022. California Institute of Technology: Materials re: Division of Chemistry and Chemical Engineering, 1957–1964.
o   Box 1.023. California Institute of Technology: Materials re: Applications for positions, Division of Chemistry and Chemical Engineering, 1932–1962.
o   Box 1.024. California Institute of Technology: Materials re: Division of Chemistry and Chemical Engineering, 1933–1963, No Date.
o   Box 1.025. California Institute of Technology: Committee on Sponsored Research, 1958–1960.
o   Box 1.026. California Institute of Technology: Committee on Sponsored Research, 1960–1961.

- o Box 1.027. California Institute of Technology: Committee on Sponsored Research, 1962; other research-related materials, 1945.
- o Box 1.028. California Institute of Technology: Research-related materials, 1944–1956.
- o Box 1.029. California Institute of Technology: Administrative Files, 1922–1968.
- o Box 1.030. California Institute of Technology: Administrative Files, 1938–1971.
- o Box 1.031. California Institute of Technology: Assorted Financial Materials, 1930–1950.
- o Box 1.032. California Institute of Technology: Assorted Financial Materials, 1945–1965.
- o Box 1.033. California Institute of Technology: General, 1922–1962, No Date.

15. Introduction

- Blog Post
  - o **Pauling as Administrator: Becoming Division Chair** (January 9, 2019), by Andy Hahn
- Additional Chapter Sourcing
  - o Image: Linus Pauling in 1937, cropped from an original photo of Pauling and J.A.A. Ketelaar. (MSS Pauling, 1937i.15)
  - o MSS Pauling, Series 1: Correspondence. Box 278. Noyes, A. A.

16. Becoming Division Chair

- Blog Posts
  - o **Becoming Division Chair: The Division Council, Pauling's Demur, and Weaver's Promise** (January 16, 2019), by Andy Hahn
  - o **Becoming Division Chair: Staffing a New Laboratory, Noyes' Death, and a Conversation with Harvard** (January 23, 2019), by Andy Hahn
  - o **Becoming Division Chair: Pauling Takes the Reins** (January 30, 2019), by Andy Hahn

- Additional Chapter Sourcing
    - Image: Linus Pauling in the laboratory, 1935. (MSS Pauling, 1935i.1)
    - MSS Pauling, Series 1: Correspondence. Folder 66.15. Conant, James B.; Folder 181.3. Ingold, Christopher K.

17. First Years in Charge

- Blog Posts
    - **The Origins of the Crellin Laboratory** (May 15, 2013), by Adam LaMascus
    - **Building the Crellin Lab (and keeping it standing)** (May 22, 2013), by Adam LaMascus
    - **The Legacy of the Crellin Laboratory** (May 29, 2013), by Adam LaMascus
    - **First Years as Division Chair: Responsibilities Large and Small** (February 6, 2019), by Andy Hahn
    - **First Years as Division Chair: Implementing the Rockefeller Grant and Dedicating Crellin** (February 13, 2019), by Andy Hahn
    - **First Years as Division Chair: Making Recommendations, Coping with an Explosion, and Bringing Aboard Zechmeister** (February 20, 2019), by Andy Hahn
- Additional Chapter Sourcing
    - Image: "Work Rewarded," *Pasadena Post*, May 5, 1937. (Cropped from a scrapbook leaf held in MSS Pauling, Series 13: Biographical, Sub-Series 6, Item 6.003.2)
    - MSS Pauling, Series 1: Correspondence. Folder 68.10, Crellin, E.W.; Folder 464.4. Zechmeister, Laszlo.
    - MSS Pauling, Series 4: Speeches. 1938s.5. "The Future of the Crellin Laboratory," dedication of the Crellin Laboratory, California Institute of Technology, Pasadena, May 16, 1938.

18. Leading the Division During World War II

- Blog Posts
    - **Chairing the Division During the War: Staffing** (April 3, 2019), by Andy Hahn

- o **Chairing the Division During the War: A Balance of Interests** (April 10, 2019), by Andy Hahn
- o **Chairing the Division During the War: Maintaining Security and Revising the Curriculum** (April 17, 2019), by Andy Hahn
- o **Taking the Division Beyond the War** (May 1, 2019), by Andy Hahn
- Additional Chapter Sourcing
  - o Image: Pauling's National Defense Research Corporation authorization papers, 1944. (MSS Pauling, Series 11: Science, Sub-Series 13: Scientific War Work. Box 13.005, Folder 3)
  - o MSS Pauling, Series 4: Speeches. 1943s.4. "Talk at Memorial Services for Elizabeth Swingle," California Institute of Technology, September 27, 1943.
  - o "Alumni News," *CIT News*, December 1943. Available online: https://calteches.library.caltech.edu/3860/1/Alumni.pdf.
  - o For more on Pauling's scientific war work, see "The Scientific War Work of Linus C. Pauling: A Documentary History," https://scarc.library.oregonstate.edu/coll/pauling/war/index.html.

19. Chairing the Division After the War

- Blog Posts
  - o **Chairing the Division After the War: Organizing the Peace** (May 8, 2019), by Andy Hahn
  - o **Chairing the Division After the War: Progress Toward Pauling's Post-War Plan** (May 15, 2019), by Andy Hahn
  - o **Chairing the Division After the War: Pauling Shifts His Focus to the Big Picture** (May 22, 2019), by Andy Hahn
- Additional Chapter Sourcing
  - o Image: Portrait of Linus Pauling, 1946. Photo by Ray Huff Studios. (MSS Pauling, 1946i.10)
  - o MSS Pauling, Series 2: Publications. 1949p.3. "Our job ahead." *Chem. Eng. News* 27 (January 1949): 9.

20. A Cold War Division Chair

- Blog Posts
  - o **The Story of Sidney Weinbaum** (May 4, 2010), by Will Clark

- o **A Cold War Division Chair: Political Activism and Institutional Pressure** (May 29, 2019), by Andy Hahn
- o **A Cold War Division Chair: Pauling Under Investigation** (June 5, 2019), by Andy Hahn
- o **A Cold War Division Chair: The Church Laboratory and a Big Step Toward Pauling's Post-War Plan** (June 12, 2019), by Andy Hahn
- Additional Chapter Sourcing
  - o Images: Portrait of Linus Pauling, 1950. Photo by Maryland Studios. (1950i.10); Newspaper photograph of Sidney Weinbaum, 1950. (MSS Pauling, Series 1: Correspondence, Folder 433.5, Weinbaum, Sidney, 1949–1950)
  - o MSS Pauling, Series 1: Correspondence. Box 433. Weinbaum, Sidney and Lina Litinskaya (Weinbaum).
  - o MSS Pauling, Series 3: Manuscripts of Articles. 1950a.5. "Statement by Linus Pauling," [re: meeting with Lee DuBridge, LP tenure review committee and Sidney Weinbaum Committee] July 18, 1950.
  - o MSS Pauling, Series 13: Biographical, Sub-Series 2: Political Issues. Box 2.010. Industrial Employment Review Board/Industrial Personnel Security Board.
  - o MSS Pauling, Series 16: Personal Safe. Folder 2.008. Collapsible file carrying case marked as "Sidney Weinbaum" — Correspondence and notes related to Weinbaum.

21. Final Years as Division Chair

- Blog Posts
  - o **The Road to Stockholm: The Appalling Life of Linus Pauling** (June 24, 2008), by Trevor Sandgathe
  - o **Louis Budenz, Informant** (May 6, 2010), by Will Clark
  - o **Pauling's Nobel Chemistry Prize** (January 13, 2016), by Anna Dvorak
  - o **Final Years as Division Chair: The Admission of Women** (June 19, 2019), by Andy Hahn
  - o **Final Years as Division Chair: Progress Within, Trouble Without** (June 26, 2019), by Andy Hahn

- o **Final Years as Division Chair: The End of Pauling's Chairmanship** (July 3, 2019), by Andy Hahn
- Additional Chapter Sourcing
  - o Image: "Dr. Pauling Steps Down to Teach," *Pasadena Independent*, June 1958. (Cropped from a scrapbook leaf held in MSS Pauling, Series 13: Biographical, Sub-Series 6: Scrapbooks, Item 6.007.158)
  - o MSS Pauling, Series 3: Manuscripts of Articles. 1951a2.14. No Title, Letter to the Editor of *The American Legion Magazine*, [re: article "Do Colleges Have to Hire Red Professors," by Louis Budenz] November 25, 1951; 1953a.1. "Statement by Linus Pauling," [re: testimony of Louis Budenz] January 4, 1953.
  - o MSS Pauling, Series 6: Research Notebooks. RNB 21 [*Meet the Press*], 1958.
  - o MSS Pauling, Series 7: Newspaper Clippings. 1950n.18. "Budenz Names 30 as Commies," Publication Unknown, 1950.
  - o MSS Pauling, Series 8: Awards. Boxes 1954h2 and 1954h3. Nobel Prize for Chemistry, 1954.
  - o MSS Pauling, Series 10: Audio/Visual. 1954v.1. "The Road to Stockholm: The Appalling Life of Dr. Pauling." The Chemistry Biology Stock Company, producer.
  - o MSS Pauling, Series 13: Biographical, Sub-Series 5: Personal Materials and Family Correspondence. Box 5.030, Folder 30.1. Oral history interview conducted by John L. Greenberg, California Institute of Technology Archives, May 10, 1984; Sub-Series 6: Scrapbooks. Item 6.006.28. "Ex-Red Tabs Pauling in House Quiz," *Pasadena Independent*, December 24, 1952; "Former Red Names 30 As Communists," Publication Unknown, December 24, 1952; "Denials Fly at Budenz Red List," *Los Angeles Herald & Express*, December 24, 1952. Item 6.006.29 "Budenz Lies, Says Caltech's Dr. Pauling," *Pasadena Star-News*, December 24, 1952; "Ex-Red Branded 'Liar' by Noted Caltech Savant," *Los Angeles Mirror*, December 24, 1952; "Pauling Denies Budenz Accusation," *Chemical and Engineering News*, January 12, 1953. Item 6.006.113. "Faculty produces a lively musical comedy to celebrate Dr. Pauling's Nobel laureate," *California Tech*, December 9, 1954.

ıMSS Pauling, Series 16: Personal Safe, Sub-Series 1: Linus Pauling's Safe — Drawer 1. Item 32.56. "Do Colleges Have to Hire Red Professors?," *American Legion*, November 1951. Item 32.57. Newspaper Clipping: "...Red Profs Well Organized, Budenz Asserts," December 10, 1951.

- o The Nobel Prize Nomination Archive (online resource): https://www.nobelprize.org/nomination/archive/search_people.php.

Part III: Period of Wandering

22. Center for the Study of Democratic Institutions

- Blog Posts
  - o **Linus Pauling and the Search for UFOs** (May 11, 2009), by Trevor Sandgathe
  - o **Lawrence Badash, 1934–2010** (August 27, 2010), by Chris Petersen
  - o **Irwin Stone: An Influential Man** (April 9, 2014), by Jessica Newgard
  - o **The Center for the Study of Democratic Institutions** (January 21, 2015), by Luis Marquez Loza
  - o **CSDI: A Platform for Action Merging Science and Peace** (January 28, 2015), by Luis Marquez Loza
  - o **Pacem in Terris** (February 4, 2015), by Luis Marquez Loza
  - o **The Triple Revolution** (February 11, 2015), by Luis Marquez Loza
  - o **A Study of Unidentified Flying Objects** (February 18, 2015), by Luis Marquez Loza
  - o **A Return to Scientific Theory** (February 25, 2015), by Luis Marquez Loza
- Major Sourcing
  - o Image: "Award," *Los Angeles Times*, August 31, 1967. (Cropped from a scrapbook leaf held in MSS Pauling, Series 13: Biographical, Sub-Series 6: Scrapbooks, Item 6.009.37)
  - o MSS Pauling, Series 1: Correspondence. Folder 369.1. Stone, Irwin: Correspondence; Folder 369.2. Stone, Irwin: Reprints. (SCARC is also home to the Irwin Stone Papers, though these materials were not in hand at the time that the relevant blog posts were written); Folder 420.10. University of California, Santa Barbara.

- MSS Pauling, Series 2: Publications. 1964p.7. "The Triple Revolution." Santa Barbara, California: The Ad Hoc Committee on the Triple Revolution, 1964; 1965p.8. "Structural significance of the principal quantum number of nucleonic orbital wave functions." *Phys. Rev. Lett.* 15 (September 1965): 499; 1965p.9. "Structural basis of neutron and proton magic numbers in atomic nuclei." *Nature* 208 (October 1965): 174; 1965p.10. "The close-packed-spheron model of atomic nuclei and its relation to the shell model." *Proc. Natl. Acad. Sci.* 54 (October 1965): 989-994; 1965p.14. "The close-packed-spheron theory and nuclear fission." *Science* 150 (October 1965): 297–305; 1965p2.1. "The nature of the problem." *Pacem in Terris — Peace on Earth* (The Proceedings of an International Convocation on the Requirements of Peace Sponsored by the Center for the Study of Democratic Institutions), Edward Reed, ed. New York: Pocket Books, Inc. 1965, 35-42.
- MSS Pauling, Series 3: Manuscripts of Articles. 1964a.22. "The Triple Revolution," 1964.
- MSS Pauling, Series 4: Speeches. 1965s.3. No Title, [re: remarks on "The Requirement of Peace: the Nature of the Problem"] Pacem in Terris Convocation, New York, February 18, 1965.
- MSS Pauling, Series 11: Science, Sub-Series 10: LP Patents; LP Notes to Self; Other Fields of Science. Box 10.009, Folder 1. LP Confidential Note to Self: "A Study of Unidentified Flying Objects", July 16, 1966.
- MSS Pauling, Series 13: Biographical, Sub-Series 1: Academia. Box 1.036. Center for the Study of Democratic Institutions.
- MSS Pauling, Series 17: Pauling Personal Library. Fuller, John G. *Incident at Exeter; The Story of Unidentified Flying Objects over America Today*. New York: Putnam, 1966; Trench, Brinsley, *The Flying Saucer Story*. New York: Ace, 1966; Gillmor, Daniel S., ed. *Final Report of the Scientific Study of Unidentified Flying Objects*. New York: Dutton, 1969.

23. University of California — San Diego

- Blog Posts
  - **Orthomolecular Psychiatry** (January 22, 2009), by Ben Jager
  - **Pauling's Eulogy for Martin Luther King, Jr.** (January 19, 2010), by Chris Petersen

- o **Pauling at UC-San Diego** (August 30, 2017), by Dani Tellvik
- o **Pauling at UCSD: Season of Tumult** (September 6, 2017), by Dani Tellvik
- o **Leaving La Jolla** (September 13, 2017), by Dani Tellvik
- Major Sourcing
  - o Images: Portrait of Linus Pauling, 1967. Photo by Jones Studio. (MSS Pauling, 1967i.17); "Pauling Speaks Out," *San Francisco Chronicle*, May 27, 1969. (Cropped from a scrapbook leaf held in MSS Pauling, Series 13: Biographical, Sub-Series 6: Scrapbooks, Item 6.009.109)
  - o MSS Pauling, Series 1: Correspondence. Folder 447.1. W: Correspondence, 1966. [Letter from Pauling to Mrs. S. Leonard Wadler, August 15, 1966]
  - o MSS Pauling, Series 2: Publications. 1968p.4. "Orthomolecular psychiatry: varying the concentrations of substances normally present in the human body may control mental disease." *Science* 160 (April 1968): 265-271.
  - o MSS Pauling, Series 4: Speeches. 1968s.5. "The Scientific Revolution," Centennial Lecture Series, University of California, Irvine, March 3, 1968; 1968s.7. "Dr. Martin Luther King, Jr.," Amherst, Massachusetts, April 9, 1968; 1969s.6. No Title, [re: support for strike] University of California, San Diego, May 21, 1969; 1969s.7. No Title, student rally protesting the presence of National Guard troops at the University of California, Berkeley, Sacramento, California, May 26, 1969.
  - o MSS Pauling, Series 13: Biographical, Sub-Series 1: Academia. 1.037. University of California, San Diego.
  - o For a thorough treatment of Pauling's ideas on eugenics, see "It's in the Blood: The Varieties of Linus Pauling's Work on Hemoglobin and Sickle Cell Anemia," by Melinda Gormley. Oregon State University master's thesis in the History of Science submitted October 22, 2003. Available online: https://ir.library.oregonstate.edu/concern/graduate_thesis_or_dissertations/p8418q54t.

24. Stanford University

- Blog Posts
  - o **Pauling at Stanford: Prelude** (April 25, 2018), by Andy Hahn
  - o **Pauling at Stanford: Settling In** (May 2, 2018), by Andy Hahn

- o **Pauling at Stanford: The Finals Years and Beyond** (May 9, 2018), by Andy Hahn
- o **Pauling, Stanford and Research — Part 1** (May 16, 2018), by Andy Hahn
- o **Pauling, Stanford and Research — Part 2** (May 23, 2018), by Andy Hahn
- o **Pauling, Stanford and Activism — Part 1** (May 30, 2018), by Andy Hahn
- o **Pauling, Stanford and Activism — Part 2** (June 6, 2018), by Andy Hahn
- Major Sourcing
  - o Image: Linus Pauling in lecture at Stanford University, 1969. Photo by George Feigen. (MSS Pauling, 1969i.37b)
  - o MSS Pauling, Series 1: Correspondence. Folder 245.2. McConnell, Harden; Folder 376.7. Stanford Research Institute; Folder 376.8. Stanford University; Folder 426.2. Vivonex Corporation.
  - o MSS Pauling, Series 4: Speeches. 1971s.14. "The Basis for Decisions," Commencement Address, University of California, Berkeley, June 5, 1971.
  - o MSS Pauling, Series 6: Research Notebooks. Details about the Paulings' personal intake of Vivonex as well as Linus Pauling's research interest in the product are available in research notebooks 24, 30, 33 and 37.
  - o MSS Pauling, Series 7: Newspaper Clippings. 1979n.48. "Linus Pauling," *Stanford Magazine*, Spring/Summer 1979.
  - o MSS Pauling, Series 13: Biographical, Sub-Series 1: Academia. Box 1.038. Stanford University; Box 1.039. Stanford University.
  - o *The Stanford Daily* Archives (online resource): https://archives.stanforddaily.com/.
  - o Yu, Kristina. "Reconstructing Linus Pauling: Scientist and Peacemaker. His Life on the Farm, a Life in the Limelight," *The Stanford Scientific Review*, Volume 1, Spring 2003. Available online: https://exhibits.stanford.edu/stanford-pubs/catalog/rs174rz8278.

Part IV: Travels

25. The Guggenheim Trip

- Blog Posts
  - **A Classic of Twentieth-Century Science: The Nature of the Chemical Bond** (May 15, 2008), by Trevor Sandgathe
  - **Linus Pauling and the Birth of Quantum Mechanics** (May 20, 2008), by Trevor Sandgathe
  - **Our Newest Addition: Pauling-Goudsmit Letters** (June 3, 2008), by Trevor Sandgathe
  - **The Guggenheim Trip, Part I: Touring in Southern Europe** (June 5, 2008), by Trevor Sandgathe
  - **The Guggenheim Trip, Part II: The Growth of a Scientist** (June 10, 2008), by Trevor Sandgathe
  - **The Guggenheim Trip, Part III: Unexpected Colleagues** (June 12, 2008), by Trevor Sandgathe
  - **Pauling Amidst the Titans of Quantum Mechanics: Europe, 1926** (May 18, 2010), by Carly Dougher
  - **Pauling and Wilson** (May 20, 2010), by Carly Dougher
  - **Pauling's Nobel Chemistry Prize** (January 13, 2016), by Anna Dvorak
  - **"The Best Work I've Ever Done"** (April 11, 2018), by Dani Tellvik
  - **Pauling's First Paper on the Nature of the Chemical Bond** (April 18, 2018), by Dani Tellvik
  - **Becoming An Asset for the Guggenheim Foundation** (January 15, 2020), by Andy Hahn
  - **The Theoretical Prediction of the Physical Properties of Many-Electron Atoms and Ions. Mole Refraction, Diamagnetic Susceptibility, and Extension in Space, 1927** (October 13, 2021), by Miriam Lipton
- Major Sourcing
  - Image: Linus Pauling at the Temple of Neptune, Paestum, Italy, April 1926. (MSS Pauling, 1926i.62)

- MSS Pauling, Series 1: Correspondence. Folder 137.4. Goudsmit, Sam; Box 278. Noyes, A.A.; Box 305. Paradowski, Robert J.; Folder 438.5. Wilson, E. Bright (Jr.).
- MSS Pauling, Series 2: Publications. 1927p.5. "The theoretical prediction of the physical properties of many-electron atoms and ions. Mole refraction, diamagnetic susceptibility, and extension in space." *Proc. Roy. Soc. A* (London) 114 (1927): 181–211;1931p.3. "The nature of the chemical bond. Application of results obtained from the quantum mechanics and from a theory of paramagnetic susceptibility to the structure of molecules." *J. Am. Chem. Soc.* 53 (April 1931): 1367–1400.
- MSS Pauling, Series 3: Manuscripts of Articles. 1976a2.3. "The Birth of Quantum Mechanics," June 28, 1976.
- MSS Pauling, Series 4: Speeches. 1929s.1. "The Development of the Quantum Mechanics," First Lecture, Berkeley Lectures on Applications of Quantum Mechanics, University of California, Berkeley, February 19, 1929.
- MSS Pauling, Series 5: Manuscripts of Books. Box 4.001, Folder 1.2. Correspondence, Reprints re: *The Nature of the Chemical Bond and the Structure of Molecules and Crystals.*
- MSS Pauling, Series 8: Awards. Box 1954h2. Nobel Prize for Chemistry, 1954.
- MSS Pauling, Series 10: Audio/Visual. 1977v.66. "NOVA audiocassettes 1–9: Interview with Linus Pauling in Office, Interview with Ava Helen Pauling, Interview with Linus Pauling in Lab." WGBH Boston, producer.
- MSS Pauling, Series 11: Science. Sub-Series 14: Scientific, Research and Grant-Funding Organizations. Box 14.013. John Simon Guggenheim Memorial Foundation, 1925–1945.
- MSS Pauling, Series 15: Ava Helen Pauling, Sub-Series 3: Biographical. Folder 3.008.1 AHP travel diary, 1926 trip to Zurich.
- MSS Pauling, Series 16: Personal Safe. Item 4.001. Diary: "Navigazione Generale Italiana Board-Day Book", during a trip from NY to Naples, Italy, March 11–March 17, 1926; Item 4.002. Diary: annotated "Memoranda — 1927–28 I think."

o Carson, Mina. *Ava Helen Pauling: Partner, Activist, Visionary*, Corvallis: Oregon State University Press, 2013.

o Nobel Prize Nomination Archive (online resource): https://www.nobelprize.org/nomination/archive/.

o Paradowski, Robert J. "An American in Munich: Truth and Controversy in the Life and Work of Linus Pauling during the Golden Years of Physics." Presentation delivered at *A Liking for the Truth: Truth and Controversy in the Work of Linus Pauling (1901–1994)*, [Pauling centenary conference], Oregon State University, Corvallis, Oregon, February 28, 2001. Available online: https://scarc.library.oregonstate.edu/events/2001paulingconference/video-s2-2-paradowski.html.

26. The Paulings Go to England

- Blog Posts
  - **Pauling's Theory of Sickle Cell Anemia** (December 11, 2008), by Chris Petersen
  - **The Importance of the Concept of Molecular Disease** (December 16, 2008), by Ben Jager
  - **Chargaff's Rules** (August 25, 2009), by Ben Jager
  - **The Paulings Go to England, 1947–1948** (February 9, 2011), by Audrey Riessman
  - **A Royal Welcome** (February 16, 2011), by Audrey Riessman
  - **New Insights into Metals and More** (February 23, 2011), by Audrey Riessman
  - **An Era of Discovery in Protein Structure** (March 2, 2011), by Audrey Riessman
  - **The Alpha Helix** (March 9, 2011), by Audrey Riessman
  - **Caltech, Cambridge and Coiled-Coils** (November 27, 2013), by Michael Mehringer
  - **Pauling and Perutz in the Golden Age of Protein Research** (June 4, 2014), by Andy Hahn
  - **The Public Response to The Nature of the Chemical Bond** (August 27, 2014), by Andy Hahn
  - **A Resonating-Valence-Bond Theory of Metals and Intermetallic Compounds, 1949** (November 3, 2021), by Miriam Lipton

- Major Sourcing
  - Image: The Pauling family at Magdalen Great Tower, Oxford, England, 1948. (MSS Pauling, 1948i.53)
  - MSS Hager, Series 2: Interviews. Box 2.002. Interviews I-L. Contains four interviews with Linda Pauling Kamb, 1990–1992.
  - MSS Pauling, Series 1: Correspondence. Box 2. Addis, Thomas; Folder 6.21. Aydelotte, Frank; Folder 67.5. Corey, Robert B.; Folder 68.11. Crick, Francis; Folder 96.10. Donohue, Jerry; Folder 163.1. Hughes, Edward; Folder 277.5. Niemann, Carl; Folder 299.8. Oxford University, [re: Eastman professorship and residency in Oxford]; Box 307. Perutz, Max; Folder 370.3. Sturdivant, James H. (Holmes).
  - MSS Pauling, Series 2: Publications. 1949p.6. "A resonating-valence-bond theory of metals and intermetallic compounds." *Proc. Roy. Soc.* 196 (1949): 343–362; 1960p.1. "Molecular structure and disease." *Disease and the Advancement of Basic Science* (1958 Lowell Institute Lectures at Harvard University), Henry K. Beecher, ed. Cambridge: Harvard University Press, 1960, 1-7.
  - MSS Pauling, Series 3: Manuscripts of Articles. 1982a2.10. "The Discovery of the Alpha Helix," September 1982.
  - MSS Pauling, Series 4: Speeches. Boxes 1947s and 1948s contain the many lectures delivered by Pauling during his residencies in England in 1947 and 1948 and are also useful in charting his travel while overseas.
  - MSS Pauling, Series 10: Audio-Visual. 1976v.7. "Life & Structure of Hemoglobin." Produced by the American Institute of Physics.
  - MSS Pauling, Series 11: Science. Sub-Series 5: Box 5.002. Materials re: Electron Theory and the Structure of Metals and Intermetallic Compounds.
  - MSS Pauling, Series 12: Peace. Sub-Series 6: Other Peace Activism. Box 6.012, Folder 1. Assorted Pauling Peace Research Notes, 1930s-1940s.
  - MSS Pauling, Series 16: Personal Safe. Item 4.033. Diary labeled as "France and England 1952 Also Toronto."
  - Pauling, Crellin. "The Personal View of Linus Pauling and His Work" and Shoemaker, David P. "My Memories and Impressions

of Linus Pauling." *The Life and Work of Linus Pauling (1901–1994): A Discourse on the Art of Biography*, conference held at Oregon State University, Corvallis, Oregon, March 1, 1995. Available online: https://scarc.library.oregonstate.edu/events/1995paulingconference/video-s3.html.
- ○ Perutz, Max. "Linus Pauling (1901–1994)," *Nature Structural Biology*, 1, 10, October 1994.
- ○ Perutz, Max. *I Wish I'd Made You Angry Earlier: Essays on Science, Scientists, and Humanity*, Plainview, New York: Cold Spring Harbor Laboratory Press, 1998.

27. Visiting Albert Schweitzer

- Blog Post
  - ○ **Visiting Albert Schweitzer** (March 14, 2008), by Trevor Sandgathe
  - ○ The material presented for this chapter expands significantly on what was originally published on the Pauling Blog.
- Major Sourcing
  - ○ Image: Albert Schweitzer and Linus Pauling in Lambéréne, Gabon, 1959. (MSS Pauling, 1959i.35)
  - ○ MSS Pauling, Series 1: Correspondence. Box 360. Schweitzer, Albert.
  - ○ MSS Pauling, Series 3: Manuscripts of Articles. 1959a.13. "Statement by Professor Pauling," [re: nuclear armament and meeting with Albert Schweitzer] August 27, 1959; 1964a.11. "Albert Schweitzer: Physician and Humanitarian," May 20, 1964.
  - ○ MSS Pauling, Series 4: Speeches. 1975s.15. "Reverence for Life — We Must Throw Off the Yoke of Militarism to Achieve Albert Schweitzer's Goal for the World," Keynote Address, Dr. Albert Schweitzer's Centennial Birth Anniversary Commemoration Symposium, Tokyo, Japan, September 25, 1975.
  - ○ MSS Pauling, Series 15: Ava Helen Pauling. Item 3.008.4 AHP diary of trip to Lambarènè, French Equatorial Africa and Europe, 1959.
  - ○ Catchpool, Frank. "Personal Reminiscences about Linus Pauling." *The Life and Work of Linus Pauling (1901–1994): A Discourse on the Art of Biography*, conference held at Oregon State University,

Corvallis, Oregon, March 1, 1995. Available online: https://scarc.library.oregonstate.edu/events/1995paulingconference/video-s3-6-catchpool.html.

28. Pauling and the Soviet Union

- Blog Posts
  - **Travels in the Soviet Union: Some Background** (August 1, 2012), by Desiree Gorham
  - **The First Two Soviet Trips** (August 7, 2012), by Desiree Gorham
  - **Back in the USSR** (August 15, 2012), by Desiree Gorham
  - **Pauling's Induction into the Soviet Academy of Sciences** (November 14, 2018), by Sarah Litwin
  - **The Lomonosov Gold Medal** (November 28, 2018), by Sarah Litwin
  - **Pauling's Receipt of the Lenin Peace Prize** (September 4, 2019), by Miriam Lipton
  - **The Lenin Peace Prize: Aftermath** (September 11, 2019), by Miriam Lipton
  - **Pauling and Sakharov** (December 11, 2019), by Miriam Lipton
  - **Pauling, the State Department, and the Right to Travel** (December 18, 2019), by Miriam Lipton
- Major Sourcing
  - Image: Linus Pauling lecturing after receiving the Lomonosov Gold Medal, Moscow, September 25, 1978. (MSS Pauling, 1978i.14)
  - MSS Pauling, Series 1: Correspondence. Folder 7.5 Academy of Sciences, U.S.S.R.; Folder 295.7. Oparin, Aleksandr Ivanovich; Folder 355.4 Sakharov, Andrei.
  - MSS Pauling, Series 2: Publications. 1970p.16. "Ortomoleculiarnye metody v meditsine." [Russian: "Orthomolecular methods in medicine"] in *Funktsionalnaia Biokhimiia Kletochnykh Struktur.* [N.M. Sisakian Memorial Volume] A. I. Oparin, ed. Moscow: Nauk, 1970, 427–432; 1978p.10. "Pauling and Sakharov." *Phys. Today* 31, no. 12 (December 1978): 81.
  - MSS Pauling, Series 3: Manuscripts of Articles. 1983a2.9. "Support Notice for Give Peace a Chance: Soviet Peace Proposals and U.S. Responses; a Panel Discussion," sponsored by the Women's

International League for Peace and Freedom, Peninsula Branch, October 1983.

o MSS Pauling, Series 4: Speeches. One excellent method for studying Pauling's travel is to engage with the Series 4 folders that correspond with lectures that he gave while overseas and that are referenced extensively in this chapter. Relevant boxes for the USSR are as follows: 1957s2, 1961s3, 1975s, 1978s3, 1983s4 and 1984s2. See also 1945s.6. No Title, [re: Soviet-American political tensions] speech at Banquet of Russian-American Club, Los Angeles, California, November 5, 1945; 1958s.24. "Science as a Cultural Subject; Chemistry and Medicine in the Future; The Place of Science in the Modern World," Pre-Commencement Alumni Seminar and Commencement Address, Antioch College, Yellow Springs, Ohio, June 19–21, 1958; 1969s.10. "The Possibilities for Social Progress," [discussion of Pauling's well-being index] Second International Congress of Social Psychiatry, London, England, August 4, 1969.

o MSS Pauling, Series 8: Awards. 1958h.4. Soviet Academy of Sciences; 1970h.1. Lenin Peace Prize Medal, U.S.S.R; 1978h.1. Lomonosov Medal.

o MSS Pauling, Series 12: Peace. Sub-Series 4: Peace Groups. Box 4.013, Folder 13.3. National Council of American-Soviet Friendship; Box 4.014, Folder 14.3. Progressive Citizens of America; Sub-Series 6: Other Peace Activism. Box 6.017. Issues of International Diplomacy and Human Rights. (So-Yu), Folder 17.6 USSR.

o MSS Pauling, Series 13: Biographical. Sub-Series 2: Political Issues. Box 2.009. California Senate Investigating Committee on Education; Boxes 2.044 and 2.045. Civil Liberties — The Cases of Julius Rosenberg, Ethel Rosenberg and Morton Sobell; Sub-series 4: Business and Financial. Box 4.003. Materials re: Taxes — Federal, State and Property, 1965–1972. See also Chapter 21 sourcing notes for references to Louis Budenz.

o MSS Pauling, Series 15: Ava Helen Pauling. Sub-Series 3: Biographical. Box 3.008, Folder 8.3. AHP travel diary, 1957 trip to Europe; Folder 8.6. Diary kept by Ava Helen Pauling for trip to USSR, 1963.

- o MSS Pauling, Series 16: Personal Safe. Folder 4.053. Fourteen Diaries, 1976–1992.
- o For a closer look at the Soviet resonance controversy, see *Visions of Linus Pauling*, Chapter 5.

29. Travels in Latin America

- Blog Posts
  - o **A Voyage on the Peace Ship** (July 23, 2009), by Chris Petersen
  - o **The Paulings in Latin America, 1940s — 1950s** (January 26, 2012), by Olga Rodriguez-Walmisley
  - o **The 1960s: The Nuclear-Free Zone, Oppression in Argentina and Molecules in Mexico** (February 1, 2012), by Olga Rodriguez-Walmisley
  - o **Science and the Future of Humanity: Chile, 1970** (February 8, 2012), by Olga Rodriguez-Walmisley
  - o **Women's Liberation, a Cruise to Acapulco and a Visit to Cuba: The 1970s** (February 15, 2012), by Olga Rodriguez-Walmisley
  - o **Symposia and the Peace Ship: Pauling in Latin America, 1980s** (February 21, 2012), by Olga Rodriguez-Walmisley
- Major Sourcing
  - o Image: Ava Helen and Linus Pauling dancing the samba in Brazil, September 1980. (MSS Pauling, 1980i.37)
  - o MSS Pauling, Series 1: Correspondence. Folder 328.2. Reagan, Ronald
  - o MSS Pauling, Series 3: Manuscripts of Articles. 1980a3.8. "Incidence of Squamous Cell Carcinoma in Hairless Mice Irradiated with Ultraviolet Light in Relation to Intake of Ascorbic Acid (Vitamin C) and of D-L--tocopheryl Acetate (Vitamin E)," September 8-10, 1980.
  - o MSS Pauling, Series 4: Speeches. As with the USSR chapter, this survey of Pauling's relationship with Latin America relied heavily on records held in the Speeches series. Relevant boxes are as follows: 1949s, 1955s, 1962s, 1963s, 1964s, 1970s, 1972s, 1977s, 1978s, 1978s2, 1980s2, 1981s, 1982s4, 1984s2.
  - o MSS Pauling, Series 7: Newspaper Clippings. 1972n.8. "Paz, Control Natal y fin a la Contaminacion, Clama Pauling," *El Mexicano*, March 6, 1972.

o MSS Pauling, Series 10: Audio/Visual. 1984v.1. "El Barco de La Paz (The Peace Ship)." Alternative Media Network/Media Alliance, producer; 1991v.9. "Dr. Linus Pauling — National Public Radio," profile on *All Things Considered*, February 28, 1991.

o MSS Pauling, Series 12: Peace. Sub-Series 6: Other Peace Activism. Box 6.009. Assorted Peace Appeals, Folder 9.6. "To People of Conscience, From the Peace Ship," [re: peace in Nicaragua] 1984; Box 6.013. Issues of International Diplomacy and Human Rights. (Af-Co); Box 6.016. Issues of International Diplomacy and Human Rights. (Ni-Ro).

o MSS Pauling: Series 13: Biographical. Box 5.039. Family Correspondence: Linus Carl Pauling, Jr., Folder 39.2. Correspondence: Linus Pauling, Jr., 1980–1996.

o MSS Pauling, Series 15: Ava Helen Pauling. Sub-Series 3: Biographical. Box 3.007. Calendars, Engagement Books, and Notebooks, Folder 7.2. "Chile 1970" engagement book; Sub-Series 5: Materials re: Political Issues and Civil Liberties. Box 5.013. Assorted Political Materials, Folder 13.21. Notebook: Liga Internacional de Mujeres por paz y libertad, Seccion Colombiana, Bogotá, July 17–22, 1970.

o MSS Pauling, Series 16: Personal Safe. Item 4.047e. Diary: labeled "1970"; Item 4.047g. Diary: labeled "Member Federation of American Scientists, Linus Pauling, Sponsor," 1985. Item 4.053_9. Diary: labeled "Member, Federation of American Scientists," 1984.

30. Japan and the Japanese

- Blog Posts
  o **Anti-Japanese Sentiment and the Rise of Pauling the Peace Activist** (March 27, 2008), by Trevor Sandgathe
  o **The Hiroshima Appeal** (August 6, 2009), by Chris Petersen
  o **Pauling's Early Development as a Peace Activist** (April 1, 2010), by Will Clark
  o **The Paulings and Japan: Roots of a Fruitful Relationship** (November 30, 2011), by Desiree Gorham
  o **Later Japan** (December 7, 2011), by Desiree Gorham
  o **Fukumi Morishige** (December 14, 2011), by Desiree Gorham

- o **Akira Murata** (December 21, 2011), by Desiree Gorham
- o **Ryoichi Sasakawa** (December 29, 2011), by Desiree Gorham
- o **Pauling and Ikeda** (May 3, 2017), by Matt McConnell
- o **A Lifelong Quest for Peace** (May 10, 2017), by Matt McConnell
- o **Carol Ikeda and Miyoshi Ikawa** (June 28, 2017), by Dani Tellvik
- Major Sourcing
  - o Image: Linus Pauling sitting outside, next to a Buddhist monk, Hiroshima, Japan, August 6, 1959. (MSS Pauling, 1959i.3)
  - o MSS Pauling, Series 1: Correspondence. Folder 136.15. Glockler, George; Folder 153.3. Heidelberger, Michael; Folder 247.1. Millikan, Robert; Folder 249.1. Morishige, Fukumi; Folder 252.3. Murata, Akira; Folder 357.4. Sasakawa, Ryoichi & Yohei; Folder 376.2. Soka Gakkai International.
  - o MSS Pauling, Series 2: Publications. 1983p.2. "Enhancement of antitumor activity of ascorbate against Ehrlich ascites tumor cells by the copper: glycylglycylhistidine complex." *Cancer Research* 43 (February 1983): 824–828.
  - o MSS Pauling, Series 3: Manuscripts of Articles. NDa.23. "An Episode that Changed My Life," No Date. [1980s]
  - o MSS Pauling, Series 4: Speeches. Boxes 1959s, 1975s, 1980s, 1981s2.
  - o MSS Pauling, Series 5: Manuscripts of Books. Box 17.001. *A Lifelong Quest for Peace: A Dialogue Between Linus Pauling and Daisaku Ikeda.*
  - o MSS Pauling, Series 7: Newspaper Clippings. 1945n.4. "Vandals Daub Paint on Scientist's Home Where Returned Jap Employed," *Pasadena Independent*, March 7, 1945; 1998n.3. "Linus Pauling Exhibition to Open in San Francisco," *The Messenger* (Oregon State University Libraries), Fall 1998; 2001n.4. "Linus Pauling Exhibit Arrives at OMSI," *Chemistry Alumni News*, vol. 20, January 2001.
  - o MSS Pauling, Series 8: Awards. 1954h2.18. Materials re: Nobel Chemistry Prize World Tour, Japan; 1980h.3. Federation of All Japan Karate-Do Organizations, Certificate of Dan: Honorary Title of 7th Dan. April 9, 1980.
  - o MSS Pauling, Series 12: Peace. Sub-Series 4, Peace Groups. Box 4.008. Japan Council Against Atomic and Hydrogen Bombs.

- o MSS Pauling, Series 13: Biographical, Sub-Series 1: Academia. Box 1.024. California Institute of Technology: Materials re: Division of Chemistry and Chemical Engineering, Folder 24.9. Interdepartmental Correspondence, C.I.T., 1941–1942; Sub-Series 2: Political Issues. Box 2.001. Materials re: "Japanese Gardener Incident"; Sub-Series 6: LP Scrapbooks. Item 6.147. "Dr. Linus Pauling...," [re: arrival in Tokyo] *Mainichi*, February 22, 1955.
- o MSS Pauling, Series 16: Personal Safe. Folder 1.017, Item 17.31. Letter from Linus Pauling to Ava Helen Pauling, September 4, 1945.
- o Agawa, Hiroyuki. *The Reluctant Admiral: Yamamoto and the Imperial Navy*, Tokyo: Kodansha International, 1979.
- o Daventry, Paula. *Sasakawa, the Warrior for Peace, the Global Philanthropist*, Oxford: Pergamon, 1981.
- o Kirkup, James. "Obituary: Ryoichi Sasakawa." *The Independent*, July 20, 1995 (online resource accessed September 26, 2011).

31. Two Trips to China

- Blog Posts
  - o **Travels in China, 1973** (September 21, 2011), by Olga Rodriguez-Walmisley
  - o **A Somber Return to China, 1981** (September 28, 2011), by Olga Rodriguez-Walmisley
  - o **Vitamin C and Cancer: Raising the Stakes** (December 2, 2015), by Matt McConnell
- Major Sourcing
  - o Image: Ava Helen and Linus Pauling at the Great Wall of China, 1973. (MSS Pauling, 1973i.23)
  - o MSS Hager, Series 2: Interviews. Box 2.002. Interviews with Linda Pauling Kamb.
  - o MSS Pauling, Series 1: Correspondence. Folder 218.7. Lu, Chia-Si, [Lu, Jiaxi]
  - o MSS Pauling, Series 3: Manuscripts of Articles. 1973a2.3. "Trip to the People's Republic of China," September–October 1973.
  - o MSS Pauling, Series 4: Speeches. Box 1981s2.

o   MSS Pauling, Series 11: Science. Sub-Series 13: Scientific War Work. Box 13.004, Folder 4.4. Correspondence and Non-Pauling Type-scripts re: Oxypolygelatin, 1972–1974.

o   MSS Pauling, Series 12: Sub-Series 6: Other Peace Activism. Box 6.013. Issues of International Diplomacy and Human Rights. (Af-Co).

o   MSS Pauling, Series 15: Ava Helen Pauling, Sub-Series 3: Biographical. Box 3.019, Folder 19.5. Materials re: AHP's medical condition and death.

o   MSS Pauling, Series 16: Item 4.053_6. Diary: labeled "Member, Federation of American Scientists," 1981.

Other Images

o   Cover: Linus Pauling during a meeting with the Department of Chemistry, Moscow State University, USSR, June 18, 1984 (MSS Pauling, 1984i.83).

o   About the Editor photo of Chris Petersen: courtesy of Sue Petersen.

o   Biographical Sketch: Linus Pauling with infant Peter Pauling outside their Pasadena home (MSS Pauling, 1931i.6).

Index

Basov, Nikolay, 299
Baumgarten, Werner, 133
Beadle, George, 155–157, 172–174
Beaver (Oregon State University), 51, 55, 56, 64
Beckman, Arnold, 122
Behnke-Walker Business School, 23
Benson, Andrew A., 134
Bergeson, Maida, 211
beta sheet (protein structure), 279
Betancur, Belisario, 316
Bethe, Hans, 153
Bible, xv, xvi, 11
Big Sur, California, 99, 201, 207, 217, 221, 340
biochemistry, 107, 112, 125, 126, 130, 135, 137–139, 155, 162, 171–174, 178
biological specificity, xxiii, 270, 285, 314
biophysics, 155
Black Panthers, 94
Black Student Union (Oregon State University), 90–94, 96
Black, William, 52
Blaine, James, 13
Blanken, Adelheit, 4
Blethen, John, 211, 226
Blethen, Margaret, 211, 226
Blinks, Lawrence, 221, 222
Boggs, James, 198
Bohr, Niels, 53, 252–255
Bolivia, 309
Bolonkin, Alexander, 300
Bonner, James, 149
Born, Max, 255
Bragg, William Lawrence, 72, 277, 280
Brahms, Johannes, 286
Branson, Herman, 279–281
Brazil, 167, 168, 243, 305, 309, 315
Brewer, Leo, 129, 130

Brezhnev, Leonid, 302, 303
Bridges, Sam, 239, 240
Briehl, Robin W., 319
British Undergraduate School of Medicine, 267
Bronk, Detlev, 296
Brooklyn College, 127
Brown, Pat, 206
Bryden, William, 18, 19
Buchman, Edwin R., 113, 114, 144, 149
Buddhism - Nichiren, 339, 341
Budenz, Louis, 176, 292
Bureau of Ordnance (United States Navy), 151

Caeser, Julius, 351
calcium boride, 164
California Institute of Technology — admission of women, 135, 176–178
California Institute of Technology — Board of Trustees, x, xxii, 105, 119, 125, 137, 138, 145, 154, 162, 175, 177, 178, 183, 186
California Institute of Technology — Division Council, 111–113, 115, 119, 121, 122
California Institute of Technology — Division of Biology, 112, 123, 124, 154, 168, 172, 187
California Institute of Technology — Division of Chemistry and Chemical Engineering, xiv, xxv, 103, 109, 110, 112, 113, 116–118, 124, 127, 132, 135, 139, 150, 153, 178, 186
California Institute of Technology — Division of Humanities, 145
California Institute of Technology — Division of Physics, Mathematics and Electrical Engineering, 153
California Institute of Technology — Executive Council, 105, 107,

eugenics, xiii, 215, 375
evolution, 52
explosives, 132, 133

Face the Nation, 319
Fair Play for Cuba, 309
Farrand, George, 173
fascism, 337
Federal Bureau of Investigation
    (United States), 166, 176, 292
Federal Unionist Club, 194
Federation of All Japan Karate-Do
    Organizations, 335
Ferry, W.H. "Ping", 198
Fetter, Alexander, 223
Fields, W.C., 36
Findeisen, Walter, 229
fires, 10, 129, 130
Flonzaley String Quartet, 62
Flory, Paul John, 227
fluid mechanics, 159
The Flying Saucer Story, 203
*Force of Nature: The Life of Linus
    Pauling*, xvi, xvii
Ford Foundation, 181, 192
Fordham University, 176
Foundation for Nutritional
    Advancement, 348, 349
Fowler, Estelle Maxine, 178
framboesia, 284
France, 172
Franklin High School (Portland,
    Oregon), 23
Franklin, Bruce, 233–240
Franklin, Rosalind, 265
Frederick, William, 4
Freedom Appeals, 301
freedom of speech, 238, 240
Freemasonry, 20
French Equatorial Africa, 283, 328
Friedrich, Christoph, 3, 4

Fritchman, Stephen, 185
Fry, D.J., 7
Fukuoka Torikai Hospital, 331
Fukuoka University, 334
Fuller, John G., 203
Fulton, John, 45, 46, 48, 53
*Functional Biochemistry of Cell
    Structures*, 298
Fund for the Republic (Ford
    Foundation), 192

Gabon, 283
Gaines, Claire, 59
Galbraith, John, 210
Galtung, Johan, 343
Gamma Tau Beta (fraternity), 42, 46,
    50
Gard, John Jay, 58
Garwood, Donald, 179
Gates and Crellin Laboratory
    (California Institute of Technology),
    103–105, 112, 113, 141, 180, 181
Gazette-Times (Corvallis, Oregon), 76
General Chemistry, xxiii, 224, 262, 270
General Petroleum, 159
genetic testing, 215
genetics, 82, 83, 154–156
Georgia Institute of Technology, 176
Germany, 3, 160, 285, 328–330, 348,
    349
Gesellschaft Deutscher Chemiker, 267
Gilfillan, Francois, 128
Gitlin, Todd, 198
*Give Peace a Chance: Soviet Peace
    Proposals and U.S. Responses*, 304
Glass, I.I., 301
Globe (Condon, Oregon), 5, 8, 10
glomerulonephritis, 213, 226, 263
Gomberg, Moses, 113, 114
Good Hope Hospital, 138
Gorbachev, Mikhail, 302, 303

Gormley, Melinda, 375
Goss, Arthur, 329
Goudsmit, Samuel, 255
graduate education, 226, 273
Graf, Samuel, 53, 67
Graham, Martha, 248
Gratcheva, Angella, 297
Great Depression, xii, 76
Gregory, Leland, 45
Gross, Paul, 127
Grundmann, Alan, 225
Guggenheim Foundation, xii, xxii, 76, 167, 168, 247–260, 281
Guggenheim, John Simon, 247
Guggenheim, Olga, 247
Guggenheim, Simon, 247, 248
Gurin, Samuel, 114
Gvishiani, Dzermen, 299

Haagen-Smit, Arie, 142
Hager, Thomas, xvi–xviii, 10, 37, 58, 59, 70, 98, 192, 249, 252, 268, 307, 353
Hahn, Andy, xiii, xiv, 366
Hale, George Ellery, 110, 115
Hansen, Thomas D., 42
Hanson, Frank Blair, 136–138, 147
Harold, Ted, 180
Harr, Rich, 90, 93
Harvard University, 68, 69, 113, 115, 116, 157, 274, 294
Harvey Mudd College, 186
Harvey, Paul, 37
Hawkins, David, 224
heart disease, xxvi, 25, 205, 224, 269, 316
heat transfer, 159
Hedberg, Ken, 93
Heidelberger, Michael, 324, 325
Heisenberg, Werner, 251, 255, 256
Heitler, Walter, 255, 256

Helmholz, Lindsay, 127, 128
Helms, Karl, 94
hemoglobin, xxiii, 107, 122, 268, 274–277, 279, 280, 305, 308, 310, 328, 347, 349
hemophilia, 276
Henry, David C., 74, 76
Henson, Fred, 124
hepatitis, 332–334
Heraclitean Fire: Sketches from a Life before Nature, 265
Herodotus, 11
Herzberg, Gerhard, 302
Hester, Hugh B., 198
Hill, Archibald, 117
Hill, Arthur, 125
Hirano, Tomosaburo, 340
Hiroshima Appeal, 329, 330
Hiroshima Fifth World Conference against Hydrogen and Atomic Bombs, 328, 329
Hiroshima University, 194, 328, 330
Hiroshima, Japan, xxiii
History of Rome, 11
Hitch, Charles, 216
Hitler, Adolf, 342
Hobson, J.E., 220
Hodgkin, Dorothy (Crowfoot), 277, 304
Hoffer, Abram, 213, 349
Hogness, Thorfin, 113
Hong Kong, 345, 346
honorary doctorates, xxvi, 31, 74–77, 262, 282
Hoover Institute (Stanford University), 234
Hopfield, John, 276
Hopkins Marine Station (Stanford University), 221, 222
hormones, 146
Houston, W.V., 133

magnetism, 53

Magnolia Petroleum, 159

malaria, 284

Manhattan Project, 128, 153, 326, 327

Many Worlds Theory, xi

Mao, Zedong, 347

March of Dimes, 155

marine biotoxins, 323

Marshall Plan, 270

Martin, Jack, 242

Massachusetts Institute of Technology, 106, 145, 148, 152, 157, 170, 177

Matkovsky, N., 309

McBain, J.W., 124

McCarthy, Joseph, 167, 301

McConnell, Harden, 222, 223, 240

McCormick, Cyrus Hall, 263

McDannald, Mahala, 57

McDonald, Lawrence, 317

McKerrow, James, 223

Mead, Clifford, xvii, xviii

measles, 334

medical research, 146, 155, 158, 173, 204, 225

Medical Tribune, 232

Meerwein, Hans Lebrecht, 259

Meet the Press, 185, 186, 319

Meloche, Villiers W., 133

meningitis, 334

mental illness, 181, 211–213, 222, 228, 349

Merrill, Frederick, 295

metabolism, 156

metallography, 54

metals, xxvi, 149, 156, 270–272

methane, 139

Mexico, 127, 306, 309–311, 313, 314, 316, 319

microbiology, 154

military draft, 133, 134

Miller, Clay, 60

Miller, George Richard, 58

Miller, Lillian, 249

Miller, Mary, 60

Miller, Milton, 60

Miller, Nora (Gard), 58, 60, 249

Miller, William F., 231, 232, 240

Millikan, Robert, 105, 110, 115, 117–119, 123, 126, 324

Milner, Gerald, 349

Milton, Fred, 91–93, 96

minerals, 28, 29

minimization of suffering, xiii, 197, 216, 238, 284, 287, 311, 341, 342

Mitchell, Lawrence, 294, 295

molecular biology, 160, 268, 286

molecular disease, xxiii, 277, 310, 311, 349

molecular orbital theory, 271

Molecule Structure Fund, 148

Montalva, Eduardo Frei, 311

Mor Roig, Arturo, 310

Morgan, Thomas Hunt, 112, 123, 126

Morishige, Fukumi, 331–335

Morton-Norwich Products, Inc., 229

Moscow State University, 293

Mount Holyoke College, 178

Muller, H.J., 82

Mulliken, Robert, 271

mumps, 334

Munro, Dorothy, 340

Munro, William Bennett, 119

Murata, Akira, 333, 334

muscular dystrophy, 276

Mussolini, Benito, 250, 251, 336

myoglobin, 268, 279

Nagasaki, Japan, xxiii

Nanjing University, 348

The Nation, 85

National Academy of Medicine, 310

National Academy of Sciences, 75, 76,
179, 295
National Broadcasting Corporation
(NBC), 185
National Commission on Educational
Reconstruction, 85
National Coordinating Center in
Solidarity with Chile, 312
National Council of American-Soviet
Friendship, 291
National Council of Atomic Scientists,
80
National Defense Research Committee,
132, 342
National Foundation for Infantile
Paralysis, 154–156
National Guard (United States), 218,
219
National Institutes of Health
(United States), 224, 230, 231
National Medal of Science, 160
National Research Council, 86, 114, 249
National School of Anthropology, 311
National Science Foundation
(United States), 158, 207, 212, 224,
231
Native Americans, 9
Nature, 281
*The Nature of the Chemical Bond
and the Structure of Molecules and
Crystals: An Introduction to Modern
Structural Chemistry*, xxiii, 79, 258
Neal, Alcy, 13, 14
Neale, Marvin Gorden, 74, 76
Netherlands, 200, 255, 270
New Mexico Institute of Mining and
Technology, 203
New York Times, 75, 295
Newton, Charles, 167
Newton, Isaac, 86, 251
Ney, Michael, 23

Ney, Thomas, 23
niacin, 349
Nicaragua, 317–320
nicotinamide, 213
nicotinic acid (B_3), 212, 213
Niemann, Carl, 114, 118, 123–125, 130,
148, 157, 159, 171, 177–179, 267
Nimaki, George, 325
The 9th Studio, 299
Nippon Foundation, 338
Nixon, Richard, 233, 345, 348
No More War!, 328
Nobel, Alfred, 248, 258, 259
Nobel Prizes, xxiii–xxvi, 69, 75, 77,
105, 124, 142, 153, 180, 187, 191,
193, 204, 242, 259, 271, 277, 283,
284, 292, 300, 302, 317, 319, 339, 341
norepinephrine, 179
North American Aviation, 159
Northern Pacific Railroad, 7
Norway, 285, 317, 319
Norwegian Helsinki Committee, 303
Noyes, A.A., 46, 54, 72, 75, 103–106,
109–111, 113–116, 118, 121, 131,
145, 147, 181, 249–251, 256
nuclear disarmament, 309
nuclear fallout, 287, 309
nuclear physics, 201
nuclear proliferation, 284
nuclear shell theory, 207
nuclear structure, xxvi, 207, 241
nuclear testing, 182, 183, 192, 284, 286,
287, 290, 297, 309, 341
nuclear testing treaty, 329
nuclear war, 193, 197–199, 286, 287, 289
nuclear weapons, xxiv, xxvi, 80, 99,
162, 175, 194, 224, 299, 305, 326,
328, 329
nucleosides, 346
nucleotides, 346
numismatics, 23